COLLOQUIA MATHEMATICA
SOCIETATIS JÁNOS BOLYAI, 15.

DIFFERENTIAL EQUATIONS

Edited by:

M. FARKAS

NORTH-HOLLAND PUBLISHING COMPANY
AMSTERDAM – OXFORD – NEW YORK

Budapest, Hungary, 1977

ISBN Bolyai: 963 8021 19 5
ISBN North-Holland: 0 7204 0496 7

Joint edition published by

JÁNOS BOLYAI MATHEMATICAL SOCIETY

and

NORTH-HOLLAND PUBLISHING COMPANY
Amsterdam–Oxford–New York

In the U.S.A. and Canada:

NORTH-HOLLAND PUBLISHING COMPANY

52 Vanderbilt Avenue

New York, N.Y. 10017

Technical Editor:
A. SZÉP
Printed in Hungary
ÁFÉSZ, VÁC
Sokszorosító üzeme

PREFACE

A Colloquium on Differential Equations was organized by the Bolyai János Mathematical Society at Keszthely (Hungary) in 1974. Research in the field of differential equations has become nowadays so extensive that it is impossible even to touch upon the main directions of development in the limitations of such a colloquium. While in the past in Hungary only a few mathematicians cultivated the subject (among whom the name of J. Egerváry is to be mentioned), recently the situation has changed and considerable interest has been shown towards this field of study also in this country. The aim of the organizing committee was to gather the mathematicians in Hungary as well as several colleagues from abroad dealing with, or interested in, differential equations. It was an honour to greet several excellent experts of this matter at Keszthely.

This volume contains 33 papers presented at the colloquium. Most of the papers deal with ordinary and functional differential equations. Topics treated in one or more papers are: stability, oscillation, local equivalence, asymptotic behaviour, boundary value problems, existence, uniqueness and continuation problems. The few papers on partial differential equations are of classical nature.

Differential equations have always been closely related to applied mathematics. Accordingly, some papers originate from, or are connected with, Hamiltonian mechanics, hydrodynamics, gyroscope problems etc.

Im some papers algorithms are described for solving different boundary or initial value problems; a part of them involves experiences on high speed computers.

A number of papers deal with functional differential equations. We refer to S.D. Norkin's paper where a brief account of the results presented at the colloquium in this branch is given.

We wish to thank Miss M. Meszlényi and Mrs. Zs. Monostori for their excellent typing of the manuscript and for the enormous amount of greatly diverse work that had to be done in preparing this volume.

The Editor

CONTENTS

SCIENTIFIC PROGRAM

September 2. Monday

Session A

15.00 – 16.00	M.L. Cartwright – H.P.F. Swinnerton-Dyer: Boundedness theorems for some second order differential equations
16.15 – 16.35	J. Kurzweil – J. Jarník: On Ryabov's special solutions of functional differential equations
16.40 – 17.00	M. Tvrdý: Boundary value type problems for functional differential equations and their adjoints
17.30 – 17.50	P.S. Gromova: Conditions for the stability of solutions of differential-difference equations in some singular case (in Russian)
17.55 – 18.15	J. Terjéki: Remarks on the existence of the solutions of functional-differential equations
18.20 – 18.40	M.M. Konstantinov – D.D. Bainov: On a boundary value problem with generalized boundary conditions for systems of differential equations of neutral type (in Russian)

Session B

16.15 – 16.35	I. Bihari: Behaviour of the 0 paths in the neighbourhood of a saddle point
16.40 – 17.00	L. Hatvani: Application of the method of Ljapunov-functions for investigating bounded stability (in Russian)
17.30 – 17.50	G. Sonnevend: On a class of nonlinear control processes
18.20 – 18.40	L. Reizin's: Local topological equivalence of differential equations (in Russian)

September 3. Tuesday

Session A

9.30 – 10.30	J. Szarski: Strong maximum principle for non linear parabolic differencial-functional inequalities
10.45 – 11.45	D. Greenspan: Conservative discrete models and a new particle theory of fluids
15.30 – 15.50	J. Walter: Methodological remarks on the notion of the saturated solutions of the differential equation $x' = f(t, x)$
15.55 – 16.15	K. Karták: On modified iterations and existence theorems
16.20 – 16.40	A. Schmidt: Über gewisse lineare Differentialgleichungen
16.45 – 17.05	G. Jank: Funktionentheoretische Untersuchungen von Lösungen gewisser elliptischer Differentialgleichungen
17.30 – 17.50	D. Schultz – D. Greenspan: Simplification and improvement of a numerical method for Navier – Stokes problems
17.55 – 18.15	J. Moravčik: Some oscillatoric and asymptotic properties of linear differential equations of order n ($n \geqslant 3$) of 3-rd kind (in Russian)
18.20 – 18.40	L.N. Fadeeva: On a certain class of non-linear boundary-value problems

Session B

15.30 – 15.50	I. Győri: On asymptotically ordinary functional-differential equations (in Russian)
15.55 – 16.15	R.J. Grimm: Some extensions of Lettenmeyer's theorem on singularities
16.20 – 16.40	P. Marušiak: Oscillation of solutions of delay differential equations

16.45 – 17.05	E. Kozakiewicz: Über das asymptotische Verhalten der positiven Lösungen einiger Differentialgleichungen mit nacheilendem Argument
17.30 – 17.50	S.D. Milusheva – D.D. Bainov: Application of the averaging method for systems of integro-differential equations of standard type with discontinuous right-hand side (in Russian)
17.55 – 18.15	J. Hegedűs: On two-sided estimates for the solutions of certain ordinary differential equations (in Russian)
18.20 – 18.40	Ju.A. Ved': Some asymptotic problems for differential equations with delayed argument (in Russian)

September 5. Thursday

Session A

9.30 – 10.30	A. Pleijel: Limit types of symmetric ordinary differential equations
10.45 – 11.45	K. Deimling: On existence and uniqueness for ordinary differential equations in Banach spaces
15.30 – 15.50	K. Schmitt – P. Volkmann: Randwert-Probleme für gewöhnliche Differentialgleichungen zweiter Ordnung in konvexen Teilmengen eines Banachraumes
15.55 – 16.15	L. Pintér: Oscillatory properties of certain ordinary second order nonlinear differential equations
16.20 – 16.40	Á. Elbert: On the distance between zeros of solutions of certain second order linear differential equations
16.45 – 17.05	B. Mehri: On the conditions for the oscillation and nonoscillation of solutions of nonlinear third order differential equations
17.30 – 17.50	I. Fenyő: On the smallest roots of solutions of fourth order differential equations

17.55 – 18.15 I. Bartsch: Über die Differentialgleichung

$$\sum_{k=1}^{n} c_k (D_1 + aD_2)^k U = 0$$

18.20 – 18.40 P. Soltész: On differential equations of infinite order

Session B

15.30 – 15.50 L. Simon: On approximation of solutions of elliptic boundary problems in unbounded domains

15.55 – 16.15 E.M. Bruins: The differential equations of quantum mechanics

16.20 – 16.40 J. Gergely: Numerical solutions of a nonlinear boundary value problem

16.45 – 17.05 J. Kačur: Application of Rothe's method to nonlinear evolution equations

17.30 – 17.50 V.M. Petkov: Sur les systemes hyperboliques a caractéristiques de multiplicité constants

17.55 – 18.15 I.F. Dorofeev: On the stable algorithms for numerical solutions of some inverse problems for an equation with delay (in Russian)

18.20 – 18.40 L. Gerencsér: A second order optimization technique for constrained variational problems

September 6. Friday

Session A

9.30 – 10.30 F.V. Atkinson: The limit-point/limit-circle classification of second-order operators

10.45 – 11.45 S.B. Norkin: On *k*-oscillation of systems with one degree of freedom (in Russian)

LIST OF PARTICIPANTS

ATKINSON, F.V., 35 Green Ridge, Brighton, Sussex, England

BAINOV, D., Ul. Oboriste 23, Sofia-4, Bulgaria

BALLA, KATALIN, MTA SzTAKI, 1052 Budapest, 112, Pf. 63, Hungary

BÁRÁNY, I., BME Gépészmérnöki Kar, Matematika Tsz., 1111 Budapest, Stoczek u. 2, Hungary

BARTSCH, INGEBORG, 25 Rostock, Universitätsplatz 1, GDR

BENKÓ, S., MTA SzTAKI, 1502 Budapest, 112, Pf. 63, Hungary

BIHARI, I., MTA Matematikai Kutató Intézet, 1053 Budapest, Reáltanoda u. 13-15, Hungary

BLEYER, A., BME Villamosmérnöki Kar, Matematika Tsz., 1111 Budapest, Stoczek u. 2, Hungary

BRUINS, E.M., Joh. Verhulststraat 185, Amsterdam-Z 1, The Netherlands

CARTWRIGHT, MARY LUCY, 38 Sherlock Close, Cambridge CB3 OHP, England

DEIMLING, K., Math. Seminar der Universität, D-23 Kiel, Olshausen Str. 40-60, GFR

DETKI, J., Szabadka (24000), J.E. Tanitya 66, Yugoslavia

DÉVÉNYI, D., 1036 Budapest, Korvin Ottó u. 63, Hungary

DOROFEEV, I.F., Univ. Druzby Nar. im. Patr. Lumumbu, Moscow B-302, Ul. Ordzonikidze, USSR

DRASNY, JÓZSEFNÉ, MTA SzTAKI, 1052 Budapest, 112, Pf. 63, Hungary

DURIKOVIČ, V., Dept. of Math., Analysis I A Komensky Univ., Matematicky pavilon, Mlinská dolina 816 31, Bratislava, Czechoslovakia

ELBERT, Á., MTA Matematikai Kutató Intézet, 1053 Budapest, Reáltanoda u. 13-15, Hungary

FARKAS, ÉVA, MTA KFKI, 1121 Budapest, Konkoly Thege út 6, Hungary

FARKAS, MIKLÓSNÉ, 1123 Budapest, Alkotás u. 11, Hungary

FAZZÁN, H.R., 1121 Budapest, Konkoly Thege u. 6, Hungary

FENYŐ, I., 1125 Budapest, Istenhegyi út 48/a, Hungary

FRIGERIO, A., via Donatello 16, 35100 Padova, Italy

FRITZ, JÓZSEFNÉ, BME Gépészmérnöki Kar, Matematika Tsz., H. ép., 1111 Budapest, Stoczek u. 2, Hungary

FRIVALDSZKY, S., KGM-ISzSzI, 1394 Budapest, Pf. 356, Hungary

FUČÍK, S., Dept. of Math., Ana. Faculty of Math-Ph. Charles Univ., 83 Sokolovská, 18600 Prague, Czechoslovakia

GERENCSÉR, L., MTA SzTAKI, 1052 Budapest, 112, Pf. 63, Hungary

GERGELY, J., MTA SzTAKI, 1052 Budapest, 112, Pf. 63, Hungary

GREENSPAN, D., c/o Dept. Computer Sciences, Univ. Wisconsin, Madison, Wisconsin 53706, USA

GRIMM, L.J.E., Department of Mathematics, University of Missouri, Rolla, Missouri 65401, USA

GROMOVA, P.S., Univ. Druzby Nar. im. Patr. Lumumbu, Moscow B-302, ul. Ordzonikidze, USSR

GYŐRI, I., 6723 Szeged, Tarján 620 ép. B. I/5, Hungary

GYURKOVICS, ÉVA, 6720 Szeged, Somogyi Béla u. 7, Hungary

HATVANI, L., 6720 Szeged, Aradi vértanúk tere 1, Bolyai Intézet, Hungary

HEGEDŰS, CS., MTA KFKI, 1121 Budapest, Konkoly Thege út 6, Hungary

HEGEDŰS, J., 6720 Szeged, Aradi vértanúk tere 1, Bolyai Intézet, Hungary

HIBBEY, L., Energiagazdálkodási Intézet, 1251 Budapest, Pf. 15, Hungary

JANK, G., A-8010 Graz, Brockmanng. 9/III, Austria

JARNÍK, J., Žitná 25, 11567 Praha 1, Czechoslovakia

JELITAI, Á., 1043 Budapest, Munkásotthon u. 34, Hungary

JUHÁSZ, F., MTA SzTAKI, 1052 Budapest, 112, Pf. 63, Hungary

KAČUR, J., Bratislava 16, Department of Mathematics, Czechoslovakia

KARTÁK, V., VSCHT, Math. Dept., Praha – Dejvice 1905, Czechoslovakia

KOZAKIEWICZ, E., Humboldt Univ. zu Berlin, 108 Berlin, Unter den Linden 6, GDR

KRISZTINKOVICS, F., 2100 Gödöllő, Nyitor tér 3/c, Hungary

LŐKÖS, ÁGNES, 1111 Budapest, Lágymányosi u. 14/a, V. em. 17, Hungary

LUSZCZKI, Z., 61860 Poznan, ul. Za Grobla 6/8, Poland

MAKAI, E., MTA Matematikai Kutató Intézet, 1053 Budapest, Reáltanoda u. 13-15, Hungary

MÁLYUSZ, K., 1011 Budapest, Batthyány u. 26, Hungary

MARUŠIAK, P., Katedra Matematiky, Marxa a Engelsa 15, 010 88 Zilina, Czechoslovakia

MEHRI, B., Dept. of Mathematics, University of Arya-Mehr, Tehran, Iran

MESKÓ, ATTILÁNÉ, MTA SzTAKI, 1052 Budapest, 112, Pf. 63, Hungary

MILE, KÁROLYNÉ, 1042 Budapest, Viola u. 3, Hungary

MILUSHEVA, S., Dede-Agacs No. 11, Sofia, Bulgaria

MOLNÁRKA, GY., ELTE TTK Num. és Gépi Mat. Tsz., 1088 Budapest, Múzeum krt. 6-8, Hungary

MORAVCIK, J., Marxa a Engelsa 25, 010 88 Zilina, Czechoslovakia

MOSON, P., 1036 Budapest, Árpád fejedelem u. 52, Hungary

OLAS, A., Inst. Fund. Techn. Research, Warsaw, ul. Swietokrzyska 21, ZUM, Poland

PETKOV, V.M., P.O. Box 373, Sofia, Bulgaria

PINTÉR, L., 6720 Szeged, Aradi vértanúk tere 1, Bolyai Intézet, Hungary

PLEIJEL, A.V.C., Stora Mossens Backe 14, 161 37 Bromma, Sweden

PÁGER, T., 1123 Budapest, Alkotás u. 39/a, Hungary

RADZISZEWSKI, B., Inst. Fund. Techn. Research, PAN, Warsaw, Swietokrzyska 21, ZUM, Poland

RENNER, G., MTA SzTAKI, 1502 Budapest, 112, Pf. 63, Hungary

SCHMIDT, A., 25 Rostock, Schliemannstr. 40, GDR

SCHNEIDER, M., 1 Berlin 33, Rohffsstr. 6, West-Berlin GDR

SCHULTZ, D.H., The University of Wisconsin-Milwaukee, Dept. of Mathematics, Milwaukee, Wisconsin 53201, USA

SCHÜTTE, ILSE, Humboldt Univ. zu Berlin Sekt. Math., 108 Berlin, Unter den Linden 6, GDR

SIMON, L., 1117 Budapest, Baranyai u. 17, V. 4, Hungary

SOLTÉSZ, P., 1083 Budapest, Tömő u. P/1, Hungary

SONNEVEND, G., MTA SzTAKI, 1502 Budapest, 112, Pf. 63, Hungary

SZADKOWSKI, A., Inst. Fund. Techn. Research PAN, Warsaw, ul. Swietokrzyska 21, ZUM, Poland

SZARSKI, J., ul. Kochanowskiego 1115, Kraków, Poland

SZEPESVÁRI, I., MTA SzTAKI, 1052 Budapest, 112, Pf. 63, Hungary

SZILÁRD, K., MTA Matematikai Kutató Intézet, 1053 Budapest, Reáltanoda u. 13-15, Hungary

SZÉP, A., MTA Matematikai Kutató Intézet, 1053 Budapest, Reáltanoda u. 13-15, Hungary

TARNAY, GY., MTA SzTAKI, 1052 Budapest, 113, Pf. 63, Hungary

TERJÉKI, J., 5094 Tiszajenő, Külsőjenő 21, Hungary

TVRDÝ, M., Matematicky ustav ČSAV, Zitná 25, 11567 Praha 1, Czechoslovakia

VILLARI, G., Istituto di Matematica Applicata, Universita di Firenze, viale Morgagni 44, 50134 Firenze, Italy

VIGASSY, J., MTA KFKI, 1525 Budapest, Konkoly Thege út 6, Hungary

VOLKMANN, P., Universität Karlsruhe (TH), Mathematisches Institut I, 75 Karlsruhe 1, GFR

WALTER, J., 51 Aachen, Templergraben 55, GFR

ZIBOLEN, E., BME Gépészmérnöki Kar, Matematika Tsz., 1111 Budapest, Stoczek u. 2, Hungary

ZOCCARATO, L., Collegio Universitario Antoniarum, via Donatello 16, 35100 Padova, Italy

COLLOQUIA MATHEMATICA SOCIETATIS JÁNOS BOLYAI
15. DIFFERENTIAL EQUATIONS, KESZTHELY (HUNGARY), 1975.

NUMERICAL SEGREGATION OF THE BOUNDED SOLUTIONS FOR SYSTEMS OF ORDINARY DIFFERENTIAL EQUATIONS

A.A. ABRAMOV – E.S. BIRGER – N.B. KONYUKHOVA – V.I. ULYANOVA

1°. Several problems in various fields of mathematical physics occur lead to systems of ordinary differential equations having singularities or being defined on an infinite interval. The role of boundary conditions at the singular point, or at infinity, is played by the condition of boundedness. As a typical example, we mention quantum mechanics where a lot of problems have the following form: Let us find the eigenvalues and eigenfunctions of the Schrödinger operator

(1) $$H\psi = \lambda\psi;$$

here $\psi = \psi(r)$, $0 < r < \infty$, $H = -\dfrac{d^2}{dr^2} + U(r)$, $U(r) \to \infty$ as $r \to 0$ and $U(r)$ has prescribed behaviour as $r \to \infty$. Let us find those λ for which there exists a non-trivial solution $\psi(r)$ such that $|\psi(r)|$ is bounded for $r \to 0$ and $r \to \infty$. When solving numerically a problem of this kind it is often useful to choose the following way. By using the expansion* of $U(r)$ in a neighbourhood of $r = 0$, we look for the expansion of the general solution and choose that solution which turns out to be bounded. In this way we obtain the required solution in the neighbourhood of $r = 0$. In similar way we look for the expansion of the required solution at $r \to \infty$. Secondly we substitute $(0, \infty)$ with a finite interval (r_0, r_∞) and solve the numerical equation (1), whlle for boundary conditions at r_0 and r_∞ we use the above expansions in the neighbourhood of $r = 0$, and at infinity, respectively. In simple cases this method really works. Unfortunately, for more involved problems this method turns out to be inapplicable.

For example consider the linear homogeneous system

$$(2) \qquad y' = A(t)y, \qquad y = \begin{Vmatrix} y_1 \\ \vdots \\ y_n \end{Vmatrix}, \qquad 0 < t < T$$

where in a neighbourhood of $t = 0$ the matrix $A(t)$ has the expansion

$$A(t) = t^{-r}(A_0 + A_1 t + A_2 t^2 + \ldots),$$

where r is a non-negative integer. Suppose that we look for the solution of (2) which is bounded at $t \to 0$ and satisfies some given boundary conditions at $t \to T$. Trying to apply the above method we need expansions of solutions bounded for $t \to 0$. This is very inconvenient from the point of view of computations. First of all, these expansions of various solutions are large and intricate. Secondly, the form of the expansions essentially depends on the values of certain parameters, which themselves have to be calculated. (For example, the expansions at $r = 1$, as a rule, have the form $y(t) = t^\sigma(y_0 + ty_1 + t^2 y_2 + \ldots)$ but there may appear logarithms the matrix A_0 admits of pairs of eigenvalues whose difference is integer). Obviously it is often impossible to tell whether numbers, calculated by computer, are actually integers or they are only near to integers. Consequently, such a method is numerically unstable. For (2) the various solutions in a

*For sake of simplicity we make no distinction between asymptotic expansions and convergent series. In order to have exact formulations one also has to define conditions on continuity or smoothness of the functions considered.

neighbourhood of the singular point behave very capriciously and do not easily abide to numerical methods. Still more difficulty we have, of course, with systems of non-linear equations, when the construction of different solutions which are bounded in near the singular point, itself proves to be a very tiresome and unstable computational problem.

Therefore it would be desirable to explore methods being free of these weaknesses.

In what follows we intend to give a report on some results which were obtained by the authors and published in [1]-[26]*. For a large class of systems one has the following picture: Instead of the individual solutions considering the family of all solutions which are bounder near the singularity (at infinity), this family can be described by relation of simpler form, moreover this relations can be derived on a numerically stable way. Let us fix some point near to the singularity. The family of all solutions which are bounded in the neighbourhood of the singular point can be described in terms of certain boundary conditions at the fixed point. We get these boundary conditions by using expansions for the family of all bounded solutions. Hence, the boundedness condition at the singular point has been transferred to an other point. Applying the same method on the other and of the interval (whenever the system is singular on both ends) we can reduce the original problem to a boundary problem for the original system on a finite interval without singularities. We emphasize once again that the behaviour of the individual solutions within the family of all bounded solutions has no importance for this second problem.

2°. We will show how to transfer the boundary condition out of a singular point.

Let us consider on $T < t < \infty$ the non-linear system

$$t^{-r} y' = f(t, y), \qquad y = \begin{Vmatrix} y_1 \\ \vdots \\ y_n \end{Vmatrix}, \quad f = \begin{Vmatrix} f_1 \\ \vdots \\ f_n \end{Vmatrix} \tag{3}$$

*We do not consider here other methods, which were worked out by other authors for more special problems.

where r is a non-negative integer.

Let $f(t, y)$ have the following behaviour at t large and $|y|$ small:

$$f(t, y) = Ay + g(t, y),$$

$$g(t, y) = \sum_{\substack{p_1 \geqslant 0, \ldots, p_n \geqslant 0,\, q \geqslant 0 \\ p_1 + \ldots + p_n + 2q \geqslant 2}} C_{q, p_1, \ldots, p_n} \frac{y_1^{p_1} \ldots y_n^{p_n}}{t^q}.$$

Let us assume that the matrix A has no eigenvalues on the imaginary axis. We take decompose A in to the quasidiagonal form

$$\left\| \begin{array}{c|c} A_+ & 0 \\ \hline 0 & A_- \end{array} \right\|$$

where the eigenvalues of A_+ and A_- lie on the right and left half-plane, respectively. In the corresponding coordinate system we write

$$y = \left\| \begin{array}{c} y_+ \\ \hline y_- \end{array} \right\| \begin{array}{l} \updownarrow k \\ \updownarrow n-k \end{array}, \quad f = \left\| \begin{array}{c} f_+ \\ \hline f_- \end{array} \right\|.$$

Then for t sufficiently large the condition

$$|y| \to 0 \ \ as \ \ t \to \infty$$

is equivalent to

$$y_+ = S(t, y_-), \qquad S = \left\| \begin{array}{c} s_1 \\ \vdots \\ s_k \end{array} \right\|.$$

Here the function $S(t, z)$ satisfies the following partial differential equation

$$(4) \qquad t^{-r}\frac{\partial S}{\partial t} + \frac{\partial S}{\partial z} f_-(t, S(t, z), z) = f_+(t, S(t, z), z)$$

and the condition

$$S(t, z) \to \beta(z) \quad \text{as} \quad t \to \infty,$$

where $\beta(z)$ satisfies the equation

$$\frac{\partial \beta}{\partial z} f_-(\infty, \beta(z), z) = f_+(\infty, \beta(z), z)$$

and can be represented by a series $\beta(z) = \sum \beta_{p_1,\ldots,p_{n-k}} z_1^{p_1} \ldots z_{n-k}^{p_{n-k}}$, the sum taken for $p_1 \geqslant 0, \ldots, p_{n-k} \geqslant 0,\ p_1 + \ldots + p_{n-k} \geqslant 2$.

The function $S(t, z)$ can be expanded into a series of integral powers of z and $\frac{1}{t}$

$$S(t, z) = \sum S_{q,p_1,\ldots,p_{n-k}} \frac{z_1^{p_1} \ldots z_{n-k}^{p_{n-k}}}{t^q},$$

the sum taken for $p_1 \geqslant 0, \ldots, p_{n-k} \geqslant 0,\ q \geqslant 0$ and $p_1 + p_2 + \ldots \ldots + p_{n-k} + 2q \geqslant 2$.

In order to evaluate the coefficients $S_{q,p_1,\ldots,p_{n-k}}$ we substitute the series formally into (4). We obtain a system of linear algebraic equations

$$(5) \qquad L\hat{S} = \varphi$$

where $\hat{S}$ is the column of $S_{q,p_2,\ldots,p_k}$ with the fixed q and the same $\sum_{j=1}^{n-k} p_j$, φ is a polynomial of the previous coefficients, $\lambda(L) = \lambda(A_+) - \sum_{j=1}^{n-k} l_j \lambda_j(A_-)$ and the l_j's are non-negative integers whence L is a non-singular.

This expansion is convenient in many aspects of computational practice. Let us mention some of them.

(1) In the above case the condition at infinity is $|y| \to 0$ as $t \to \infty$. Therefore, for t large enough, the relation

$$y_+ = S(t, y_-)$$

has only to be applied for small $|y_-|$. Consequently, for practical computation we may conveniently use the expansion of $S(t, z)$.

(2) We have expansion by integral powers independently of such particular factors as, for example, the Jordan structure of A.

(3) The coefficients of the expansion of S are determined by using algebraic equations which can be solved in stable way. The computations require only to represent A in quasi-diagonal form.

3°. If equation (3) is linear, our formulae become very simple. In particular, $S(t, z)$ takes the form

$$S(t, z) = M(t)z + N(t)$$

and (4) reduces to a system of ordinary differential equations for $M(t)$ and $N(t)$. Formulae (5) also have simplified form, accordingly.

4°. We assumed in 2° that every eigenvalue of A lies either in the right or in the left half-plane. In the case of linear system this assumption implies that every particular solution either increases or decreases not slowlier as the exponential function. If A has purely imaginary eigenvalues, the situation becomes much mroe complicated. A final result has been derived for linear systems in particular for the case if the point $t = \infty$ is a regular singularity of the given linear system.

5°. We mention some further results related to the problems treated 2°-4°.

(1) We consider the system

$$y'' + P(t)y = 0, \qquad y = \left\| \begin{matrix} y_1 \\ \vdots \\ y_n \end{matrix} \right\|$$

on the interval (α, β) where $\alpha < 0 < \beta$ and $P(t)$ has the following expansion at $t = 0$

$$P(t) = \frac{1}{t}(P_0 + tP_1 + t^2P_2 + \ldots).$$

One has to separate the set of solutions admitting analytic continuation into the lower half-plane. This can be done by transferring the boundary conditions from α to β along the real axis. At $t = 0$ the transfer can be carried out by using some special expansions having same convenient properties as the series used in 2°.

(2) Instead of solving systems of the form (2), some problems require to find homogeneous functionals of the form

$$\left(\int_0^T y^*(t)Q(t)y(t)\,dt\right)\left(\int_0^T y^*(t)R(t)y(t)\,dt\right)^{-1}$$

where $Q(t)$ and $R(t)$ are given quadratic matrices. There exists a method for computing such functionals without determining $y(t)$. This leads to a Cauchy problem for some auxiliary system of differential equations and stable in that very direction in which the equations has to be solved.

(3) The algebraic problem we met in 2°, occurs in various numerical problems. Given a quadratic matrix A without pure imaginary eigenvalues. We are required to decompose it into quasi-diagonal form

$$\left\| \begin{matrix} A_+ & 0 \\ 0 & A_- \end{matrix} \right\|$$

where the eigenvalues of A_+ and A_- are in the right and left half-plane, respectively.

This is possible without determining the individual eigenvalues and eigenvectors, the Jordan structure of A, without not using complex numbers if A is real.

6°. The authors applied the above methods to many problems of quantum mechanics, radiophysics and hydromechanics. The results proved to be satisfactory.

REFERENCES

[1] A. A. Abramov, On transferring boundedness condition for some system of ordinary linear differential equations. *Ž. V. Mat. i Mat. Fiz.*, 1 (1961), 733-737 (in Russian).

[2] A. A. Abramov, On boundary values in a singular point for linear ordinary differential equations. *Ž. V. Mat. i Mat. Fiz.*, 11 (1971), 275-278 (in Russian).

[3] A. A. Abramov – B. A. Tareev – V. I. Ulyanova, Baroclynic instability of the frontal double-layer model of Kočina on the β-plane. *Izv. AN SSSR, Physics of Atmosphere and Ocean,* 8, 2 (1972), 131-141 (in Russian).

[4] A. A. Abramov – B. A. Tareev – V. I. Ulyanova, Long waves and baroclynic instability on inclined boundary surface, in: *Proc. Soviet-French Symp. "Inner waves in the Ocean", Novosibirsk,* (1972), 244-257 (in Russian).

[5] A. A. Abramov – B. A. Tareev – V. I. Ulyanova, Instability of double-layer geostrophic flow with antisymmetric velocity-shave in the upper layer, *Izv. A.N. SSSR, Physics of Atmosphere and Ocean,* 10 (1972), 1017-1028 (in Russian).

[6] A. A. Abramov – V. I. Ulyanova, On the solution of equations for energy levels of ionized Hydrogene molecula. *Ž. V. Mat. i Mat. Fiz.*, 1 (1961), 351-354 (in Russian).

[7] A. A. Abramov – V. I. Ulyanova, Computation of energy levels for systems of two kernels and one electron. *T.E.Kh.*, 6 (1970), 384-386.

[8] E. S. Birger, Error estimation for transfer of boundedness condition for linear differential equation on infinite interval. *Ž. V. Mat. i Mat. Fiz.*, 8 (1968), 674-678 (in Russian).

[9] E. S. Birger, On stable computation of some functionals of eigenfunctions of the Sturm – Liouville problem on an infinite interval. *Ž. V. Mat. i Mat. Fiz.*, 8 (1968), 1126-1133 (in Russian).

[10] E.S. Birger, Computation of functionals of eigenfunctions of boundary value problems for systems of linear ordinary differential equations. *Ž. V. Mat. i Mat. Fiz.*, 13 (1973), 227-233 (in Russian).

[11-12] E.S. Birger – N.B. Lyalikova, Solutions with a given condition at infinity for certain systems of ordinary differential equations I, II. *Ž. V. Mat. i Mat. Fiz.*, 5 (1965), 979-990, 6 (1966), 446-453 (in Russian).

[13] E.S. Birger – N.B. Konyukhova, On molecula calculations by approximation with one electron and one centre. *T.E.Kh.*, 4 (1968), 29-36 (in Russian).

[14] E.S. Birger – N.B. Konyukhova, Numerical computation of radiowaves in vertically-inhomogeneous troposphere. *Radiotekhnika i Elektronika,* 14 (1969), 1147-1156 (in Russian).

[15] E.S. Birger – B.O. Kerbikov – N.B. Konyukhova – I.S. Shapiro, On connected quasi-nuclear states of the system $2N\,2\bar{N}$. *Yadernaya Fiz.*, 17 (1973), 178-185 (in Russian).

[16] N.B. Konyukhova, Numerical segregation of solutions which converge to zero at infinity for some two-dimensional non-linear systems of ordinary differential equations. *Ž. V. Mat. i Mat. Fiz.*, 10 (1970), 74-87 (in Russian).

[17] N.B. Konyukhova, Behaviour inside and outside of a stable manifold of solutions of some two-dimensional non-linear systems of ordinary differential equations. *Mat. Zametki,* 8 (1970), 285-295 (in Russian).

[18] N.B. Konyukhova, On solution of boundary value problems in an infinite interval for some non-linear systems of ordinary differential equations with a singularity. *Ž. V. Mat. i Mat. Fiz.*, 10 (1970), 1150-1163 (in Russian).

[19] N.B. Konyukhova, Segregation of stable manifolds for some non-linear systems of ordinary differential equations with a singularity. *Ž. V. Mat. i Mat. Fiz.*, 13 (1973), 609-626 (in Russian).

[20] V.I. Ulyanova, Transfer of boundary conditions through some singular points. *Ž.V. Mat. i Mat. Fiz.*, 12 (1972), 528-532 (in Russian).

[21] N.B. Konyukhova, On iterative solution of nonlinear boundary value problems segregating small solutions of some systems of ordinary differential equations with a singularity. *Ž.V. Mat. i Mat. Fiz.*, 14 (1974), 1221-1231 (in Russian).

[22] E.S. Birger – N.B. Konyukhova, On stable computation of eigenvalues and homogeneous functionals of eigenfunctions of some singular Sturm – Liouville problems, occurring in nuclear physics. OIYAI, D10-7707, Dubna, (1974), 253-259 (in Russian).

[23] V.I. Ulyanova, Method of transfer of boundary conditions through some singular points and of it application to solution of some physical problems. OIYAI, D10-7707, Dubna (1974), 260-165.

[24] A.M. Badalyan – E.S. Birger – N.B. Konyukhova, The search of resonance in a system of three neutrons, *Yadernaya Fiz.*, 20 (1974), 1147-1157 (in Russian).

[25] N.A. Voronov – I.Yu. Kobzarev – N.B. Konyukhova, On a possibility of existence of the mesons of new type. *Letters in Ž.E.T. Fiz.*, 22 (1975), 590-594 (in Russian).

[26] E.S. Birger – L.A. Wainstein – N.B. Konyukhova, Diffraction of a wave bunch on a plasma cylinder. *Ž.V. Mat. i Mat. Fiz.*, 16, 6 (1976) (in Russian).

A.A. Abramov – E.S. Birger – N.B. Konyukhova – V.I. Ulyanova

Acad. Sci. USSR, Computing Centre, Moscow, USSR.

COLLOQUIA MATHEMATICA SOCIETATIS JÁNOS BOLYAI
15. DIFFERENTIAL EQUATIONS, KESZTHELY (HUNGARY), 1975.

CONSTRUCTION OF PERIODIC SOLUTIONS FOR THE EULER – POISSON EQUATIONS BY MEANS OF POWER SERIES EXPANSION CONTAINING A SMALL PARAMETER

YU.A. ARKHANGELSKII

It is well known that the motion of a rigid body about a fixed point can be described by the nonlinear differential equations of Euler and Poisson

$$
\begin{aligned}
&A\frac{dp}{dt} + (C-B)qr = \mathrm{Mg}\,(y^0\gamma'' - z^0\gamma'),\\
&B\frac{dq}{dt} + (A-C)pr = \mathrm{Mg}\,(z^0\gamma - x^0\gamma''),\\
&C\frac{dr}{dt} + (B-A)pq = \mathrm{Mg}\,(x^0\gamma' - y^0\gamma),\\
&\frac{d\gamma}{dt} = r\gamma' - q\gamma'', \quad \frac{d\gamma'}{dt} = p\gamma'' - r\gamma, \quad \frac{d\gamma''}{dt} = q\gamma - p\gamma',
\end{aligned}
\tag{1.1}
$$

possessing three first integrals

$$
\begin{aligned}
Ap^2 + Bq^2 + Cr^2 - 2\,\mathrm{Mg}\,(x^0\gamma + y^0\gamma' + z^0\gamma'') =\\
= Ap_0^2 + Bq_0^2 + Cr_0^2 - 2\,\mathrm{Mg}\,(x^0\gamma_0 + y^0\gamma_0' + z^0\gamma_0'')
\end{aligned}
\tag{1.2}
$$

$$Ap\gamma + Bq\gamma' + Cr\gamma'' = Ap_0\gamma_0 + Bq_0\gamma_0' + Cr_0\gamma_0'',$$

$$\gamma^2 + \gamma'^2 + \gamma''^2 = 1.$$

Here $p, q, r, \gamma, \gamma', \gamma''$ denote the unknown functions; A, B, C, x^0, y^0, z^0 are constant parameters characterizing the mass-distribution within the rigid body; $p_0, q_0, r_0, \gamma_0, \gamma_0', \gamma_0''$ are initial values of the corresponding variables.

One can show that in order to obtain the general solution of (1.1) it would be sufficient to find one further time-independent first integral. However, this fourth integral has not been found yet in full generality, despite all the efforts by many outstanding mathematicians during the last twohundred years. We know this integral only in three special cases when the constant parameters satisfy certain relations. These are the cases of Euler, Lagrange and Kovalevskaya; for them we really have the general solution of (1.1). Furthermore, particular solutions by quadratures can be obtained under certain restrictions on the initial values $p_0, q_0, r_0, \gamma_0, \gamma_0', \gamma_0''$. Some other particular solutions of (1.1) have been found in the form of power series involving a small parameter, cf. [1]-[4]. In [5] there were investigated some periodic solutions, near to the initial values of the parameters p_0, q_0, r_0.

By the mere amount of scholarship sacrificed to it and the plenty of results during the last two centuries it is obvious that the system of Euler – Poisson is a unique phenomenon among nonlinear differential equations. Therefore every new step forward demands special efforts and deserves more and more interest as we are getting nearer to the final solution of the Euler – Poisson system.

In the present paper we construct new exact periodic solutions in terms of power series of a small parameter, assuming that the rigid body has large angular speed r_0 at the beginning.

We make the following assumptions

(1.3) $\gamma_0 \geqslant 0, \quad 0 < \gamma_0'' < 1.$

Let us introduce the notations

$$a = \frac{A}{C}, \quad b = \frac{B}{C}, \quad c^2 = \frac{\mathrm{Mg}\, l}{C}, \quad \mu = c\,\frac{\sqrt{\gamma_0''}}{r_0}$$

(1.4)

$$x^0 = l x_1^0, \quad y^0 = l y_1^0, \quad z^0 = l z_1^0, \quad l^2 = x^{02} + y^{02} + z^{02}.$$

Then μ will be a small parameter in respect to r_0. We introduce the variables without dimension $p_1, q_1, r_1, \gamma_1, \gamma_1', \gamma_1'', \tau$.

$$p = c\sqrt{\gamma_0''}\, p_1, \quad q = c\sqrt{\gamma_0''}\, q_1, \quad r = r_0 r_1$$

(1.5)

$$\gamma = \gamma_0'' \gamma_1, \quad \gamma' = \gamma_0'' \gamma_1', \quad \gamma'' = \gamma_0'' \gamma_1'', \quad t = \frac{\tau}{r_0}.$$

Equations (1.1) and their integrals (1.2) obtain the following forms

$$\dot{p}_1 + A_1 q_1 r_1 = \mu a^{-1}(y_1^0 \gamma_1'' - z_1^0 \gamma_1'), \quad \dot{\gamma}_1 = r_1 \gamma_1' - \mu_1 \gamma_1''$$

$$\dot{q}_1 + B_1 p_1 r_1 = \mu b^{-1}(z_1^0 \gamma_1 - x_1^0 \gamma_1''), \quad \dot{\gamma}_1' = \mu p_1 \gamma_1'' - r_1 \gamma_1$$

(1.6)

$$\dot{r}_1 = \mu^2(- C_1 p_1 q_1 + x_1^0 \gamma_1' - y_1^0 \gamma_1), \quad \dot{\gamma}_1'' = \mu(q_1 \gamma_1 - p_1 \gamma_1')$$

where $\dot{u} = \dfrac{du}{d\tau}$, $A_1 = \dfrac{C - B}{A}$, $B_1 = \dfrac{A - C}{B}$, $C_1 = \dfrac{B - A}{C}$.

(1.7)

$$r_1^2 = 1 + \mu^2 S_1; \ \{S_1 = a(p_{10}^2 - p_1^2) + b(q_{10}^2 - q_1^2) -$$

$$- 2[x_1^0(\gamma_{10} - \gamma_1) + y_1^0(\gamma_{10}' - \gamma_1') + z_1^0(1 - \gamma_1'')\}$$

(1.8) $$r_1 \gamma_1'' = 1 + \mu S_2; \ \{S_2 = a(p_{10}\gamma_{10} - p_1\gamma_1) + b(q_{10}\gamma_{10}' - q_1\gamma_1')\}$$

(1.9) $$\gamma_1^2 + \gamma_1'^2 + \gamma_1''^2 = (\gamma_0'')^{-2}.$$

We may use the first integrals (1.7) and (1.8) to express the variables r_1 and γ_1'' through the remaining variables $p_1, q_1, \gamma_1, \gamma_1'$, their initial values $p_{10}, q_{10}, \gamma_{10}, \gamma_{10}'$ and the small parameter μ

$$r_1 = 1 + \frac{1}{2}\mu^2[S_1 + 2z_1^0(1 - \gamma_1'')] + \dots$$

(1.10)

$$\gamma_1'' = 1 + \mu S_2 - \frac{1}{2}\mu^2[S_1 + 2z_1^0(1 - \gamma_1'')] + \dots .$$

Let us reduce the remaining four equations of (1.6) to two equations of second order

(1.11)
$$\begin{aligned}\ddot{p}_1 + \omega^2 p_1 &= \mu[z_1^0(a^{-1} - A_1 b^{-1})\gamma_1 + A_1 b^{-1} x_1^0] + \\ &+ \mu^2\{-\omega^2 p_1[S_1 + 2z_1^0(1-\gamma_1'')] + \\ &+ A_1 b^{-1} x_1^0 \cdot S_2 + A_1 C_1 p_1 q_1^2 - A_1 x_1^0 q_1 \gamma_1' - \\ &- y_1^0 a^{-1} p_1 \gamma_1' + y_1^0(A_1 + a^{-1}) q_1 \gamma_1 - z_1^0 a^{-1} p_1\} + \\ &+ \mu^3 z_1^0\Big\{\frac{1}{2}(a^{-1} - b^{-1}A_1)[S_1 + 2z_1^0(1-\gamma_1'')]\gamma_1 - \\ &- (2\omega^2 + a^{-1}) p_1 S_2\Big\} + \dots,\end{aligned}$$

(1.12)
$$\begin{aligned}\ddot{\gamma}_1 + \gamma_1 &= \mu(1+B_1)p_1 + \mu^2\{-\gamma_1[S_1 + 2z_1^0(1-\gamma_1'')] + \\ &+ (1+B_1)p_1 S_2 + (1+C_1)p_1 q_1 \gamma_1' + x_1^0 \gamma_1'^2 - y_1^0 \gamma_1 \gamma_1' - \\ &- b^{-1} z_1^0 \gamma_1 + b^{-1} x_1^0 - q_1^2 \gamma_1\} + \mu^3[-z_1^0(2+b^{-1})\gamma_1 S_2 + \\ &+ 2b^{-1} x_1^0 S_2] + \dots,\end{aligned}$$

(1.13)
$$\omega^2 = -A_1 B_1 = \frac{(a-1)(b-1)}{ab} = \frac{(A-C)(B-C)}{AB}.$$

From the first and the fourth equations of (1.6) we express the variables q_1 and γ_1' explicitly

(1.14)
$$\begin{aligned}q_1 &= (A_1 r_1)^{-1}[-\dot{p}_1 + \mu a^{-1}(y_1^0 \gamma_1'' - z_1^0 \gamma_1')], \\ \gamma_1' &= (r_1)^{-1}[\dot{\gamma}_1 + \mu q_1 \gamma_1'']\end{aligned}$$

where r_1 and γ_1'' are replaced by their expressions (1.10). Further, we put the expressions (1.14) into the right hand side of the equations (1.11) and (1.12). Then we have a quasilinear autonomous system with degree of freedom 2 whose right hand sides depend on $p_1, \dot{p}_1, \gamma_1, \dot{\gamma}_1, p_{10}, \dot{p}_{10}, \gamma_{10}, \dot{\gamma}_{10}$.

We shall look for periodical solutions of this system, assuming that $A > B > C$ or $A < B < C$ (ω^2 is a positive number).

Let p_2 and γ_2 be new variables defined by

(1.15) $$p_1 = p_2 + \mu\kappa + \mu\kappa_1\gamma_2,$$

(1.16) $$\gamma_1 = \gamma_2 + \mu\nu p_2,$$

(1.17) $$\kappa = \frac{x_1^0 A_1}{b\omega^2}, \quad \kappa_1 = -\frac{z_1^0}{1-\omega^2}\left(\frac{1}{a} - \frac{A_1}{b}\right), \quad \nu = \frac{1-B_1}{1-\omega^2}.$$

Applying (1.15), (1.16), (1.10) and (1.14) one can get the following expansion according to powers of μ

$$S_i = S_{i1} + 2^{2-i}\mu S_{i2} + \ldots \quad (i = 1, 2),$$

(1.18) $$r_1 = 1 + \frac{1}{2}\mu^2 S_{11} + \ldots,$$

$$\gamma_1'' = 1 + \mu S_{21} + \mu^2\left(S_{22} - \frac{1}{2} S_{11}\right) + \ldots,$$

$$q_1 = -A_1^{-1}\dot{p}_2 + \mu A_1^{-1}(y_1^0 a^{-1} - \kappa_2\dot{\gamma}_2) + \ldots,$$

(1.19) $$\gamma_1' = \dot{\gamma}_2 + \mu\nu_2\dot{p}_2 + \ldots,$$

$$\kappa_2 = \kappa_1 + a^{-1}z_1^0, \quad \nu_2 = \nu - A_1^{-1},$$

where

$$S_{11} = a(p_{20}^2 - p_2^2) + \frac{b(p_{20}^2 - p_2^2)}{A_1^2} -$$
$$- 2[x_1^0(\gamma_{20} - \gamma_2) + y_1^0(\dot{\gamma}_{20} - \dot{\gamma}_2)],$$

$$S_{12} = a[\kappa(p_{20} - p_2) + \kappa_1(p_{20}\gamma_{20} - p_2\gamma_2)] -$$

(1.20) $$- bA_1^{-2}[y_1^0 a^{-1}(\dot{p}_{20} - \dot{p}_2) - \kappa_2(\dot{\gamma}_{20}\dot{p}_{20} - \dot{\gamma}_2\dot{p}_2)] -$$
$$- \kappa_1^0\nu(p_{20} - p_2) - y_1^0\nu_2(\dot{p}_{20} - \dot{p}_2) + z_1^0 S_{21},$$

$$S_{21} = a(p_{20}\gamma_{20} - p_2\gamma_2) - bA_1^{-1}(\dot{p}_{20}\dot{\gamma}_{20} - \dot{p}_2\dot{\gamma}_2),$$

$$S_{22} = a[\nu(p_{20}^2 - p_2^2) + \kappa(\gamma_{20} - \gamma_2) + \kappa_1(\gamma_{20}^2 - \gamma_2^2) +$$
$$+ bA_1^{-1}[-\nu_2(\dot{p}_{20}^2 - \dot{p}_2^2) + a^{-1}y_1^0(\dot{\gamma}_{20} - \dot{\gamma}_2) - \kappa_2(\dot{\gamma}_{20}^2 - \dot{\gamma}_2^2)].$$

Putting (1.15), (1.16) and (1.18) into equations (1.11) and (1.12) we get

$$\ddot{p}_2 + \omega^2 p_2 = \mu^2 F(p_2, \dot{p}_2, \gamma_2, \dot{\gamma}_2, \mu), \qquad F = F_2 + \mu F_3 + \dots,$$

$$(1.21) \qquad \ddot{\gamma}_2 + \gamma_2 = \mu^2 \Phi(p_2, \dot{p}_2, \gamma_2, \dot{\gamma}_2, \mu), \qquad \Phi = \Phi_2 + \mu\Phi_3 + \dots,$$

$$F_2 = f_2 - \nu\kappa_1(1-\omega^2)p_2,$$

$$F_3 = f_3 - \kappa_1\varphi_2 - \nu\kappa_1\kappa(1-\omega^2) - \nu\kappa_1^2(1-\omega^2)\gamma_2,$$

$$(1.22) \qquad \Phi_2 = \varphi_2 + \nu\kappa(1-\omega^2) + \nu\kappa_1(1-\omega^2)\gamma_2,$$

$$\Phi_3 = \varphi_3 - \nu f_2 + \nu^2\kappa_1(1-\omega^2)p_2,$$

where

$$(1.23) \qquad f_2 = -\omega^2 p_2 S_{11} + A_1 b^{-1} x_1^0 S_{21} + C_1 A_1^{-1} p_2 p_2^{-2} + x_1^0 p_2^0 \dot{\gamma}_2 -$$

$$- y_1^0 a^{-1} p_2 \dot{\gamma}_2 - y_1^0 A_1^{-1}(A_1 + a^{-1})\gamma_2 \dot{p}_2 - z_1^0 a^{-1} p_2$$

$$f_3 = -\omega^2(\kappa S_{11} + \kappa_1\gamma_2 S_{11} + 2p_2 S_{12}) + A_1 b^{-1} x_1^0 S_{22} +$$

$$+ C_1 A_1^{-1}[\kappa\dot{p}_2^2 + \kappa_1\gamma_2\dot{p}_2^2 - 2p_2\dot{p}_2(y_1^0 a^{-1} - \kappa_2\dot{\gamma}_2)] -$$

$$- x_1^0[-\nu_2\dot{p}_2^2 + \dot{\gamma}_2(y_1^0 a^{-1} - \kappa_2\dot{\gamma}_2)] - y_1^0 a^{-1}[\dot{\gamma}_2(\kappa + \kappa_1\gamma_2) +$$

$$+ \nu_2 p_2 \dot{p}_2] + y_1^0 A_1^{-1}(A_1 + a^{-1})[\gamma_2(y_1^0 a^{-1} - \kappa_2\dot{\gamma}_2) -$$

$$- \nu p_2 \dot{p}_2] - z_1^0 a^{-1}(\kappa + \kappa_1\gamma_2) + \frac{1}{2} z_1^0(a^{-1} + A_1 b^{-1})\gamma_2 S_{11} +$$

$$+ z_1^0 a^{-1} p_2 S_{21},$$

$$\varphi_2 = -\gamma_2 S_{11} + (1 + B_1)p_2 S_{21} - (1 - C_1)A_1^{-1} p_2 \dot{p}_2 \dot{\gamma}_2 +$$

$$+ x_1^0 \dot{\gamma}_2^2 - y_1^0 \gamma_2 \dot{\gamma}_2 - z_1^0 b^{-1}\gamma_2 + x_1^0 b^{-1} - A_1^{-2}\gamma_2\dot{p}_2^2,$$

$$\varphi_3 = -\nu p_2 S_{11} - 2\gamma_2 S_{21} + (1 + B_1)p_2 S_{22} +$$

$$+ (1 + B_1)(\kappa + \kappa_1\gamma_2)S_{21} +$$

$$+ (1 + C_1)A_1^{-1}[-(\kappa + \kappa_1\gamma_2)\dot{p}_2\dot{\gamma}_2 -$$

$$- \nu_2 p_2 \dot{p}_2^2 + p_2\dot{\gamma}_2(y_1^0 a^{-1} - \kappa_2\dot{\gamma}_2)] + 2x_1^0\nu_2\dot{p}_2\dot{\gamma}_2 -$$

$$- y_1^0(\nu p_2\dot\gamma_2 + \nu_2\gamma_2\dot p_2) - z_1^0 b^{-1}\nu p_2 - z_1^0 b^{-1}\gamma_2 S_{21} +$$

$$+ 2x_1^0 b^{-1} S_{21} - A_1^{-2}[- 2\gamma_2\dot p_2(y_1^0 a^{-1} - \kappa_2\dot\gamma_2) + \nu p_2\dot p_2^2].$$

By (1.9) this system has the first integral

$$\gamma_2^2 + \dot\gamma_2^2 + 2\mu(\nu p_2\gamma_2 + \nu_2\dot p_2\dot\gamma_2 + S_{21}) + \mu^2(\ldots) = (\gamma_0'')^{-2} - 1. \tag{1.24}$$

Let us look for periodic solutions $p_2(\tau, \mu), \dot p_2(\tau, \mu), \gamma_2(\tau, \mu), \dot\gamma_2(\tau, \mu)$ of (1.21) satisfying the condition

$$\dot\gamma_2(0, \mu) = 0. \tag{1.25}$$

Since the system (1.21) is autonomous this condition does not restrict the generality of the solution to be obtained (cf. [6]).

Since the generating systems

$$\ddot p_2 + \omega^2 p_2 = 0, \qquad \ddot\gamma_2 + \gamma_2 = 0 \tag{1.26}$$

have frequencies ω and 1 respectively, there are three possibilities for periodic solutions of (1.21):

1°. The frequencies are different and commensurable (i.e. $\omega = \frac{m}{n}$, where m and n are integers, relatively-prime to each other).

2°. The frequencies are equal ($\omega = 1$);

3°. The frequencies are not commensurable (ω irrational).

2. At first let us consider the case $\frac{m}{n}$. Now the generating system (1.26) admits of periodic solutions with period $T_0 = 2\pi n$.

$$p_2^{(0)} = M_1 \cos\omega\tau + M_2 \sin\omega\tau, \quad \gamma_2^{(0)} = M_3 \cos\tau. \tag{2.1}$$

We will assume that the original autonomous system (1.21) has periodic solutions with periods $T_0 + \alpha$, reducing to the generating solution (2.1) for $\mu = 0$. We express the initial conditions by means of the relations

$$\begin{aligned} &p_2(0, \mu) = M_1 + \beta_1, \quad \dot p_2(0, \mu) = \omega(M_2 + \beta_2) \\ &\gamma_2(0, \mu) = M_3 + \beta_3, \quad \dot\gamma_2(0, \mu) = 0. \end{aligned} \tag{2.2}$$

Let us define the operator

$$U = u + \frac{\partial u}{\partial M_1}\beta_1 + \frac{\partial u}{\partial M_2}\beta_2 + \frac{\partial u}{\partial M_3}\beta_3 + \frac{1}{2}\frac{\partial^2 u}{\partial M_1^2}\beta_1^2 + \ldots,$$

$$\begin{pmatrix} U = G_k, H_k \\ u = g_k, \ h_k \end{pmatrix}$$

and look for a periodical solution in the following form [7]

$$(2.3) \qquad \begin{aligned} p_2(\tau, \mu) &= (M_1 + \beta_1)\cos\omega\tau + (M_2 + \beta_2)\sin\omega\tau + \sum_{k=2}^{\infty} G_k(\tau)\mu^k \\ \gamma_2(\tau, \mu) &= (M_3 + \beta_3)\cos\tau + \sum_{k=2}^{\infty} H_k(\tau)\mu^k \end{aligned}$$

$$g_k(\tau) = \frac{1}{\omega_0}\int_0^\tau F'_k(t_1)\sin\omega(\tau - t_1)\,dt_1,$$

$$(2.4) \qquad \begin{aligned} h_k(\tau) &= \int_0^\tau \Phi'_k(t_1)\sin(\tau - t_1)\,dt_1, \\ F'_k(\tau) &= \frac{1}{(k-2)!}\left(\frac{d^{k-2}F}{d\mu^{k-2}}\right)_{\beta=\mu=0}, \\ \Phi'_k(\tau) &= \frac{1}{(k-2)!}\left(\frac{d^{k-2}\Phi}{d\mu^{k-2}}\right)_{\beta=\mu=0}. \end{aligned}$$

Here $\beta_1, \omega\beta_2, \beta_3$ denote the deviations of the initial values of $p_2, \dot{p}_2, \gamma_2$ of the periodic solutions of (1.21) we look for, from the initial values of these same quantities in the generating solution (2.1); these deviations are functions of μ vanishing at $\mu = 0$.

We observe that the right hand sides of the system (1.21) begin with terms of order μ^2 and therefore we have

$$(2.5) \qquad \begin{aligned} F'_k(\tau) &= F_k(p_2^{(0)}, \dot{p}_2^{(0)}, \gamma_2^{(0)}, \dot{\gamma}_2^{(0)}) \equiv F_k^{(0)} \qquad (k = 2, 3) \\ \Phi'_k(\tau) &= \Phi_k(p_2^{(0)}, \dot{p}_2^{(0)}, \gamma_2^{(0)}, \dot{\gamma}_2^{(0)}) \equiv \Phi_k^{(0)}. \end{aligned}$$

Let us find expression for the functions $F_2^{(0)}$ and $\Phi_2^{(0)}$. Put

$$E = \sqrt{M_1^2 + M_2^2}, \quad \cos \epsilon = \frac{M_1}{E}, \quad \sin \epsilon = \frac{M_2}{E}. \tag{2.6}$$

Then formula (2.1) can be written as

$$p_2^{(0)}(\tau) = E \cos(\omega\tau - \epsilon), \quad \gamma_2^{(0)}(\tau) = M_3 \cos \tau. \tag{2.7}$$

By applying (2.7) for the formulae

$$S_{11}^{(0)} = S_{11}(p_2^{(0)}, \dot{p}_2^{(0)}, \gamma_2^{(0)}, \dot{\gamma}_2^{(0)}),$$
$$S_{21}^{(0)} = S_{21}(p_2^{(0)}, \dot{p}_2^{(0)}, \gamma_2^{(0)}, \dot{\gamma}_2^{(0)})$$

we get from (1.20)

$$\begin{aligned} S_{11}^{(0)} &= E^2\Big\{\Big[a\Big(\cos^2 \epsilon - \frac{1}{2}\Big) + b\omega^2 A_1^{-2}\Big(\sin^2 \epsilon - \frac{1}{2}\Big)\Big] + \\ &+ \frac{1}{2}(b\omega^2 A_1^{-2} - a)\cos 2(\omega\tau - \epsilon)\Big\} - \\ &- 2M_3[x_1^0(1 - \cos\tau) + y_1^0 \sin\tau] \\ S_{21}^{(0)} &= M_3 E\Big\{a \cos\epsilon + \frac{1}{2}(b\omega A_1^{-1} - a)\cos[(\omega - 1)\tau - \epsilon] - \\ &- \frac{1}{2}(b\omega A_1^{-1} + a)\cos[(\omega + 1)\tau - \epsilon]\Big\}. \end{aligned} \tag{2.8}$$

Now we substitute expressions (2.7) and (2.8) into (1.22) and (1.23), respectively, and get the following expressions for all values of ω which were taken into consideration (except $\omega = \frac{1}{2}$):

$$\begin{aligned} F_2^{(0)} &= M_1 L(\omega) \cos \omega\tau + M_2 L(\omega) \sin \omega\tau + \dots, \\ \Phi_2^{(0)} &= M_3 N(\omega) \cos \tau + \dots \end{aligned} \tag{2.9}$$

where

$$\begin{aligned} L(\omega) &= \omega^2\Big[-(aM_1^2 + b\omega^2 A_1^{-2} M_2^2) + \\ &+ \frac{1}{4}(M_1^2 + M_2^2)(C_1 A_1^{-1} + 3a + b\omega^2 A_1^{-2})\Big] + 2M_3\omega^2 x_1^0 - \\ &- [z_1^0 a^{-1} + \kappa_1 \nu(1 - \omega^2)], \\ N(\omega) &= -(aM_1^2 + b\omega^2 A_1^{-2} M_2^2) + 2M_3 x_1^0 - 2z_1^0. \end{aligned} \tag{2.10}$$

From (2.4) and (2.5) we get

$$
(2.11)\qquad
\begin{aligned}
g_k(T_0) &= -\frac{1}{\omega}\int_0^{T_0} F_k^{(0)}(t_1)\sin\omega t_1\,dt_1,\\
\dot{g}_k(T_0) &= \int_0^{T_0} F_k^{(0)}(t_1)\cos\omega t_1\,dt_1\\
h_k(T_0) &= -\int_0^{T_0} \Phi_k^{(0)}(t_1)\sin t_1\,dt_1,\\
\dot{h}_k(T_0) &= \int_0^{T_0} \Phi_k^{(0)}(t_1)\cos t_1\,dt_1, \qquad \begin{pmatrix} T_0 = 2\pi n\\ k = 2, 3\end{pmatrix}.
\end{aligned}
$$

Hence by (2.9) we will obtain

$$
(2.12)\qquad
\begin{aligned}
&g_2(T_0) = -\pi n\omega^{-1}M_2L(\omega), \qquad \dot{g}_2(T_0) = \pi nM_1L(\omega)\\
&h_2(T_0) = 0, \qquad \dot{h}_2(T_0) = \pi nM_3N(\omega).
\end{aligned}
$$

We have to determine the constants $M_1, \omega M_2, M_3$ representing the initial values of the generating solutions (2.1) and also the deviations $\beta_1(\mu), \omega\beta_2(\mu), \beta_3(\mu)$ and the correction of the period α. To this end we consider the periodicity conditions on the solutions and their first derivatives:

$$
(2.13)\qquad
\begin{aligned}
\Psi_1 &= p_2(T_0+\alpha,\mu) - p_2(0,\mu) = 0,\\
\Psi_2 &= \dot{p}_2(T_0+\alpha,\mu) - \dot{p}_2(0,\mu) = 0,\\
\Psi_3 &= \gamma_2(T_0+\alpha,\mu) - \gamma_2(0,\mu) = 0,\\
\Psi_4 &= \dot{\gamma}_2(T_0+\alpha,\mu) - \dot{\gamma}_2(0,\mu) = 0.
\end{aligned}
$$

We observe that since the system (1.21) has the first integral (1.24) the condition $\Psi_3 = 0$ is not independent [8]. In fact, writing (1.24) in the form

$$
\begin{aligned}
&\gamma_2^2(T_0+\alpha,\mu) + \dot{\gamma}_2^2(T_0+\alpha,\mu) + \mu(\ldots) =\\
&\qquad = \gamma_2^2(0,\mu) + \dot{\gamma}_2^2(0,\mu) + \mu(\ldots)
\end{aligned}
$$

and using the condition (2.2) we get from (2.13)

(2.14) $$2(M_3 + \beta_3)\Psi_3 + \Psi_3^2 + \mu\varphi_1(\Psi_1, \Psi_2, \Psi_3, \Psi_4, \mu) = 0$$

where φ_1 is an entire function of all of its variables, and $\varphi_1(0,0,0,0,\mu) = 0$. If $M_3 \neq 0$, and we will later show this is always the case, it follows from (2.14) that $\Psi_3 = f(\Psi_1, \Psi_2, \Psi_4, \mu)$ where f is an entire function of all of its arguments and $f(0, 0, 0, \mu) = 0$. Then it follows immediately that condition $\Psi_3 = 0$ in (2.13) is a consequence of the remaining ones,

(2.15) $$\Psi_1 = \Psi_2 = \Psi_4 = 0.$$

Substituting the initial values (2.2) into the integral (1.24) evaluated at $\tau = 0$ we can determine M_3 and β_3 from the equations

$$M_3^2 + 2M_3\beta_3 + \beta_3^2 + 2\mu\nu M_3(M_1 + \beta_1) + \ldots = (\gamma_0'')^{-2} - 1.$$

Supposing that γ_0'' does not depend on μ we get

(2.16) $$M_3^2 = (\gamma_0'')^{-2} - 1, \quad \beta_3^2 + 2M_3\beta_3 + 2\mu\nu M_3(M_1 + \beta_1) + \ldots = 0.$$

From the equations (2.16) and condition (1.3) it follows that

(2.17) $$0 < M_3 = (1 - \gamma_0''^2)^{\frac{1}{2}}(\gamma_0'')^{-1} < \infty, \quad \beta_3 = -\mu\nu(M_1 + \beta_1) + \ldots$$

and since γ_0'' is an arbitrary parameter, M_3 may be any positive constant.

This means that a periodic solution (2.3) depends on an arbitrary constant M_3 and on a function $\beta_3(\mu)$ vanishing at $\mu = 0$. We observe that this is independent of the choice of ω and holds for every case.

Expanding the right hand sides of the equations (2.13) according to powers of α and writing out the terms of order not larger than one (and consequently also omitting the terms $\mu^2\alpha$) we obtain independent conditions for the periodicity of (2.15):

$$p_2(T_0, \mu) - M_1 - \beta_1 + \alpha\omega(M_2 + \beta_2) = 0$$

(2.18) $$\dot{p}_2(T_0, \mu) - \omega(M_2 + \beta_2) - \alpha\omega^2(M_1 + \beta_1) = 0$$

$$\dot{\gamma}_2(T_0, \mu) - \alpha(M_3 + \beta_3) = 0.$$

From the last equation of (2.18), from (2.17) and (2.3) we define the function

$$(2.19) \qquad \alpha = \frac{\mu^2[\dot{H}_2(T_0) + \mu\dot{H}_3(T_0) + \ldots]}{M_3 + \beta_3}, \qquad (T_0 = 2\pi n).$$

It follows then that by omitting the terms of order α^2 and $\mu^2\alpha$ in (2.18) we also omit the terms of order μ^4.

Applying (2.19) and (2.3) for two equations of (2.18) and reducing by μ^2 we obtain for β_1 and β_2 the system

$$(2.20) \qquad \begin{gathered} G_2(T_0) + \mu G_3(T_0) + \omega(M_2 + \beta_2)\frac{[\dot{H}_2(T_0) + \mu\dot{H}_3(T_0) + \ldots]}{M_3 + \beta_3} + \\ + \mu^2(\ldots) = 0, \\ \dot{G}_2(T_0) + \mu\dot{G}_3(T_0) - \omega^2(M_1 + \beta_1)\frac{[\dot{H}_2(T_0) + \mu\dot{H}_3(T_0) + \ldots]}{M_3 + \beta_3} + \\ + \mu^2(\ldots) = 0. \end{gathered}$$

By virtue of (2.12) this system can be written in the following form:

$$(2.21) \qquad \begin{gathered} -\widetilde{K}(\omega)\pi n\frac{\widetilde{M}_2}{\omega} + \mu[G(T_0) + \ldots] = 0 \\ \widetilde{K}(\omega)\pi n\widetilde{M}_1 + \mu[\dot{G}_3(T_0) + \ldots] = 0. \end{gathered}$$

Here, and further on, the $\sim$ denotes the result of substitution

$$(2.22) \qquad M_i \to \widetilde{M}_i = M_i + \beta_i \qquad (i = 1, 2, 3).$$

Then from (1.13), (1.17), (1.19) and (2.10) we have

$$(2.23) \qquad K(\omega) = L(\omega) - \omega^2 N(\omega) = W(\omega)(M_1^2 + M_2^2) - z_1^0 W_1(\omega)$$

$$(2.24) \qquad W(\omega) = \frac{(a-1)(a+b-2)}{2b}, \quad W_1 = \frac{3a + 3b - 4ab - 2}{ab}.$$

From the equations (2.21) taken at $\mu = 0$ we obtain equations for M_1 and M_2. We remark that zero values for the basic amplitudes

$$(2.25) \qquad M_1 = 0, \qquad M_2 = 0$$

satisfy these equations.

In [9] there were constructed, for arbitrary ω, periodic solutions of the Euler – Poisson equations corresponding to the zero basic amplitudes. As it follows from the construction, conditions (2.25) are necessary only in the case when ω is irrational and in the case of the disk ($\omega = 1$, $A + B = C$, $z_1^0 = 0$). Otherwise, by using the above expansion (up to μ^3, including) one can show that there are only three values of ω: $\omega = 2$, $\omega = \frac{1}{2}$ and $\omega = \frac{1}{3}$ for which there exist periodic solutions of the Euler – Poisson equations with non-zero amplitudes

$$M_1^2 + M_2^2 \neq 0. \tag{2.26}$$

3. Let us consider periodic solutions of the Euler – Poisson equations for $\omega = 2$.

If (2.26) holds we can write formula (1.20) in the form

$$\begin{aligned} S_{12}^{(0)} &= E\Big\langle\Big\{[\kappa a + \kappa_1 a M_3 - x_1^0 \nu] \cos \epsilon - \\ &- y_1^0 \omega\Big[\frac{b}{aA_1^2} + \nu_2\Big] \sin \epsilon\Big\} - \omega y_1^0 \Big(\frac{b}{aA_1^2} + \nu_2\Big) \sin(\omega\tau - \epsilon) - \\ &- \frac{1}{2} M_3 (a\kappa_1 + b\omega\kappa_2 A_1^{-2}) \cos[(\omega - 1)\tau - \epsilon] + \\ &+ (a\kappa + x_1^0 \nu) \cos(\omega\tau - \epsilon) - \\ &- \frac{1}{2} M_3 (a\kappa_1 - b \cos \kappa_2 A_2^{-2}) \cos[(\omega + 1)\tau - \epsilon]\Big\rangle + z_1^0 S_{21}^{(0)}, \\ S_{22}^{(0)} &= E^2\big[(a\nu \cos^2 \epsilon - b\omega^2 \nu_2 A_1^{-1} \sin^2 \epsilon) - \\ &- \frac{1}{2} (\nu a - b\omega^2 \nu_2 A_1^{-1})\big] + a\kappa M_3 + \\ &+ \frac{1}{2} M_3^2 (a\kappa_1 + b\kappa_2 A_1^{-1}) + b y_1^0 a^{-1} A_1^{-1} M_3 \sin \tau - \\ &- a\kappa M_3 \cos \tau - \frac{1}{2} M_3^2 (a\kappa_1 + b\kappa_2 A_1^{-1}) \cos 2\tau - \\ &- \frac{1}{2} E^2 (a\nu - b\omega^2 \nu_2 A_1^{-1}) \cos^2(\omega\tau - \epsilon), \end{aligned} \tag{3.1}$$

where $S_{11}^{(0)}$ and $S_{21}^{(0)}$ are taken from (2.8).

Substituting the expressions (3.1), (2.6) and (2.7) into (1.22) and (1.23) and considering only the terms with $\sin 2\tau$ and $\cos 2\tau$ we obtain

$$F_3^{(0)} = - V_1 \cos 2\tau - V_2 \sin 2\tau - EP(2) \cos (2\tau - \epsilon) + \dots . \tag{3.2}$$

For $\Phi_3^{(0)}$ we get

$$\Phi_3^{(0)} = M_3 Q(2) \cos \tau + \dots . \tag{3.3}$$

Here

$$\begin{aligned} V_1 &= -\frac{x_1^0 z_1^0 M_3^2}{6ab^2}(12b^2 - b - 1), \\ V_2 &= -\frac{y_1^0 z_1^0 M_3^2}{6a^2 b(1 - b)}(9ab^2 - 17ab + 2b^2 + 4a - 3b + 1), \end{aligned} \tag{3.4}$$

$$\begin{aligned} P(\omega) = M_1\{M_3[2\omega^2 a\kappa_1 + 2\omega^2 az_1^0 + z_1^0 + a\kappa_1(1 + B_1)] + \\ + 2\omega^2(\kappa a - x_1^0 \nu)\} - 2\omega^3 y_1^0 (ba^{-1} A_1^{-2} + \nu_2) M_2, \\ Q(\omega) = M_1\{M_3 a[\kappa_1(1 + B_1) - z_1^0(2 + b^{-1}) - 2\kappa_1] + \\ + 2(x_1^0 \nu - \kappa a)\} + 2\omega^3 y_1^0(ba^{-1} A_1^{-2} + \nu_2) M_2. \end{aligned}$$

By virtue of (3.2), (3.3) and (2.11) we have

$$\begin{aligned} g_3(T_0) &= \frac{1}{2}\pi[V_2 + EP(2)\sin\epsilon], \quad h_3(T_0) = 0 \qquad (T = 2\pi), \\ \dot{g}_3(T_0) &= -\pi[V_1 + EP(2)\cos\epsilon], \quad \dot{h}_3(T_0) = \pi M_3 Q(2). \end{aligned} \tag{3.5}$$

Substituting the expression (3.5) into the equation of periodicity (2.20) we get

$$\begin{aligned} \widetilde{M}_2 \widetilde{K}(2) - \mu[\widetilde{M}_2 \widetilde{K}_1(2) + \widetilde{V}_2] + \mu^2(\dots) = 0, \\ \widetilde{M}_1 \widetilde{K}(2) - \mu[\widetilde{M}_1 \widetilde{K}_1(2) + \widetilde{V}_1] + \mu^2(\dots) = 0, \end{aligned} \tag{3.6}$$

where $K(\omega)$ is defined by (2.23) and $K_1(\omega)$ is

$$K_1(\omega) = P(\omega) + \omega^2 Q(\omega) = M_1 M_3 az_1^0 W_1(\omega). \tag{3.7}$$

Let us now write the equations of periodicity in the following new form, as given in [10]

$$\widetilde{M}_1 \widetilde{V}_2 - \widetilde{M}_2 \widetilde{V}_1 + \mu(\ldots) = 0, \quad \widetilde{M}_2 \widetilde{K}(2) + \mu(\ldots) = 0. \tag{3.8}$$

Using the equations

$$M_1 V_2 - M_2 V_1 = 0, \quad M_2[M_1^2 + M_2^2 - z_1^0 W_2(2)] = 0,$$

$$W_2(\omega) = \frac{W_1(\omega)}{W(\omega)} \tag{3.9}$$

we can determine M_1 and M_2, assuming that

$$z_1^0 W_1(2) > 0, \quad V_1 V_2 \neq 0. \tag{3.10}$$

Thus we have

$$M_1 = \pm V_1 \sqrt{\chi}, \quad M_2 = \pm V_2 \sqrt{\chi}, \quad \chi = \frac{z_1^0 W_2(2)}{V_1^2 + V_2^2}. \tag{3.11}$$

The corresponding values β_1 and β_2 can be obtained from (3.8).

If the conditions (3.10) hold, by (1.16), (1.18), (2.3) and (2.6) we have the following formula for the periodic solutions of the Euler – Poisson equations and for the period:

$$\begin{aligned}
p_1 &= \widetilde{E} \cos(\omega\tau - \widetilde{\epsilon}) + \mu\kappa + \mu\kappa_1 M_3 \cos\tau + \mu^2(\ldots), \\
q_1 &= \omega \widetilde{E} A_1^{-1} \sin(\omega\tau - \widetilde{\epsilon}) + \mu A_1^{-1}(y_1^0 a^{-1} + \kappa_2 M_3 \sin\tau) + \\
&\quad + \mu^2(\ldots), \\
r_1 &= 1 + \mu^2(\ldots), \\
\gamma_1 &= \widetilde{M}_3 \cos\tau + \mu\nu E \cos(\omega\tau - \epsilon) + \mu^2(\ldots), \\
\gamma_1' &= -\widetilde{M}_3 \sin\tau - \mu\nu_2 \omega E \sin(\omega\tau - \epsilon) + \mu^2(\ldots), \\
\gamma_1'' &= 1 + \frac{1}{2}\mu M_3 E\{2a\cos\epsilon + \sigma_1(\omega)\cos[(\omega - 1)\tau - \epsilon] - \\
&\quad - \sigma_2(\omega)\cos[(\omega + 1)\tau - \epsilon]\} + \mu^2(\ldots),
\end{aligned} \tag{3.12}$$

$$\alpha = \mu^2 \pi n[-(aM_1^2 + b\omega^2 A_1^{-2} M_2^2) + 2M_3 x_1^0 - 2z_1^0] + \mu^3(\ldots)$$

(3.13)

$$\sigma_1(\omega) = \frac{1 - b - \omega a}{(\omega - 1)(1 - b)}, \qquad \sigma_2(\omega) = \frac{1 - b + \omega a}{(1 + \omega)(1 - b)}.$$

In these formulae we have to substitute 2 for ω, the values (3.11) for M_1 and M_2 and the unity for n.

4. Let us now consider the periodic solutions for $\omega = \frac{1}{3}$.

We substitute into (1.22) and (1.23) the expressions (3.1) (taken at $\omega = \frac{1}{3}$), (2.6) and (2.7) and restrict ourselves to the terms with $\sin \frac{1}{3} \tau$ and $\cos \frac{1}{3} \tau$. Then we have

(4.1)

$$F_3^{(0)} = \left[EP\left(\tfrac{1}{3}\right) \cos \epsilon + E^2 V \cos 2\epsilon\right] \cos \frac{1}{3} \tau + \\ + \left[EP\left(\tfrac{1}{3}\right) \sin \epsilon - E^2 V \sin 2\epsilon\right] \sin \frac{1}{3} \tau + \ldots,$$

(4.2)

$$V = M_3 \Big\{ \frac{1}{4}(-a + b\omega^2 A_1^{-1}) \times \\ \times \left[-\omega^2 \kappa_1 + \frac{1}{2} z_1^0 (a^{-1} - A_1 b^{-1}) + \kappa_1\right] + \\ + \frac{1}{2} \omega^2 (a\kappa_1 + b\omega\kappa_2 A_1^{-2}) - \\ - \frac{1}{4} (-a + b\omega A_1^{-1})[z_1^0(2\omega^2 + a^{-1}) + \\ + \kappa_1(1 + B_1)] - \frac{1}{4} \omega^2 \kappa_1 A_1^{-1}(C_1 - A_1^{-1}) + \\ + \frac{1}{4} \omega(2C_1 A_1^{-1} \kappa_2 + (1 - C_1) A_1^{-1} \kappa_1) \Big\}.$$

We observe that we have terms containing V in (4.1) only in the case $\omega = \frac{1}{3}$. Taking (4.1), (4.2), (2.6), (2.11) and (3.3) at $\omega = \frac{1}{3}$ we get

(4.3)

$$g_3(T_0) = 9\pi\left[M_2 P\left(\tfrac{1}{3}\right) - 2M_1 M_2 V\right], \quad h_3(T_0) = 0, \quad (T_0 = 6\pi), \\ \dot{g}_3(T_0) = 3\pi\left[M_1 P\left(\tfrac{1}{3}\right) + (M_1^2 - M_2^2) V\right], \quad \dot{h}_3(T_0) = 3\pi M_3 Q\left(\tfrac{1}{3}\right)$$

whence, by virtue of the equations of periodicity (2.20) we have the following equations for M_1 and M_2:

$$(4.4)\qquad \begin{aligned} &\widetilde{M}_2\left[\widetilde{K}\left(\tfrac{1}{3}\right) - \mu\widetilde{K}_1\left(\tfrac{1}{3}\right)\right] - 2\mu\widetilde{M}_1\widetilde{M}_2\widetilde{V} + \mu^2(\ldots) = 0,\\ &\widetilde{M}_2\left[\widetilde{K}\left(\tfrac{1}{3}\right) - \mu\widetilde{K}_1\left(\tfrac{1}{3}\right)\right] + \mu(\widetilde{M}_1^2 - \widetilde{M}_2^2)V + \mu^2(\ldots) = 0. \end{aligned}$$

These equations of periodicity (4.4) correspond to the form considered in [10]. Subtracting from the first equation, multiplied by $\widetilde{M}_1$, the second equation of (4.4) multiplied by $\widetilde{M}_2$, and dividing by μ we have

$$\widetilde{M}_2\widetilde{V}_2(3\widetilde{M}_1^2 - \widetilde{M}_2^2) + \mu(\ldots) = 0.$$

This relation together with any (under the condition $M_1M_2 \neq 0$) of the equations (4.4), for example, with the first one, will represent a new form of the periodicity equations (after multiplication by suitable nonzero multipliers)

$$(4.5)\qquad \begin{aligned} &3\widetilde{M}_1^2 - \widetilde{M}_2^2 + \mu(\ldots) = 0\\ &\widetilde{K}\left(\tfrac{1}{3}\right) + \mu(\ldots) = 0. \end{aligned}$$

We observe that the expression for V, which takes the form

$$V = -\frac{7z_1^0M_3}{12b\left(\frac{9}{8} - b\right)(1 - b)}\left(b - \frac{3}{2}\right)\left(b - \frac{3}{4}\right)\left(b - \frac{33}{28}\right)$$

does not vanish, provided the following condition holds (and we assume it throughout)

$$(4.6)\qquad z_1^0(A - B)\left(B - \frac{33}{28}C\right) \neq 0$$

since from $\omega = \dfrac{1}{3}$ it follows that either $a = b = \dfrac{3}{2}$ or $a = b = \dfrac{3}{4}$.

From the equations for the basic amplitudes

$$(4.7)\qquad 3M_1^2 - M_2^2 = 0, \quad M_1^2 + M_2^2 - z_1^0W_2\left(\tfrac{1}{3}\right) = 0$$

where $W_2(\omega) = \dfrac{W_1(\omega)}{W(\omega)}$ one obtains the values M_1 and M_2 which are different from zero

$$(4.8) \qquad M_1 = \pm \frac{1}{2}\sqrt{z_1^0 W_2\left(\tfrac{1}{3}\right)}, \quad M_2 = \pm \frac{\sqrt{3}}{2}\sqrt{z_1^0 W_2\left(\tfrac{1}{3}\right)}$$

and from (4.5) the appropriate values β_1 and β_2.

In the case $M_1 M_2 = 0$ the equations for the basic amplitudes, derived from (4.4)

$$(4.9) \qquad \begin{aligned} M_1\left[M_1^2 + M_2^2 - z_1^0 W_2\left(\tfrac{1}{3}\right)\right] &= 0, \\ M_2\left[M_1^2 + M_2^2 - z_1^0 W_2\left(\tfrac{1}{3}\right)\right] &= 0 \end{aligned}$$

are satisfied by two groups of solutions

$$(4.10) \qquad 1.\ M_1 = 0, \quad M_2 = \pm\sqrt{z_1^0 W_2\left(\tfrac{1}{3}\right)}$$

$$(4.11) \qquad 2.\ M_2 = 0, \quad M_1 = \pm\sqrt{z_1^0 W_2\left(\tfrac{1}{3}\right)}.$$

However, for the first group of solutions, equations (4.4) do not allow any values β_1 and β_2.

Hence the only solutions of (4.9), satisfying the condition $M_1 M_2 = 0$ $(M_1^2 + M_2^2 \neq 0)$ are the solutions (4.11); the corresponding β_1 and β_2 are determined by (4.4).

By virtue of the relation

$$\operatorname{sign} W_1(\omega) = \operatorname{sign}(C - A) \qquad (\omega < 1)$$

we may write

$$(4.12) \qquad \begin{aligned} W_2(\omega) &= (C - A) R^2(\omega), \\ R^2(\omega) &= \frac{2C[(3(A + B)C - 4AB - 2C^2]}{A(C - A)^2(2C - A - B)} \end{aligned}$$

and conclude that M_1 and M_2 are real under the condition

(4.13) $z_1^0(C - A) > 0.$

In the case of rational ω $\left(\omega \neq 1, \frac{1}{2}, \frac{1}{3}, 2\right)$ the equations of periodicity (2.20) take the form

$$(4.14)\qquad \begin{aligned} &\widetilde{M}_2 \widetilde{K}(\omega) + \mu \widetilde{M}_2 \widetilde{K}_1(\omega) + \mu^2(\ldots) = 0,\\ &\widetilde{M}_1 \widetilde{K}(\omega) + \mu \widetilde{M}_1 \widetilde{K}_1(\omega) + \mu^2(\ldots) = 0 \end{aligned}$$

therefore it follows from the results of [10] and the above that the condition (4.13) is both necessary and sufficient for the existence of periodic solutions with nonzero basic amplitudes $M_1^2 + M_2^2 \neq 0$ in the case of any rational ω with the exception of the four cases mentioned above.

Under conditions (4.6) and (4.13) one can obtain the periodic solutions of the Euler – Poisson equations in the present case and also the value α from the general formulae (3.12)-(3.14) by taking $\omega = \frac{1}{3}$ and the corresponding values (4.8) and (4.11) for M_1 and M_2.

5. Let us consider the case $\omega = \frac{1}{2}$. From (1.23) and (1.22) we obtain

$$(5.1)\qquad \begin{aligned} F_2^{(0)} &= \left\{\left[L\left(\tfrac{1}{2}\right) + x_1^0 M_3 L_2\right] M_1 + y_1^0 M_3 L_3 M_2\right\} \cos \tfrac{1}{2}\tau +\\ &+ \left\{\left[L\left(\tfrac{1}{2}\right) - x_1^0 M_3 L_2\right] M_2 + y_1^0 M_3 L_3 M_1\right\} \sin \tfrac{1}{2}\tau + \ldots, \end{aligned}$$

$$(5.2)\qquad \begin{aligned} &\Phi_2^{(0)} = M_3 N\left(\tfrac{1}{2}\right) \cos \tau + \ldots, \qquad L_2 = -\tfrac{1}{2} A_1 \left(\tfrac{a}{b} - \tfrac{1}{2} A_1^{-1}\right),\\ &L_3 = \tfrac{1}{2}\left(1 - \tfrac{1}{2} A_1^{-1}\right) a^{-1}. \end{aligned}$$

From (2.11) and (5.1) we get the expressions

$$(5.3)\qquad \begin{aligned} g_2(T_0) &= -4\pi\left\{\left[L\left(\tfrac{1}{2}\right) + x_1^0 M_3 L_2\right] M_2 + y_1^0 M_1 M_3 L_3\right\},\\ \dot{g}_2(T_0) &= 2\pi\left\{\left[L\left(\tfrac{1}{2}\right) + x_1^0 M_3 L_2\right] M_1 + y_1^0 M_2 M_3 L_3\right\}, \end{aligned}$$

$$h_2(T_0) = 0, \quad \dot{h}_2(T_0) = 2\pi M_3 N\left(\tfrac{1}{2}\right), \quad T_0 = 4\pi$$

where, by eliminating a (since $\omega = \frac{1}{2}$) we have L_2 and L_3 in the form

$$L_2 = \frac{3}{4} M_3 b^{-1}\left(b - \frac{2}{3}\right), \quad L_3 = \frac{3}{8} M_3 (b-1)^{-1}\left(b - \frac{2}{3}\right). \tag{5.4}$$

We note that, provided $\omega = \frac{1}{2}$, the condition $a = b$ holds only in the cases $a = b = \frac{2}{3}$ and $a = b = \frac{1}{2}$.

By substituting (5.3) into the equations of basic amplitudes which we get from (2.20) taking $\mu = 0$, we obtain

$$\begin{aligned} &M_2 K\left(\tfrac{1}{2}\right) - l_2 M_2 + l_3 M_1 = 0, \quad M_1 K\left(\tfrac{1}{2}\right) + l_2 M_1 + l_3 M_2 = 0, \\ &l_2 = x_1^0 L_2, \quad l_3 = y_1^0 L_3. \end{aligned} \tag{5.5}$$

Since

$$K\left(\tfrac{1}{2}\right) = W\left(\tfrac{1}{2}\right)(M_1^2 + M_2^2) - z_1^0 W_1\left(\tfrac{1}{2}\right) \tag{5.6}$$

where $W(\omega)$ and $W_1(\omega)$ are defined by (2.24), and are independent of M_1 and M_2, equations (5.5) are nonlinear algebraic equations. Under the condition $l_3 \neq 0$ corresponding to the assumption $M_1 M_2 \neq 0$ we can proceed as follows.

Subtracting from the first equation, multiplied by M_1 the second equation, multiplied by M_2 we raise this expression to the second power. This yields the first relation

$$\frac{4 l_2^2}{l_3^2} M_1^2 M_2^2 = (M_1^2 - M_2^2)^2. \tag{5.7}$$

The product of the two equations in (5.5), reduced by $M_1 M_2$ will be chosen as the second relation

$$K^2\left(\tfrac{1}{2}\right) - l_2^2 = l_3^2,$$

which we write in the following final form, by virtue of (5.6)

$$M_1^2 + M_2^2 = s^2$$

(5.8)

$$s^2 = \frac{1}{W\left(\frac{1}{2}\right)} \left[z_1^0 W_1\left(\frac{1}{2}\right) \pm \sqrt{l_2^2 + l_3^2} \right].$$

One can determine the real numbers M_1 and M_2 from relations (5.7) and (5.8) by solving the following biquadratic equation

$$M^4 - s^2 M^2 + \frac{s^4 l_3^2}{4(l_2^2 + l_3^2)} = 0.$$

Finally we have

$$M_1 = \pm \frac{s}{\sqrt{2}} \sqrt{1 \pm \frac{l_2}{\sqrt{l_2^2 + l_3^2}}},$$

(5.9)

$$M_2 = \pm \frac{s}{\sqrt{2}} \sqrt{1 \mp \frac{l_2}{\sqrt{l_2^2 + l_3^2}}},$$

and we get from (2.20) the appropriate β_1 and β_2.

Note that (5.9) can also be used for $l_3 = 0$, $l_2 \neq 0$; this case corresponds to the condition $M_1 M_2 = 0$.

We see from (5.8) that $s^2 > 0$. This inequality can be taken to the form

(5.10) $$z_1^0 W_1\left(\frac{1}{2}\right) \pm M_3 \sqrt{x_1^0 L_2^2 + y_1^0 L_3^2} > 0.$$

From this inequality it follows that under the conditions

1. $z_1^{0^2} + (x_1^{0^2} + y_1^{0^2})\left(b - \frac{2}{3}\right)^2 = 0,$

(5.11)

2. $z_1^0(C - A) < 0, \quad (x_1^{0^2} + y_1^{0^2})\left(b - \frac{2}{3}\right) = 0$

there exist no periodic solution for $\omega = \frac{1}{2}$ satisfying (2.26).

For arbitrary numbers x_1^0, y_1^0, z_1^0, $(x_1^{0^2} + y_1^{0^2} + z_1^{0^2}) \neq 0$, A, B, C, $\left(\omega = \frac{1}{2}\right)$ which do not satisfy any of the conditions (5.11), relation (5.10) can hold for every γ_0'' (i.e., every M_3 by (2.16)) either on the whole interval $0 < \gamma_0'' < 1$ or on one of its parts $0 < \gamma_0'' < \Gamma_0''$

$$\Gamma_0'' = (l_2^2 + l_3^2)^{\frac{1}{2}} \left[l_2^2 + l_3^2 + z_1^{0^2} W_1^2 \left(\tfrac{1}{3}\right) \right]^{-\frac{1}{2}}$$

depending on the sign of $z_1^0(C - A)$ and the sign in (5.10).

The corresponding periodic solution of the Euler – Poisson equations and the value α can be obtained from the general formulae (3.12)-(3.14), by taking $\omega = \frac{1}{2}$, and M_1 and M_2 accordingly to (5.9).

As it was pointed out above, these three classes of periodic solutions with nonzero amplitudes were constructed in such a way that we needed only to consider the first terms of the expansions of the right hand sides of (1.21). One can expect that by using further terms of the expansion it is possible to construct periodic solutions for other rational values of ω such that the basic amplitudes satisfy (2.26). It seems to be difficult, however, to derive results of this kind from the symbolic information, without the use of computers.

Finally, we note that in [11]-[14] there are given some series representing periodic solutions of the Euler – Poisson equations for $\gamma_0'' = 0$, $\gamma_0'' = 1$ and in a neighbourhood of these values, and based on results of [15] there exists an estimation on the radius of convergence of these series.

Moreover a program has been worked out on the language LISP [16] and new periodic solutions obtained in a neighbourhood of $\gamma_0 = 0$.

REFERENCES

[1] R.I. Chertkov, *The method of Jacobi in the dynamics of rigid body.* Leningrad, 1960, Sudpromgiz (in Russian).

[2] V.A. Toporova, On a new case of exact integrability of rotation of a heavy rigid body about a fixed point. *Trudy Inst. Mat. AN UzSSR,* 24 (1962), (in Russian).

[3] W.D. MacMillan, *Dynamics of rigid bodies,* New York – London 1936, McGraw-Hill.

[4] M.P. Khalimanovich, On the motion of a non-completely symmetric heavy gyroscope at small angles of nutation. *Uch. Zap. Belorussk. Univ.,* 15 (1953) (in Russian).

[5] E. Mettler, Periodische und asymptotische Bewegungen des unsymmetrischen schweren Kreisels. *Math. Ztschr.,* 43 (1937).

[6] I.G. Malkin, *Some problems in the theory of nonlinear oscillation.* Moscow, 1956, Gostekhizdat (in Russian).

[7] A.P. Proskuryakov, Periodic oscillation of quasilinear autonomous systems with two degrees of freedom. *Prikl. Mat. Mekh.,* 24, 6 (1960) (in Russian).

[8] Yu.A. Arkhangelskiĭ, On the periodic solutions of quasilinear autonomous systems having first integrals. *Prikl. Mat. Mekh.,* 27, 2 (1963) (in Russian).

[9] Yu.A. Arkhangelskiĭ, On the motion of heavy rigid body set in fast rotation about a fixed point. *Prikl. Mat. Mekh.,* 27, 5 (1963) (in Russian).

[10] Yu.A. Arkhangelskiĭ, On a case of constructing of periodic solutions of quasilinear systems. *Prikl. Mat. Mekh.,* 28, 5 (1964) (in Russian).

[11] V.S. Sergeev, On the estimation of convergence domain for series representing solutions of the equations of motion of a heavy rigid body with a fixed point. *Vestnik Mosk. Univ., Mat. Mekh.,* 1969, No. 2 (in Russian).

[12] V.S. Sergeev, On periodic solutions for the equations of motion of a heavy rigid body about a fixed point. *Vestnik Mosk. Univ., Mat. Mekh.*, 1969, No. 1 (in Russian).

[13] S.P. Sedunova, On some motions of a gyroscope. *Vestnik Mosk. Univ.*, 1973, No. 1 (in Russian).

[14] S.P. Sedunova, Some motions of a heavy rigid body about a fixed point. In: *Dynamics of machines and working processes, Vol.* 129, 1973, Chelyabinsk Polytechn. Inst. (in Russian).

[15] Yu.A. Ryabov, Determination of existence domain for some implicit functions. *Trudy Vsesoyuznogo Zaochn. Energ. Inst. Mat.*, 1957, No. 2 (in Russian).

[16] S.S. Lavrov – G.S. Sigaladze, *Input language and interpreter of programming system on base of the language LISP for the computer BESM*-6, Moscow, 1969, Acad. Sci. USSR (in Russian).

Yu. A. Arkhangelskii,
Moscow 121309, Bolshaya Filevskaya ul. 13/7, kv. 86, USSR.

COLLOQUIA MATHEMATICA SOCIETATIS JÁNOS BOLYAI
15. DIFFERENTIAL EQUATIONS, KESZTHELY (HUNGARY), 1975.

ON A BOUNDARY VALUE PROBLEM WITH A GENERALIZED BOUNDARY CONDITION FOR SYSTEMS OF SUPER NEUTRAL DIFFERENTIAL EQUATIONS

D.D. BAINOV — M.M. KONSTANTINOV

1. INTRODUCTION

Boundary value problems for differential equations with deviating argument are of great importance both from theoretical and practical point of view. These problems have been studied by a number of authors under various assumptions [1]-[10].

The results concerning the boundary value problem investigated in the present paper generalize in a great extent certain results of [6]. A vector differential equation with deviating argument of superneutral type with a boundary condition in the form of a multidimensional vector-tunctional is considered. In particular, this functional may be defined by the help of a Lebesgue – Stieltjes integral. Under a special choice of the integrands the generalized boundary condition may be transformed into an ordinary nonlinear multipoint boundary condition.

To improve the growth estimation result of the corresponding multi-dimensional functions generalized norms are used.

In the sequel, unless otherwise specified, the value of the index i is equal to 0 or 1, index l runs from 1 to $r \geqslant 1$, the indices j, k, p assume all values between 1 and $n \geqslant 1$.

2. FORMULATION OF THE PROBLEM. BASIC ASSUMPTIONS

Consider the initial value problem

$$\dot{x}(t) = X(t, \lambda, x(\Delta_1^0), \ldots, x(\Delta_r^0), \dot{x}(\Delta_1'), \ldots, \dot{x}(\Delta_r^1)),$$

(1) $$t \in J_\tau = [0, \tau];$$

$$x(t) = \varphi_0(t, \lambda), \dot{x}(t) = \varphi_1(t, \lambda) = \frac{\partial}{\partial t} \varphi_0(t, \lambda), t \leqslant 0$$

with boundary condition

(2) $$F(\lambda, U_\tau^0 x(\theta), U_\tau^1 \dot{x}(\theta)) = 0,$$

where $x = \text{col}\,(x^k)$, $X = \text{col}\,(X^k)$, $\lambda = \text{col}\,(\lambda^k)$, $\varphi_i = \text{col}\,(\varphi_i^k)$, $F = \text{col}\,(F^k)$, $U_t^i z(\theta)$ are n-dimensional functionals (column vectors), being at the same time restrictions of the function $z(\theta)$ on the interval $(0, t]$.

The delays Δ_l^i are of the form $\Delta_l^i = \Delta_l^i(t, \lambda, x(t), \dot{x}(t))$.

Further, we can assume $\tau = 1$ and $\varphi_i \equiv 0$, since this can be always achieved by means of an appropriate substitution of the variables.

Let $R^{(n)}$ be the real n-dimensional space, partially ordered (with order relation $\leqslant$) defined by the non-negative oone K of all n-dimensional vectors with non-negative components. For $c = \text{col}\,(c^k) \in R^{(n)}$ let $|c| = \text{col}\,(|c^k|) \in K$.

Define the set $\Lambda = \{\lambda\colon a \leqslant \lambda \leqslant b\}$, where $a \leqslant b$; $a = \text{col}\,(a^k)$, $b = \text{col}\,(b^K) \in R^{(n)}$.

Let Ω_i be the set of all functions $z \in C$ (where C denotes the space of the continuous bounded functions $z\colon J \to R^{(n)}$, $J = I \cup J_1$, $I = (-\infty, 0]$; $z(t) = 0$, $t \in I$) such that

$$z \in \Omega_0 \Rightarrow |z(t)| \leqslant \int_0^t X_0(\theta)\, d\theta,$$

$$z \in \Omega_1 \Rightarrow |z(t)| \leqslant X_0(t),$$

where $X_0 = \operatorname{col}(X_0^K)$: $J_1 \to K$ is a continuous function.

Suppose that the following set (A) of assumptions holds:

A 1. The function $X(t, \lambda, \xi_0, \xi_1)$, $\xi_i = (\xi_{li})$, $\xi_{li} = \operatorname{col}(\xi_{li}^k)$, is defined in the domain $Q = J_1 \times D = J_1 \times \Lambda \times \omega_0^r \times \omega_1^r$, where

$$\omega_0 = \left\{\xi:\ |\xi| \leqslant \int_0^1 X_0(\theta)\, d\theta\right\} \subset R^{(n)},$$

$$\omega_1 = \{\xi:\ |\xi| \leqslant \operatorname{col}(\max\{X_0^k(t):\ t \in J_1\}) = \operatorname{col}(\beta_0^k) = \beta_0\} \subset R^{(n)}$$

and there it satisfyes the Lipschitz conditions

$$|X(t, \lambda, \xi_0, \xi_1) - X(\bar{t}, \bar{\lambda}, \bar{\xi}_0, \bar{\xi}_1)| \leqslant T|t - \bar{t}| +$$

$$+ L|\lambda - \bar{\lambda}| + \sum_{i=0}^{1} \sum_{l=1}^{r} M_{li} |\xi_{li} - \bar{\xi}_{li}|,$$

$$\bar{\xi}_i = (\bar{\xi}_{li}), \qquad \bar{\xi}_{li} = \operatorname{col}(\xi_{li}^k).$$

Here $T = \operatorname{col}(T^k)$, $L = (L^{jk})$, $M_{li} = (M_{li}^{jk})$.

A 2. The function $F(\lambda, \eta_0, \eta_1)$, $\eta_i = \operatorname{col}(\eta_i^k)$, is defined in the domain $\Lambda \times u_0 \times u_1$,

$$u_i = \bigcup_{z \in \Omega_i} \{U_1^i z(\theta)\}$$

and there it satisfies the conditions:

A 2.1.

$$|F(\lambda, \eta_0, \eta_1) - F(\bar{\lambda}, \eta_0, \eta_1)| \leqslant P|\lambda - \bar{\lambda}|, \qquad P = (P^{jk}).$$

A 2.2.

$$|F(\lambda, \eta_0, \eta_1) - F(\lambda + h, \eta_0, \eta_1)| \geqslant R|h|, \qquad R = \operatorname{diag}(R^k),$$

$$h = \operatorname{col}(h^k), \qquad \operatorname{col}(R^k) > 0,$$

and the functions $F^k(\lambda, \eta_0, \eta_1)$ are increasing with respect to λ^k when $k = 1, \ldots, m$ $(m \leqslant n)$, and decreasing with respect to λ^k when $k = m+1, \ldots, n$.

A 2.3. The functional F is continuous in the following sense: for each $\epsilon > 0$ there exists $\delta = \delta(\epsilon) > 0$, such that

$$|F(\lambda, U_1^0 x(\theta), U_1^1 \dot{x}(\theta)) - F(\lambda, U_1^0 \bar{x}(\theta), U_1^1 \dot{\bar{x}}(\theta))| < \epsilon E$$

when $x, \bar{x} \in \Omega_0$, $\dot{x}, \dot{\bar{x}} \in \Omega_1$, ($E$ is a unit n-vector) and $\|\dot{x} - \dot{\bar{x}}\| < \delta E$, where

$$\|x\| = \operatorname{col}(\max\{|x^k(t)|:\ t \in J_1\}).$$

A 3. The inequalities

$$\max\{|X^k(t, \lambda, \xi_0, \xi_1)|:\ (\lambda, \xi_0, \xi_1) \in D\} \leqslant X_0^k(t)$$

are fulfilled.

A 4. The inequalities $a \leqslant \lambda_0 \leqslant \lambda_1 \leqslant b$ hold, where $\lambda_i = \operatorname{col}(\lambda_i^k)$,

$$\lambda_0 = \inf\{\lambda - \Lambda_0^{-1} F(\lambda, \eta_0, \eta_1):\ (\lambda, \eta_0, \eta_1) \in \Lambda \times u_0 \times u_1\},$$

$$\lambda_1 = \sup\{\lambda - \Lambda_0^{-1} F(\lambda, \eta_0, \eta_1):\ (\lambda, \eta_0, \eta_1) \in \Lambda \times u_0 \times u_1\},$$

and $\Lambda_0 = \operatorname{diag}(\Lambda_0^k)$,

$$\Lambda_0^k = 2(P^{kk} + R^k)^{-1} \operatorname{sign}\left(m + \frac{1}{2} - k\right).$$

A 5. The functions $\Delta_l^i(t, \lambda, \eta_0, \eta_1)$ are defined in the domain $J_1 \times \Lambda \times \omega_0 \times \omega_1$ and there they satisfy the conditions

$$|\Delta_l^i(t, \lambda, \eta_0, \eta_1) - \Delta_l^i(\bar{t}, \bar{\lambda}, \bar{\eta}_0, \bar{\eta}_1)| \leqslant T^{li} |t - \bar{t}| +$$

$$+ L_{li} |\lambda - \bar{\lambda}| + \sum_{\alpha=0}^{1} \mu_{li\alpha} |\eta_\alpha - \bar{\eta}_\alpha|,$$

where $L_{li} = (L_{li}^k)$, $\mu_{li\alpha} = (\mu_{li\alpha}^k)$.

A 6. The compatibility condition for $t = 0$ holds identically with respect to λ.

3. EXISTENCE OF THE SOLUTIONS OF THE BOUNDARY VALUE PROBLEM

Theorem 1. *Suppose, that*

1.1. *Conditions* (A) *are satisfied.*

1.2. *The system of algebraic inequalities*

$$T^p + \sum_{l=1}^{r} \sum_{k=1}^{n} \Big(M_{l0}^{pk} \beta_0^k \Big(T^{l0} + \sum_{j=1}^{n} (\mu_{l00}^j \beta_0^j + \mu_{l01}^j \beta^j)\Big) +$$

$$+ M_{l1}^{pk} \beta^k \Big(T^{l1} + \sum_{j=1}^{r} (\mu_{l10}^j \beta_0^j + \mu_{l11}^j \beta^j)\Big)\Big) \leqslant \beta^p$$

admits a solution $\beta \in K$ $(\beta = \operatorname{col}(\beta^p))$.

Then, the boundary value problem (1), (2) *has at least one solution with a derivative satisfying a Lipschitz condition.*

Remark 1. For simplifying the computations, condition 1.2 may be substituted by some more restricting conditions. For example, if $\beta^1 = \ldots = \beta^n$ is assumed, then β^1 will be defined as $\beta^1 = \inf M$ (with $M \neq \phi$), where M is the intersection of the existence domains of the solutions of n quadratic equations.

It can be assumed that the solution $\beta \in K$ is such that there exists no other solution satisfying $\beta \geqslant \beta_* \in K$.

Proof. Let Ω be the set of all the functions $y \in \Omega_1$, satisfying the condition

(3) $\qquad |y(t) - y(\bar{t})| \leqslant \beta |t - \bar{t}|,$

where the vector $\beta \in K$ is defined in condition 1.2 of Theorem 1.

Let the operator $\Pi = (\hat{\Pi}, \check{\Pi})$ act in $W = \Omega \times \Lambda$ according to the formula

$$\hat{\Pi} y(t) = \frac{1}{2}(1 + \operatorname{sign} t) X(t, \lambda, x(\Delta_1^0), \ldots$$

$$\ldots, x(\Delta_r^0), y(\Delta_1^1), \ldots, y(\Delta_r^1)),$$

$$\check{\Pi}\lambda = -\Lambda_0^{-1} F(\lambda, U_1^0 x(\theta), U_1^1 y(\theta)) + \lambda,$$

where $x(t) = \int_0^t y(\theta)\, d\theta$. It is easy to show that the operator equation $w = \Pi w$, $w = (y, \lambda)$, is equivalent to the boundary value problem (1), (2).

For $w, \bar{w} \in W$, $\bar{w} = (\bar{y}, \bar{\lambda})$, there holds

$$\begin{aligned}
\| \hat{\Pi} y - \hat{\Pi}\bar{y} \| &\leqslant L | \lambda - \bar{\lambda} \| + \sum_{i=1}^{r} (M_{l0}(| x(\Delta_l^0) - \bar{x}(\Delta_l^0)| + \\
&+ | \bar{x}(\Delta_l^0) - \bar{x}(\bar{\Delta}_l^0)|) + M_{l1}(| y(\Delta_l^1) - \bar{y}(\Delta_l^1)| + \\
&+ | \bar{y}(\Delta_l^1) - \bar{y}(\bar{\Delta}_l^1)|)) \leqslant \qquad (4) \\
&\leqslant \Big(\sum_{i=0}^{1} \sum_{l=1}^{r} M_{li}(S + \beta_i(\mu_{li0} + \mu_{li1})) \Big) \| y - \bar{y} \| + \\
&+ \Big(L + \sum_{i=0}^{1} \sum_{l=1}^{r} M_{li}\beta_i L_{li}\Big) | \lambda - \bar{\lambda} |,
\end{aligned}$$

where

$$S = \operatorname{diag} \underbrace{(1, \ldots, 1)}_{n}, \qquad \beta_1 = \beta.$$

On the other hand, by condition A 2 the restriction $\check{\Pi}$ of the operator Π on the set Λ is continuous. Hence the operator Π is continuous on W.

Finally, we shall show that the operator Π transforms the set W into itself. For, by conditions A 1 - A 4, it is sufficient to show that for the function $\hat{\Pi} y$, $(y \in \Omega)$ condition (3) holds.

Let $t, \tilde{t} \in J_1$. Then, for the p-th component $\hat{\Pi}^p$ of the operator $\hat{\Pi}$ we have

$$\begin{aligned}
| \Pi^p y(t) - \Pi^p y(\tilde{t}) | &\leqslant T^p | t - \tilde{t} | + \\
&+ \sum_{l=1}^{r} \sum_{k=1}^{n} M_{l0}^{pk} | x^k(\Delta_l^0) - x^k(\tilde{\Delta}_l^0) | +
\end{aligned}$$

$$+ \sum_{l=1}^{r} \sum_{k=1}^{n} M_{l1}^{pk} | y^k(\Delta_l^1) - y^k(\widetilde{\Delta}_l^1) | \leqslant$$

$$\leqslant T^p | t - \widetilde{t} | + \sum_{l=1}^{r} \sum_{k=1}^{n} M_{l0}^{pk} \beta_0^k | \Delta_l^0 - \widetilde{\Delta}_l^0 | +$$

$$+ \sum_{l=1}^{r} \sum_{k=1}^{n} M_{l1}^{pk} \beta^k | \Delta_l^1 - \widetilde{\Delta}_l^1 |$$

and since

$$| \Delta_l^i - \widetilde{\Delta}_l^i | \leqslant T^{li} | t - \widetilde{t} | + \sum_{k=1}^{n} \mu_{li0}^k | x^k(t) - x^k(\widetilde{t}) | +$$

$$+ \sum_{k=1}^{n} \mu_{li1}^k | y^k(t) - y^k(\widetilde{t}) | \leqslant$$

$$\leqslant | t - \widetilde{t} | \left(T^{li} + \sum_{k=1}^{n} (\mu_{li0}^k \beta_0^k + \mu_{li1} \beta^k) \right),$$

then

$$| \Pi^p y(t) - \Pi^p y(\widetilde{t}) | \leqslant$$

$$\leqslant | t - \widetilde{t} | \left(T^p + \sum_{l=1}^{r} \sum_{k=1}^{n} M_{l0}^{pk} \left(T^{l0} + \sum_{j=1}^{n} (\mu_{l00}^j \beta_0^j + \mu_{l01}^j \beta^j) \right) + \right.$$

$$\left. + \sum_{l=1}^{r} \sum_{k=1}^{n} M_{l1}^{pk} \beta^k \left(T^{l1} + \sum_{j=1}^{n} (\mu_{l10}^j \beta_0^j + \mu_{l11}^j \beta^j) \right) \right).$$

Hence we obtain by condition 1.2 of Theorem 1, $\Pi W \subset W$. By Tychonov's theorem the assertion follows, since W is obviously convex and compact.

4. UNIQUENESS OF SOLUTION OF THE BOUNDARY VALUE PROBLEM

Theorem 2. *Suppose, that*

2.1. *The conditions of Theorem* 1 *are satisfied.*

2.2. *The functional F satisfies the Lipschitz conditions*

$$|F(\lambda, U_1^0 x(\theta), U_1^1 \dot{x}(\theta)) - F(\lambda, U_1^0 \bar{x}(\theta), U_1^1 \dot{\bar{x}}(\theta))| \leqslant$$
$$\leqslant A_0 \|x - \bar{x}\| + A_1 \|\dot{x} - \dot{\bar{x}}\|, \qquad A_i = (A_i^{jk}).$$

2.3. *The inequality* $\rho(\Phi) < 1$ *is satisfied where* $\rho(\Phi)$ *denotes the spectral radius of the matrix* Φ,

$$\Phi = \begin{pmatrix} A, & B \\ C_0, & D_0 \end{pmatrix},$$

$$A = \sum_{i=0}^{1} \sum_{l=1}^{r} M_{li}(S + \beta_i(\mu_{li0} + \mu_{li1})),$$

$$B = L + \sum_{i=0}^{1} \sum_{l=1}^{r} M_{li}\beta_i L_{li}, \qquad C_0 = \Lambda_0^{-1}(A_0 + A_1),$$

$$D_0 = (D_0^{jk}) = \left(\frac{P^{kk} - R^k}{P^{kk} + R^k} + \text{sign}^2(k-j)\left(\frac{2P^{kj} + R^k - P^{kk}}{P^{kk} + R^k}\right)\right).$$

Then, the boundary value problem (1), (2) *has a unique solution with a derivative satisfying a Lipschitz condition.*

Proof. By condition 2.1 of the theorem there exists at least one solution $w = (y, \lambda)$ of the boundary value problem (1), (2).

It will be shown that this solution is unique. Indeed, if the boundary value problem (1), (2) has a solution $\bar{w} \neq w$, $\bar{w} = (\bar{y}, \bar{\lambda})$, then for $\|y - \bar{y}\|$ the estimation of the form (3) holds.

On the other hand,

$$|\lambda - \bar{\lambda}| = |\check{\Pi}\lambda - \check{\Pi}\bar{\lambda}| \leqslant D_0 |\lambda - \bar{\lambda}| + \Lambda_0^{-1} A_0 \|x - \bar{x}\| +$$
$$+ \Lambda_0^{-1} A_1 \|y - \bar{y}\| \leqslant \Lambda_0^{-1}(A_0 + A_1)\|y - \bar{y}\| + D_0 |\lambda - \bar{\lambda}|,$$

where $D_0 = (D_0^{jk})$,

$$D_0^{jk} = \begin{cases} (P^{kk} - R^k)(P^{kk} + R^k)^{-1}, & j = k \\ 2P^{jk}(P^{kk} + R^k)^{-1}, & j \neq k \end{cases}$$

Whence, by condition 2.3 there follows that $\bar{w} = w$, i.e. the solution of the boundary value problem (1), (2) is unique, and the proof is complete.

Remark 2. In Theorem 1 the Lipschitz conditions with respect to λ for the functions F and Δ_l^i (which play role in the formulation of Theorem 2 as well) may be left out, substituting them with conditions for uniform continuity of the corresponding functions with respect to λ.

5. A PARTICULAR CASE OF THE FUNCTIONAL F

Consider the case when the functional F is given by means of Stieltjes integral, in the following form:

$$F = F\left(\lambda, \int_0^1 (dG_0(\theta))x(\theta), \int_0^1 (dG_1(\theta))\dot{x}(\theta)\right),$$

where the kernels $G_i = (G_i^{jk})$, are measurable.

Denote by $\widetilde{G}_i$, the matrix with elements

$$\widetilde{G}_i^{jk} = \bigvee_0^1 (G_i^{jk}).$$

So, if

$$|F(\lambda, \eta_0, \eta_1) - F(\lambda, \bar{\eta}_0, \bar{\eta}_1)| \leqslant \sum_{i=0}^{1} A_{i1} |\eta_i - \bar{\eta}_i|$$

then A_i can be taken as $A_i = A_{i1} \widetilde{G}_i$ (see Theorem 2).

Remark 3. With a certain particular choice of the kernels G_i, the boundary value problem (1), (2) with generalized boundary conditions may be transformed into the usual multipoint boundary problem of the first order for a vector differential equation of superneutral type.

Remark 4. The outline of investigations remains in principle the same if the kernels G_i depend on the unknown function and its derivative.

REFERENCES

[1] K.T. Ahmedov – S.V. Israilov, The Cauchy – Nicoletti multipoint boundary value problem for retarded differential equations and certain problems concerning oscillatoric properties of the solutions, *Doklady A.N. Az. SSR,* 19, 9 (1973) (in Russian).

[2] K.T. Ahmedov – N.A. Svarichevskaja – M.A. Yakubov, Approximate solution of a two point boundary value problem with a parameter, by the method of averaging of functional corrections, *Doklady A.N. Az. SSR,* 29, 8 (1973) (in Russian).

[3] H. Bensaad – S.B. Norkin, A boundary value problem with controlling initial function for a system of nonlinear differential equations with retarded argument, *Ukrain. Mat. Ž.,* 26, 1 (1974) (in Russian).

[4] I. Israilov, On a boundary value problem with parameter for differential-extremal equations with retarded argument, *Uč. Zam. Az. G.U.,* (1968) (in Russian).

[5] V.V. Mosjiagin, A boundary value problem for retarded differential equations in Banach space, *Uč. Zap. L.G.U.,* 387 (1968) (in Russian).

[6] Z.B. Seidov, A boundary value problem for retarded differential equations, *Ukrain. Mat. Ž.,* 25, 6 (1973) (in Russian).

[7] Z.B. Seidov. A boundary value problem for retarded differential equations, *Izv. Ab. N. Az. SSR,* No 2 (1973) (in Russian).

[8] M.M. Konstantinov, Existence and uniqueness of solutions of boundary value problems for differential equations of superneutral type, *Math. Balcanica,* 3 (1973) (in Russian).

[9] M.M. Konstantinov – D.D. Bainov, On a second order boundary value problem for systems of differential equations of superneutral type, *Soobš. A.N. Gruz. SSR,* 74, 2 (1974), (in Russian).

[10] K. Schmitt, On solutions of nonlinear differential equations with deviating argument, *SIAM J. Appl. Math. Vol.* 17, No 6 (1969).

D.D. Bainov

University of Plovdiv, Paissji Hilendarski, Bulgaria.

M.M. Konstantinov

Higher Institute for Machines and Electrotechnics "V.I. Lenin", Bulgaria.

COLLOQUIA MATHEMATICA SOCIETATIS JÁNOS BOLYAI
15. DIFFERENTIAL EQUATIONS, KESZTHELY (HUNGARY), 1975.

ON THE APPLICATION OF THE AVERAGING METHOD FOR CERTAIN CLASSES OF EQUATIONS WITH DELAY

D.D. BAINOV – S.D. MILUSHEVA

The averaging method for differential and integro-differential equations has been applied in papers [1]-[16].

The present paper justifies the averaging method for certain classes of ordinary differential equations and integro-differential equations of Volterra type with delay.

Consider the initial value problem

$$(1)\qquad \begin{aligned} &\dot{x}(t) = \epsilon X(t, x(t), x_\tau(t)), \qquad t \in I = (t_0, +\infty),\\ &x(t) = \kappa(t), \qquad t \in \bar{I} = (-\infty, t_0], \end{aligned}$$

where $x_\tau(t) = x(t - \tau(t, x(t)))$, $x, X \in R^n$, $\epsilon > 0$ is a small parameter, and $\kappa(t)$ is a function defined and continuous when $t \in \bar{I}$.

Suppose that the limit

$$(2) \qquad \lim_{T \to \infty} \frac{1}{T} \int_{t_0}^{t_0+T} X(t, x, u)\, dt = \bar{X}(x, u)$$

exists.

Then, to equation (1) we put in correspondence the averaged equation

$$(3) \qquad \begin{aligned} &\dot{\xi}(t) = \epsilon \bar{X}(\xi(t), \xi(t)), \qquad t \in I, \\ &\xi(t_0) = \kappa(t_0). \end{aligned}$$

Note, that if $x = (x^{(1)}, \ldots, x^{(n)})$ and $A = (a_{ij})_{n,m}$, then by definition

$$\|x\| = \Big[\sum_{i=1}^{n} (x^{(i)})^2\Big]^{\frac{1}{2}}, \qquad \|A\| = \Big[\sum_{j=1}^{m} \sum_{i=1}^{n} a_{ij}^2\Big]^{\frac{1}{2}}.$$

Theorem 1. *Let the following assumptions be fulfilled:*

1. *The function* $\tau(t, x)$ *is defined, continuous and non-negative in the domain* $\Omega(t, x) = \Omega(t) \times \Omega(x)$, *where* $\Omega(t) = I$, *and* $\Omega(x)$ *is a certain open domain in the space* R_n.

The function $X(t, x, u)$ *is defined and continuous in the domain* $\Omega(t, x, u) = \Omega(t) \times \Omega(x) \times \Omega(u)$, *where* $\Omega(u) = \Omega(x) \cup K$,
$K = \bigcup_{-\infty}^{t=t_0} \{\kappa(t)\}$.

2. *On the corresponding projections of the domain* $\Omega(t, x, u)$ *the following inequalities hold:*

$$\tau(t, x) \leqslant \Delta(t),$$

$$\|X(t, x, u)\| \leqslant M = \text{const.},$$

$$\|X(t, x, u) - X(t, x', u')\| \leqslant \lambda[\|x - x'\| + \|u - u'\|], \quad \lambda = \text{const.}$$

3. *The set*

$$I_0 = I \cap \{t\colon\ t - \Delta(t) \leqslant 0,\ t \in I\}$$

has a finite measure $\mu = \text{mes}\, I_0 < \infty$.

4. *For every* $L > 0$ *there exists the limit*

$$\lim_{\epsilon \to 0} \epsilon^2 \int_{I^*} \Delta(t)\, dt = 0, \qquad I^* = I^\epsilon \setminus I_0, \qquad I^\epsilon = [0, L\epsilon^{-1}].$$

5. *The initial value problem* (1) *has a unique and continuous solution for* $t \geqslant 0$ *belonging to the domain* $\Omega(x)$ *for* $t \geqslant 0$.

6. *For every point* (x, u) *of the domain* $\Omega(x, u)$, *the limit* (2) *exists.*

7. *The Cauchy problem* (3) *has a unique and continuous solution for* $t \geqslant 0$ *belonging to the domain* $\Omega(x)$ *for* $t \geqslant 0$.

8. *The function* $\bar{X}(x, u)$ *is continuous in the domain* $\Omega(x, u)$, *and satisfies the Lipschitz condition*

$$\| \bar{X}(x, u) - \bar{X}(x', u') \| \leqslant \nu[\| x - x' \| + \| u - u' \|], \qquad \nu = \text{const.},$$

and, moreover, on every finite interval $[t_1, t_2]$, *for a solution* $\xi(t)$ *the following inequality holds:*

$$\left\| \int_{t_1}^{t_2} \bar{X}(\xi(t), \xi(t))\, dt \right\| \leqslant \mathcal{N}(t_2 - t_1), \qquad \mathcal{N} = \text{const.}$$

So, if $x(t)$ *is a solution of initial problem* (1), $\xi(t)$ *is a solution of the Cauchy problem* (3), *then, for every* $\eta > 0$ *and* $L > 0$ *there exists an* $\epsilon_0 = \epsilon_0(\eta, L)$ *such that for* $0 < \epsilon \leqslant \epsilon_0$ *and* $t \in I^\epsilon$ *the inequality*

$$\| x(t) - \xi(t) \| < \eta$$

holds.

Proof. Let $y(t)$ be a solution of the equation

$$(4) \qquad \begin{aligned} &\dot{y}(t) = \epsilon X(t, y(t), y(t)), \qquad t \in I, \\ &y(t) = \kappa(t_0). \end{aligned}$$

It will be shown that for every $\frac{\eta}{2} > 0$ and $L > 0$, there exists an $\epsilon_1 = \epsilon_1\left(\frac{\eta}{2}, L\right) > 0$, such that for $\epsilon < \epsilon_1$ and $t \in I^\epsilon$ the following inequality holds:

(5) $$\| x(t) - y(t) \| < \frac{\eta}{2}.$$

Indeed, for the solutions of equations (1) and (4) the equalities

(6) $$x(t) = \begin{cases} \kappa(t_0) + \epsilon \int_{t_0}^{t} X(\theta, x(\theta), x_\tau(\theta))\, d\theta, & t \in I^\epsilon \\ \kappa(t), & t \in \bar{I}, \end{cases}$$

(7) $$y(t) = \kappa(t_0) + \epsilon \int_{t_0}^{t} X(\theta, y(\theta), y(\theta))\, d\theta, \qquad t \in I^\epsilon$$

hold.

Subtracting (7) from (6) and estimating the difference obtained, we get:

$$\begin{aligned} &\| x(t) - y(t) \| \leqslant \\ &\leqslant \epsilon \int_{t_0}^{t} \lambda[\| x(\theta) - y(\theta) \| + \| x(\theta - \tau(\theta, x(\theta))) - y(\theta) \|]\, d\theta \leqslant \\ &\leqslant 2\epsilon\lambda \int_{t_0}^{t} \| x(\theta) - y(\theta) \|\, d\theta + \\ &+ \epsilon\lambda \int_{t_0}^{t} \| x(\theta - \tau(\theta, x(\theta))) - x(\theta) \|\, d\theta \leqslant \\ &\leqslant 2\epsilon\lambda \int_{t_0}^{t} \| x(\theta) - y(\theta) \|\, d\theta + \epsilon\lambda\mu P + \epsilon^2 \lambda M \int_{I^*} \tau(\theta, x(\theta))\, d\theta \leqslant \\ &\leqslant 2\epsilon\lambda \int_{t_0}^{t} \| x(\theta) - y(\theta) \|\, d\theta + \epsilon\lambda\mu P + \epsilon^2 \lambda M \int_{I^*} \Delta(\theta)\, d\theta, \end{aligned}$$

where $P = \sup_{t \in I_0} \| x(t - \tau(t, x(t))) - x(t) \|$.

According to the conditions of the theorem for ϵ sufficiently small ($\epsilon < \epsilon'_1$), the inequality

$$\epsilon^2 \int_{I^*} \Delta(\theta)\, d\theta < \frac{1}{4} \eta (\lambda M e^{2\lambda L})^{-1}$$

holds.

Let $\epsilon_1 = \min(\epsilon_2', \epsilon_1'')$, where $\epsilon_1'' = \frac{1}{4}\eta(\lambda\mu P e^{2\lambda L})^{-1}$. Then it follows from the Gronwall – Bellman lemma, that for $\epsilon < \epsilon_1$ and $t \in I^\epsilon$ inequality (5) is fulfilled.

On the other hand, following Theorem III. 1 [8], it is not difficult to see that for arbitrary $\frac{\eta}{2}$ and $L > 0$ there exists an $\epsilon_2 = \epsilon_2\left(\frac{\eta}{2}, L\right)$, such that for $\epsilon < \epsilon_2$ and $t \in I^\epsilon$ the inequality

$$\| y(t) - \xi(t) \| < \frac{\eta}{2}$$

holds.

Therefore, when $\epsilon < \epsilon_0 = \min(\epsilon_1, \epsilon_2)$ and $t \in I^\epsilon$ the inequality

$$\| x(t) - \xi(t) \| \leqslant \| x(t) - y(t) \| + \| y(t) - \xi(t) \| < \eta$$

is satisfied, hence Theorem 1 is proved.

Consider the initial value problem

$$\begin{aligned} &\dot{x}(t) = \epsilon X\left(t, x(t), x_\tau(t), \int_{t_0}^{t} \varphi(t, s, x(s), x_\tau(s))\, ds\right), \qquad t \in I \\ &x(t) = \kappa(t), \qquad t \in \bar{I}, \end{aligned} \tag{8}$$

where $x, X \in R^n$, $x_\tau(t) = x(t - \tau(t, x(t)))$, $\varphi \in R^m$, $\kappa(t)$ is an initial function, defined and continuous when $t \in \bar{I}$, and $\epsilon > 0$ is a small parameter. Consider the following schemes of averaging.

First scheme of averaging. Suppose that the integral

$$\int_{t_0}^{t} \varphi(t, s, x, u)\, ds,$$

where t, x and u are parameters exists, calculated and define

$$\int_{t_0}^{t} \varphi(t, s, x, u)\, ds = \varphi_1(t_0, t, x, u).$$

Suppose that the limit

$$(9) \qquad \lim_{T\to\infty} \frac{1}{T} \int_{t_0}^{t_0+T} X(t, x, u, \varphi_1(t_0, t, x, u))\, dt = \bar{X}_1(x, u)$$

exists.

Then, the system (8) we put in to correspondence the averaged system

$$(10) \qquad \begin{aligned} &\dot{\xi}(t) = \epsilon \bar{X}_1(\xi(t), \xi(t)), \qquad t \in I, \\ &\xi(t_0) = \kappa(t_0). \end{aligned}$$

Second scheme of averaging. Suppose that the integral

$$\int_{t_0}^{\infty} \varphi(t, s, x, u)\, ds,$$

where t, x and u are parameters, exists and define

$$\int_{t_0}^{\infty} \varphi(t, s, x, u)\, ds = \varphi_2(t_0, t, x, u).$$

Suppose that the limit

$$(11) \qquad \lim_{T\to\infty} \frac{1}{T} \int_{t_0}^{t_0+T} X(t, x, u, \varphi_2(t_0, t, x, u))\, dt = \bar{X}_2(x, u)$$

exists.

Then, the system (8) we put in to correspondence the averaged system

$$(12) \qquad \begin{aligned} &\dot{\xi}(t) = \epsilon \bar{X}_2(\xi(t), \xi(t)), \qquad t \in I, \\ &\xi(t_0) = \kappa(t_0). \end{aligned}$$

Theorem 2. *Let the following assumptions be fulfilled:*

1. *The function* $\tau(t, x)$ *is defined, continuous and non-negative in the domain* $\Omega(t, x) = \Omega(t) \times \Omega(x)$, *where* $\Omega(t) \equiv I$, *and* $\Omega(x)$ *is an open domain in the space* R^n.

The function $X(t, x, u, v)$ *is defined and continuous in the domain* $\Omega(t, x, u, v) = \Omega(t) \times \Omega(x) \times \Omega(u) \times \Omega(v)$, *where* $\Omega(u) = \Omega(x) \cup K$,

$$K = \bigcup_{-\infty}^{t=t_0} \{\kappa(t)\}, \qquad \Omega(v) \equiv R_m .$$

The function $\varphi(t, s, x, u)$ *is defined and continuous in the domain* $\Omega(t, s, x, u) = \Omega(t) \times \Omega(s) \times \Omega(x) \times \Omega(u)$, *where* $\Omega(s) \equiv I$.

2. *In the corresponding projections of the domain* $\Omega(t, s, x, u, v) = \Omega(t) \times \Omega(s) \times \Omega(x) \times \Omega(u) \times \Omega(v)$ *the following inequalities hold:*

$$\tau(t, x) \leqslant \Delta(t),$$

$$\| X(t, x, u, v) \| \leqslant M,$$

$$\| X(t, x, u, v) - X(t, x', u', v') \| \leqslant$$

$$\leqslant \lambda[\| x - x' \| + \| u - u' \| + \| v - v' \|],$$

$$\| \varphi(t, s, x, u) - \varphi(t, s, x', u') \| \leqslant \sigma(t, s)[\| x - x' \| + \| u - u' \|],$$

where λ *and* M *are positive constants,* $\Delta(t)$ *is a function continuous for* $t \in I$, *and the function* $\sigma(t, s)$ *satisfies*

$$\lim_{t \to \infty} \sigma_0(t - t_0) = \lim_{t \to \infty} \frac{1}{t - t_0} \int_{t_0}^{t} d\theta \int_{t_0}^{\theta} \sigma(\theta, s)\, ds = 0.$$

3. *The set*

$$I_0 = I \cap \{t\colon\ t - \Delta(t) \leqslant 0,\ t \in I\}$$

is of finite measure $\mu = \operatorname{mes} I_0 < \infty$.

4. *There exist the limits:*

$$\lim_{\epsilon \to 0} \epsilon^2 \int_{I^*} \Delta(t)\, dt = 0, \quad I^* = I^\epsilon \setminus I_0, \quad I^\epsilon = [t_0, t_0 + L\epsilon^{-1}],$$

$$L = \text{const.},$$

$$\lim_{\epsilon \to 0} \epsilon^2 \iint_{D^*} \sigma(\theta, s) \Delta(s)\, d\theta ds, \quad D^* = D^\epsilon \setminus D_0,$$

$$D_0 = \{(s, \theta)\colon\ s - \Delta(s) \leqslant 0,\ s \in I,\ s \leqslant \theta < \infty\},$$

$$D^\epsilon = I^\epsilon(s) \times I^\epsilon(\theta), \quad I^\epsilon(s) \equiv I^\epsilon(l) \equiv I^\epsilon.$$

5. *The initial value problem* (8) *has a unique and continuous solution* $x(t)$ *for* $t \geqslant 0$, *belonging to the domain* $\Omega(x)$ *for* $t \geqslant 0$.

6. *At every point* (x, u) *of the domain* $\Omega(x, u)$ *the limit* (9) *exists.*

The function $\bar{X}_1(x, u)$ *is continuous in the domain* $\Omega(x, u)$, *and satisfies the Lipschitz condition*

$$\|\bar{X}_1(x, u) - \bar{X}_1(x', u')\| \leqslant \nu[\|x - x'\| + \|u - u'\|], \qquad \nu = \text{const.}$$

7. *The Cauchy problem* (10) *has a unique and continuous solution* $\xi(t)$ *for* $t \geqslant 0$, *belonging to the domain* $\Omega(x)$ *for* $t \geqslant 0$.

Then, if $x(t)$ *is a solution of initial problem* (8), *and* $\xi(t)$ *is a solution of the Cauchy problem* (10), *then for every* $\eta > 0$ *and* $L > 0$ *there exists an* ϵ_0 $(\epsilon_0 = \epsilon_0(\eta, L))$ *such that for* $0 < \epsilon \leqslant \epsilon_0$ *and* $t \in I^\epsilon$ *the inequality*

$$\|x(t) - \xi(t)\| < \eta$$

holds.

Proof. For the solutions of the systems (8) and (10) the following equalities are fulfilled:

$$(13) \qquad x(t) = \begin{cases} \kappa(t_0) + \epsilon \int\limits_{t_0}^{t} X\Big(\theta, x(\theta), x_\tau(\theta), \\ \qquad\qquad \int\limits_{t_0}^{t} \varphi(\theta, s, x(s), x_\tau(s))\, ds\Big)\, d\theta, & t \in I^\epsilon, \\ \kappa(t), & t \in \bar{I} \end{cases}$$

$$(14) \qquad \xi(t) = \kappa(t_0) + \epsilon \int\limits_{t_0}^{t} \bar{X}_1(\xi(\theta), \xi(\theta))\, d\theta, \qquad t \in I^\epsilon.$$

Subtracting (14) from (13) the difference will be estimated. For this purpose the following representations will be used:

$$\epsilon \int_{t_0}^{t} \Big[X\Big(\theta, x(\theta), x_\tau(\theta), \int_{t_0}^{\theta} \varphi(\theta, s, x(s), x_\tau(s))\, ds\Big) -$$

$$- \bar{X}_1(\xi(\theta), \xi(\theta))\Big]\, d\theta =$$

$$= \epsilon \int_{t_0}^{t} \Big[X\Big(\theta, x(\theta), x_\tau(\theta), \int_{t_0}^{\theta} \varphi(\theta, s, x(s), x_\tau(s))\, ds\Big) -$$

$$- X\Big(\theta, \xi(\theta), \xi(\theta), \int_{t_0}^{\theta} \varphi(\theta, s, \xi(s), \xi(s))\, ds\Big)\Big]\, d\theta +$$

$$+ \epsilon \int_{t_0}^{t} \Big[X\Big(\theta, \xi(\theta), \xi(\theta), \int_{t_0}^{\theta} \varphi(\theta, s, \xi(s), \xi(s))\, ds\Big) -$$

$$- X\Big(\theta, \xi(\theta), \xi(\theta), \int_{t_0}^{\theta} \varphi(\theta, s, \xi(\theta), \xi(\theta))\, ds\Big)\Big]\, d\theta +$$

$$+ \epsilon \int_{t_0}^{t} \Big[X\Big(\theta, \xi(\theta), \xi(\theta), \int_{t_0}^{\theta} \varphi(\theta, s, \xi(\theta), \xi(\theta))\, ds\Big) -$$

$$- \bar{X}_1(\xi(\theta), \xi(\theta))\Big]\, d\theta.$$

We obtain

$$(15) \qquad \| x(t) - \xi(t) \| \leqslant \epsilon\lambda \int_{t_0}^{t} \Big\{\| x(\theta) - \xi(\theta)\| + \| x_\tau(\theta) - \xi(\theta)\| +$$

$$+ \int_{t_0}^{\theta} \sigma(\theta, s)[\| x(s) - \xi(s)\| + \| x_\tau(s) - \xi(s)\| +$$

$$+ 2\| \xi(s) - \xi(\theta)\|]\, ds\Big\}\, d\theta +$$

$$+ \epsilon\Big\| \int_{t_0}^{t} \Big[X\Big(\theta, \xi(\theta), \xi(\theta), \int_{t_0}^{\theta} \varphi(\theta, s, \xi(\theta), \xi(\theta))\, ds\Big) -$$

$$- \bar{X}_1(\xi(\theta), \xi(\theta))\Big]\, d\theta\Big\| \leqslant$$

$$\leqslant 2\epsilon\lambda \int_{t_0}^{t} \Big[\| x(\theta) - \xi(\theta)\| + \int_{t_0}^{\theta} \sigma(\theta, s)\| x(s) - \xi(s)\|\, ds\Big]\, d\theta +$$

$$+ \epsilon\lambda\mu P + \epsilon^2 \lambda M \int_{I^*} \Delta(\theta)\, d\theta + \epsilon\lambda P \int_{t_0}^{t} d\theta \int_{t_0}^{\theta} \sigma(\theta, s)\, ds +$$

$$+ \epsilon^2 \lambda M \iint_{D^*} \sigma(\theta, s) \Delta(s)\, d\theta ds + 2\epsilon^2 \lambda M \int_{t_0}^{t} d\theta \int_{t_0}^{\theta} (\theta - s) \sigma(\theta, s)\, ds +$$

$$+ \epsilon \Big\| \int_{t_0}^{t} \Big[X\Big(\theta, \xi(\theta), \xi(\theta), \int_{t_0}^{\theta} \varphi(\theta, s, \xi(\theta), \xi(\theta))\, ds\Big) -$$

$$- \bar{X}_1(\xi(\theta), \xi(\theta))\Big]\, d\theta \Big\|,$$

where

$$P = \sup_{t \in I^0} \| x(t - \tau(t, x(t))) - x(t) \|,$$

$$I^0 = I \cap \{t\colon\ t - \tau(t, x(t)) \leqslant 0,\ t \in I\}, \qquad I^0 \subset I_0.$$

Introducing the notations

$$X\Big(\theta, \xi(\theta), \xi(\theta), \int_{t_0}^{\theta} \varphi(\theta, s, \xi(\theta), \xi(\theta))\, ds\Big) = \hat{X}(\theta, \xi(\theta)),$$

$$\hat{X}(\theta, \xi(\theta)) - \bar{X}_1(\xi(\theta), \xi(\theta)) = \psi(\theta, \xi(\theta))$$

we have

$$\epsilon \Big\| \int_{t_0}^{t} \Big[X\Big(\theta, \xi(\theta), \xi(\theta), \int_{t_0}^{\theta} \varphi(\theta, s, \xi(\theta), \xi(\theta))\, ds\Big) -$$

$$- \bar{X}_1(\xi(\theta), \xi(\theta))\Big]\, d\theta \Big\| = \epsilon \Big\| \int_{t_0}^{t} \psi(\theta, \xi(\theta))\, d\theta \Big\|.$$

It will be shown that for every $\epsilon > 0$ there exists a function $a(\epsilon, m)$, monotonously tending to zero, as $\epsilon \to 0$ and $m \to \infty$, for which the inequality

(16) $$\epsilon \Big\| \int_{t_0}^{t} \psi(\theta, \xi(\theta))\, d\theta \Big\| \leqslant a(\epsilon, m), \qquad t \in I^\epsilon$$

holds.

Indeed, divide the interval I^ϵ into m equal parts by the points

$$t_k = t_0 + \frac{kL}{\epsilon m}, \qquad k = \overline{0, m}$$

and evaluate the integral

$$\int_{t_0}^{t} \psi(\theta, \xi(\theta))\, d\theta$$

on the whole interval I^ϵ, assuming that for $\theta > t$ the function ψ vanishes. We have

$$\begin{aligned} &\epsilon \left\| \int_{t_0}^{t} \psi(\theta, \xi(\theta))\, d\theta \right\| \leqslant \\ &\leqslant \epsilon \left\| \sum_{k=0}^{m-1} \int_{t_k}^{t_{k+1}} [\psi(\theta, \xi(\theta)) - \psi(\theta, \xi_k)]\, d\theta \right\| + \\ &+ \epsilon \left\| \sum_{k=0}^{m-1} \int_{t_k}^{t_{k+1}} \psi(\theta, \xi_k)\, d\theta \right\|, \end{aligned} \tag{17}$$

where $\xi_n = \xi(t_k)$.

For the first term on the right hand side of (17) the inequalities

$$\begin{aligned} &\epsilon \left\| \sum_{k=0}^{m-1} \int_{t_k}^{t_{k+1}} [\psi(\theta, \xi(\theta)) - \psi(\theta, \xi_k)]\, d\theta \right\| \leqslant \\ &\leqslant 2\epsilon \sum_{k=0}^{m-1} \int_{t_k}^{t_{k+1}} \left(\lambda + \nu + \lambda \int_{t_0}^{\theta} \sigma(\theta, s)\, ds \right) \times \\ &\times \| \xi(\theta) - \xi_k \|\, d\theta \leqslant \\ &\leqslant 2\epsilon^2 M \sum_{k=0}^{m-1} \int_{t_k}^{t_{k+1}} \left(\lambda + \nu + \lambda \int_{t_0}^{\theta} \sigma(\theta, s)\, ds \right) | \theta - t_k |\, d\theta \leqslant \\ &\leqslant (\lambda + \nu) \frac{ML^2}{m} + \frac{2\epsilon\lambda ML}{m} \int_{t_0}^{t_0 + L\epsilon^{-1}} d\theta \int_{t_0}^{\theta} \sigma(\theta, s)\, ds \leqslant \\ &\leqslant (\lambda + \nu) \frac{ML^2}{m} + \frac{2\lambda ML}{m} \delta(\epsilon) \equiv \alpha(\epsilon, m), \end{aligned} \tag{18}$$

hold, where

$$\delta(\epsilon) = \sup_{0 \leqslant \tau \leqslant L} \tau \sigma_0(\tau \epsilon^{-1}) \to 0, \qquad \epsilon \to 0.$$

Therefore, $\alpha(\epsilon, m) \to 0$ as $m \to \infty$ and $\epsilon \to 0$.

Introduce the notation

$$\Phi(t - t_0, \xi) = \left\| \frac{1}{t - t_0} \int_{t_0}^{t} [X(\theta, \xi, \xi, \varphi_1(t_0, \theta, \xi, \xi)) - \right.$$
$$\left. - \bar{X}_1(\xi, \xi)] \, d\theta \right\|.$$

By the assumptions of Theorem 2, for every fixed ξ in the domain $\Omega(x)$, $\Phi(t - t_0, \xi) \to 0$ as $t \to \infty$. Hence, for $t \geqslant t_0$ the equality

$$\epsilon \left\| \int_{t_0}^{t} \psi(\theta, \xi) \, d\theta \right\| = \epsilon(t - t_0) \Phi(t - t_0, \xi) = \tau \Phi\left(\frac{\tau}{\epsilon}, \xi\right),$$

holds, where $\tau = \epsilon(t - t_0)$.

For k, m and ξ_k are fixed this yields

$$\epsilon \left\| \int_{t_0}^{t_1} \psi(\theta, \xi_0) \, d\theta \right\| \leqslant \sup_{0 \leqslant \tau \leqslant \frac{L}{m}} \tau \Phi\left(\frac{\tau}{\epsilon}, \xi_0\right) \to 0 \qquad \text{as} \qquad \epsilon \to 0,$$

$$\epsilon \left\| \int_{t_0}^{t_1} \psi(\theta, \xi_1) \, d\theta \right\| \leqslant \sup_{0 \leqslant \tau \leqslant \frac{L}{m}} \tau \Phi\left(\frac{\tau}{\epsilon}, \xi_1\right) \to 0 \qquad \text{as} \qquad \epsilon \to 0,$$

$$\epsilon \left\| \int_{t_0}^{t_{k+1}} \psi(\theta, \xi_k) \, d\theta \right\| \leqslant L \Phi\left(\frac{(k+1)L}{\epsilon m}, \xi_k\right) \to 0$$

$$(k = \overline{1, m-1}) \qquad \text{as} \qquad \epsilon \to 0,$$

$$\epsilon \left\| \int_{t_0}^{t_k} \psi(\theta, \xi_k) \, d\theta \right\| \leqslant L \Phi\left(\frac{kL}{\epsilon m}, \xi_k\right) \to 0$$

$$(k = \overline{1, m-2}) \qquad \text{as} \qquad \epsilon \to 0.$$

Therefore,

$$\epsilon\Big\| \sum_{k=0}^{m-1} \int_{t_k}^{t_{k+1}} \psi(\theta, \xi_k)\, d\theta \Big\| \leqslant \epsilon \Big\| \int_{t_0}^{t_1} \psi(\theta, \xi_0)\, d\theta \Big\| +$$

$$+ \epsilon \Big\| \int_{t_0}^{t_1} \psi(\theta, \xi_1)\, d\theta \Big\| + \epsilon \sum_{k=1}^{m-1} \Big\| \int_{t_k}^{t_{k+1}} \psi(\theta, \xi_k)\, d\theta \Big\| +$$

$$(19) \qquad + \epsilon \sum_{k=2}^{m-1} \Big\| \int_{t_0}^{t_k} \psi(\theta, \xi_k)\, d\theta \Big\| \leqslant$$

$$\leqslant \sup_{0 \leqslant \tau \leqslant \frac{L}{m}} \tau \Phi\Big(\frac{\tau}{\epsilon}, \xi_0\Big) + \sup_{0 \leqslant \tau \leqslant \frac{L}{m}} \tau \Phi\Big(\frac{\tau}{\epsilon}, \xi_1\Big) +$$

$$+ L \sum_{k=1}^{m-1} \Phi\Big(\frac{(k+1)L}{\epsilon m}, \xi_k\Big) + L \sum_{k=2}^{m-1} \Phi\Big(\frac{kL}{\epsilon m}, \xi_k\Big) \equiv \beta(\epsilon, m).$$

Obviously for fixed m

$$\beta(\epsilon, m) \to 0 \qquad \text{as} \qquad \epsilon \to 0.$$

Thus (17)-(19) yield inequality (20), and

$$a(\epsilon, m) = \alpha(\epsilon, m) + \beta(\epsilon, m).$$

Introduce the notations

$$\gamma_1(\epsilon) = \epsilon^2 \int_{I^*} \Delta(\theta)\, d\theta, \quad \gamma_2(\epsilon) = \epsilon \iint_{D^*} \sigma(\theta, s) \Delta(s)\, d\theta\, ds,$$

$$\gamma_i(\epsilon) \to 0 \qquad (i = 1, 2) \qquad \text{as} \qquad \epsilon \to 0.$$

From (15) and (16), when $t \in I^\epsilon$, it follows

$$\| x(t) - \xi(t) \| \leqslant \{ \epsilon\lambda\mu P + \lambda M[\gamma_1(\epsilon) + \gamma_2(\epsilon)] +$$

$$(20) \qquad + \lambda(2ML + P)\delta(\epsilon) + a(\epsilon, m)\} +$$

$$+ 2\epsilon\lambda \int_{t_0}^{t} \Big[\| x(\theta) - \xi(\theta) \| + \int_{t_0}^{\theta} \sigma(\theta, s) \| x(s) - \xi(s) \|\, ds \Big] d\theta.$$

For the remainder of the proof we shall need the following lemma.

Lemma 1 [8]. *Suppose that for* $t \geqslant \theta \geqslant t_0$ *the following assumption hold:*

1. *The functions* $u(t)$, $\nu(t)$ *and* $\mu(t, \theta)$ *are nonnegative and continuous.*

2. *The function* $u(t)$ *satisfies the inequality*

$$u(t) \leqslant \alpha + \beta \int_{t_0}^{t} [\nu(\theta) u(\theta) + \int_{t_0}^{\theta} \mu(\theta, s) u(s)\, ds]\, d\theta, \quad \alpha = \text{const.} > 0,$$

$$\beta = \text{const.} > 0.$$

Then

$$u(t) \leqslant \alpha \exp \{\beta \int_{t_0}^{t} [\nu(\theta) + \int_{t_0}^{\theta} \mu(\theta, s)\, ds]\, d\theta.$$

Applying Lemma 1 to (20), we obtain:

$$\| x(t) - \xi(t) \| \leqslant A(\epsilon, m) \exp \{2\lambda(L + \delta(\epsilon))\},$$

where

$$A(\epsilon, m) = \epsilon\lambda\mu P + \lambda M[\gamma_1(\epsilon) + \gamma_2(\epsilon)] +$$

$$+ \lambda(2ML + P)\, \delta(\epsilon) + a(\epsilon, m)$$

and $A(\epsilon, m) \to 0$ as $m \to \infty$ and $\epsilon \to 0$.

Hence Theorem 2 is proved.

Theorem 3. *Assume, that*

1. *Conditions* 1-5 *of Theorem* 2 *are satisfied.*

2. *At every point* (x, u) *of the domain* $\Omega(x, u)$ *the limit* (11) *exists.*

The function $\bar{X}_2(x, u)$ *is continuous in the domain* $\Omega(x, u)$, *and satisfies the Lipschitz condition*

$$\| \bar{X}_2(x, u) - \bar{X}_2(x', u') \| \leqslant \nu[\| x - x' \| + \| u - u' \|], \quad \nu = \text{const.}$$

3. *For* $t \geqslant 0$ *the Cauchy problem* (12) *has a unique and continuous, solution* $\xi(t)$, *belonging to the domain* $\Omega(x)$ *for* $t \geqslant 0$.

4. *For each solution* $\xi(t)$

$$\lim_{t\to\infty} \frac{1}{t-t_0} \int_{t_0}^{t} d\theta \left\| \int_{\theta}^{\infty} \varphi(\theta, s, \xi(\theta), \xi(\theta))\, ds \right\| = 0.$$

5. *The function* $\varphi_2(t_0, t, x, u)$, *in the domain* $\Omega(t, x, u)$ *satisfies the Lipschitz condition*

$$\| \varphi_2(t_0, t, x, u) - \varphi_2(t_0, t, x', u') \| \leqslant \beta[\| x - x' \| + \| u - u' \|],$$

$$\beta = \text{const.}$$

Then, if $x(t)$ *is a solution of initial problem* (8), *and* $\xi(t)$ *is a solution of the Cauchy problem* (12), *then for every* $\eta > 0$ *and* $L > 0$ *there exists an* $\epsilon_0 > 0$ $(\epsilon_0 = \epsilon_0(\eta, L))$, *such that for* $0 < \epsilon \leqslant \epsilon_0$ *and* $t \in I^\epsilon$ *the inequality*

$$\| x(t) - \xi(t) \| < \eta$$

holds.

Proof. The proof of Theorem 3 is analogous to that of Theorem 2, hence it will be omitted.

Remark 1. Replacing condition 5 and the condition

$$\lim_{t\to\infty} \frac{1}{t-t_0} \int_{t_0}^{t} d\theta \int_{t_0}^{\theta} \sigma(\theta, s)\, ds = 0$$

of Theorem 3 by the condition

$$\lim_{t\to\infty} \frac{1}{t-t_0} \int_{t_0}^{t} d\theta \int_{t_0}^{\infty} \sigma(\theta, s)\, ds = 0$$

then, the assertion of Theorem 3 still holds.

REFERENCES

[1] R.R. Ahmerov, On the averaging principle for functional differential equations of neutral type, *Ukrain. Mat. Z.*, 25, 5 (1973), 579-588 (in Russian).

[2] L.G. Fedorenko, On the averaging for integro-differential equations with delay depending on the solution, *Ukrain. Mat. Ž.*, 25, 5 (1973), 696-701 (in Russian)

[3] A.N. Filatov, Averaging methods for differential and integro-differential equations, Izd. FAN UzSSR, Tashkent, 1971 (in Russian).

[4] V.I. Fodchuk, Averaging method for difference-differential equation of neutral type, *Ukrain. Mat. Ž.*, 20, 2 (1968), 203-209 (in Russian).

[5] V.I. Fodchuk, On the continuous dependence on parameters of solutions of retarded differential equations of neutral type, *Ukrain. Mat. Ž.*, 16, 2 (1964), 273-279 (in Russian).

[6] V.I. Fodchuk, On the fundaments of the averaging method for differential equations with retarded argument, III. *Konferenz Über nichtlineare Schwingungen,* Berlin, vom 25 bis 30 May 1964, Vol. 1, Akademie-Verlag, 1965, 45-50 (in Russian).

[7] V.I. Fodchuk, Foundation of the averaging method for a class of singularly perturbed system with delay, *Trudy Sem. Teor. Differencial. Uravnenii Otklon. Argumentom Univ. Družby Narodov Patrisa Lumumby.* 4 (1967), 163-172 (in Russian).

[8] A. Halanai, Systems with delay, *Matematika – Period. Sb. Perevodov Inostran. Stateĭ.* 10 (1966), 85-102 (in Russian).

[9] A. Halanai, The averaging method for systems of differential equations with delay, *Rev. Math. pures et appl. Acad.* R. RR. 4 (1959), 467-483 (in Russian).

[10] J.K. Hale, Averaging methods for differential equations with retarded arguments and a small parameter, *Brown University, Technical Report* 1964, 64-1.

[11] J.K. Hale, Averaging methods for differential equations with retarded arguments and a small parameter, *J. Diff. Eqn.* 2 (1966), 57-73.

[12] U.A. Mitropolski – V.I. Fodchuk, Fundations of the averaging method for difference-differential equation in Hilbert-space, *Ukrain Mat. Ž.*, 23 (1971), 745-752 (in Russian).

[13] U.A. Mitropolski – V.I. Fodchuk, Applying asymptotic methods of nonlinear mechanics for nonlinear differential equations with delay, *Ukrain Mat. Ž.*, 18 (1966), 65-84 (in Russian).

[14] V.P. Rubanik, Justifying the applicability of averaging method for systems of difference-differential equations, *Nauch. Ezegod. Chernovitsk. Univ,* 1959 (in Russian).

[15] V.M. Volosov – G.M. Medvedev – B.I. Morgunov, An application of the averaging method for solving a system of differential equations with delay, *Vestnik M.G.U. Ser.* III, *Fizika, Astronomija,* 6 (1965), 89-91.

[16] V.M. Volosov – G.M. Medvedev – B.I. Morgunov, On the application of the averaging method for certain systems of differential equations with delay, *Vestink M.G.U. Ser.* III, *Fizika, Astronomija,* 1968, 251-294.

D.D. Bainov – S.D. Milusheva

University of Plovdiv, Paissji Hilendarski, Bulgaria.

COLLOQUIA MATHEMATICA SOCIETATIS JÁNOS BOLYAI
15. DIFFERENTIAL EQUATIONS, KESZTHELY (HUNGARY), 1975.

ON THE MOST GENERAL SOLUTION OF SOME BOUNDARY VALUE PROBLEMS FOR THE DIFFERENTIAL EQUATION $\sum_{k=0}^{n} c_k(D_1 + aD_2)^k U = 0$

I. BARTSCH

The aim of this work is to find a distribution U satisfying the differential equation

$$(1) \qquad \sum_{k=0}^{n} c_k(D_1 + aD_2)^k U = 0$$

and assuming prescribed values at certain points. These are not necessarily functions: distributions as boundary values are also permitted. In (1), D_1 and D_2 denote distribution derivatives with respect to the first and second independent variable, respectively. We assume $a(x, y), c_k(x, y) \in \in C^\infty(R^2)$; this assures that the left-hand side of (1) is defined for all distributions. Denote by $\mathscr{D}(R^n)$ the fundamental space of test functions in the sense of L. Schwarz, and let $\mathscr{D}'(R^n)$ be the space of distributions [5]. The value of a distribution T at φ will be denoted by $T\varphi$. The test functions and distributions will be assumed to be real. When considering boundary value problems for distributions a common difficulty arises, since the value of a distribution at a given point may not exist.

In the paper it will be shown that if a distribution U satisfies the differential equation (1) then the corresponding local values do exist (in the sense of Łojasiewicz [7]). Hence the boundary value problems considered below are meaningful for distributions too.

1. In [2] it was shown that the distribution solutions of the differential equation (1) have the general form

$$U = \sum_{j=1}^{n} \mathfrak{w}_j(Z_j) \tag{2}$$

where

$$\mathfrak{w}_j(Z_j)\varphi = Z_j \int_{R^1} Y_j(t, s) \frac{\partial(x, y)}{\partial(t, s)} \varphi(x(t, s), y(t, s))\, dt$$

$$(j = 1, 2, \ldots, n;\ \varphi \in \mathscr{D}(R^2)),$$

Z_j $(j = 1, 2, \ldots, n)$ are arbitrary distributions belonging to $\mathscr{D}'(R^1)$, and $\{Y_j(t, s)\}$ is a fundamental system of the ordinary differential equation arising from equation (1).

For deducing the general solution (2), a diffeomorphism [6] $s = s(x, y)$, $t = t(x, y)$ of the plane onto itself with the following properties is used (a)-(d):

(a) $s(x, y) \in C^\infty(R^2)$.

(b) The function $s = s(x, y)$ satisfies the differential equation $(D_1 + aD_2)U = 0$.

(c) $s = s(x, y)$ is strictly monotone as a function of x and as a function of y. Hence there exists a function $f(x, s)$, strictly monotone in x, such that

$$y = f(x, s(x, y)).$$

(d) $t = t(x, y)$ has the properties (a) and (b).

Using the notations of [2] we prove the following lemmas:

Lemma 1. *For arbitrary fixed* x *the transformation*

$$\mathfrak{v}_x^*(\psi) = \psi(f(x,s))\,\frac{\partial f(x,s)}{\partial s} \qquad (\psi \in \mathcal{D}(R^1)) \tag{3}$$

is a continuous linear map of $\mathcal{D}(R^1)$ *into* $\mathcal{D}(I_x)$, *where* I_x *denotes the range of* $f(x,s)$ *for fixed* x.

Proof. Since ψ is a test function of one variable, its support is contained in some finite interval:

$$\operatorname{supp}\psi \subset (r_1, r_2), \qquad r_1 < r_2.$$

By assumption, the map $t = t(x,y)$, $s = s(x,y)$ is a diffeomorphism, hence it carries maps compact supports into compact supports. Again by the assumption, $y = f(x,s)$ is a monotone function of s, therefore its domain is a (not necessarily finite) interval I_x. Thus for fixed x there exist s_1 and s_2 such that

$$r_1 = f(x, s_1) \quad \text{and} \quad r_2 = f(x, s_2),$$

i.e. $\operatorname{supp}\psi(f(x,s)) \subset (s_1, s_2)$.

This yields $\mathfrak{v}_x^*(\psi) \in \mathcal{D}(I_x) \subset \mathcal{D}(R^1)$. The proof of linearity and continuity is obvious.

On the boundary of I_x the function $\psi(f(x,s))$ vanishes together with all its derivatives, and it is infinitely differentiable with respect to s at any point of I_x. We extend it to the whole real line by defining $\psi(f(x,s)) \equiv 0$ for $s \notin I_x$. Let the function $\dfrac{\partial f(x,s)}{\partial s}$ also be extended in some way to the points $s \notin I_x$. Then can write $\mathfrak{w}_x^* \in \mathcal{D}(R^1)$.

We denote by $\mathfrak{v}_x$ the adjoint of the transformation $\mathfrak{v}_x^*$. By definition,

$$\mathfrak{v}_x(V)\,\psi := V\mathfrak{v}_x^*(\psi) \tag{4}$$

for all $V \in \mathcal{D}'(R^1)$ and $\psi \in \mathcal{D}(R^1)$. By Lemma 1, $\mathfrak{v}_x$ is a one-to-one, linear and continuous transformation of $\mathcal{D}'(I_x)$ into $\mathcal{D}'(R^1)$.

Lemma 2. *If* $\{Y_j(t, s)\}$ *denotes a fundamental system required for the representation of the general solution of the differential equation* (1) and Z_j *is an arbitrary distribution belonging to* $\mathscr{D}'(R^1)$, *then the functions*

$$Z_j Y_j(t(x, f(x, s)), s)\mathfrak{v}_x^*(\psi) = F_j(x) \qquad (j = 1, 2, \ldots, n) \tag{5}$$

continuously depend on x.

Proof. For any fixed $x = x_0$ and $h > 0$ we have

$$\begin{aligned} & F_j(x_0 + h) - F_j(x_0) = \\ & = Z_j \cdot Y_j(t(x_0 + h, f(x_0 + h; s)), s)\mathfrak{v}^*_{x_0+h}(\psi) - \\ & - Z_j \cdot Y_j(t(x_0, f(x_0; s)), s)\mathfrak{v}^*_{x_0}(\psi) = \\ & = Z_j \cdot \{Y_j(t(x_0 + h, f(x_0 + h; s)), s)[\mathfrak{v}^*_{x_0+h}(\psi) - \mathfrak{v}^*_{x_0}(\psi)] + \\ & + \mathfrak{v}^*_{x_0}(\psi)[Y_j(t(x_0 + h, f(x_0 + h; s)), s) - \\ & - Y_j(t(x_0, f(x_0; s)), s)] \, . \end{aligned}$$

It is easily seen that

$$\mathfrak{v}^*_{x_0+h}(\psi) - \mathfrak{v}^*_{x_0}(\psi) \xrightarrow{\mathscr{D}} 0$$

with respect to s, as $h \to 0$.

Moreover, for all $p \geqslant 0$,

$$\frac{\partial^p Y_j(t(x_0 + h, f(x_0 + h; s)), s)}{\partial s^p} - \frac{\partial^p Y_j(t(x_0, f(x_0; s)), s)}{\partial s^p} \to 0$$

uniformly for s in the support of $\mathfrak{v}^*_{x_0}(\psi)$. This completes the proof.

2. In [4] I. Fenyő introduced a transformation

$$\circ\colon \mathscr{D}'(R^n \times R^n) \times \mathscr{D}(R^n) \to \mathscr{D}'(R^n)$$

in the following way. Let $\alpha \in \mathscr{D}(R^n)$, $A \in \mathscr{D}'(R^n \times R^n)$, and $\zeta \in \mathscr{D}(R^n)$ an arbitrary test function. Then

(6) $$(A \circ \alpha)\zeta := A(\zeta \otimes \alpha).$$

(Here $\otimes$ denotes the tensor product of two functions.) If A is represented by a locally integrable function $f \in L_{\text{loc}}(R^n \times R^n)$, then

$$(f \circ \alpha)\zeta = \int_{R^n} \left(\int_{R^n} f(x, y)\alpha(y)\, dy \right) \zeta(x)\, dx,$$

i.e. $f \circ \alpha$ concides with the distribution defined by the function

$$\int_{R^n} f(x, y)\alpha(y)\, dy.$$

For an arbitrary term of the general solution (2) obtain

$$(\mathfrak{w}_j(Z_j) \circ \psi)\zeta = \mathfrak{w}_j(Z_j) \cdot \zeta \otimes \psi =$$

$$= Z_j \int_{R^1} \frac{\partial(x, y)}{\partial(t, s)} Y_j(t, s)\zeta(x(t, s))\, \psi(y(t, s))\, dt \quad (\zeta, \psi \in \mathscr{D}(R^1)).$$

Introducing the new variable x we find

(7) $$(\mathfrak{w}_j(Z_j) \circ \psi)\zeta = Z_j \int_{R^1} Y_j(t(x, f(x, s)), s)\mathfrak{v}_x^*(\psi)\zeta(x)\, dx.$$

Approximating the integral on the right-hand side by finite sums in the usual way it can be shown that the distribution Z_j as an operator commutes with the integral operator, i.e.

$$(\mathfrak{w}_j(Z_j) \circ \psi)\zeta =$$

$$= \int_{R^1} Z_j \cdot Y_j(t(x, f(x, s)), s)\mathfrak{v}_x^*(\psi)\zeta(x)\, dx = \int_{R^1} F_j(x)\zeta(x)\, dx.$$

Applying this result to the general solution (2) of equation (1) it follows that

$$(U \circ \psi) \cdot \zeta = \left(\sum_{j=1}^{n} \mathfrak{w}_j(Z_j) \circ \psi \right) \cdot \zeta = \sum_{j=1}^{n} (\mathfrak{w}_j(Z_j) \circ \psi) \cdot \zeta =$$

$$= \sum_{j=1}^{n} \int_{R^1} F_j(x)\zeta(x)\, dx.$$

Since each integrand in the last sum vanishes outside a finite interval, we may write

(8) $$\left(\sum_{j=1}^{n} \mathfrak{w}_j(Z_j) \circ \psi\right) \zeta = \int_{R^1} \sum_{j=1}^{n} F_j(x)\zeta(x)\,dx.$$

Hence, at any point x_0, the distribution $\sum_{j=1}^{n} \mathfrak{w}_j(Z_j) \circ \psi$ admits a local value (in the sense of Łojasiewicz) equal to

$$\sum_{j=1}^{n} F_j(x_0) = \sum_{j=1}^{n} Z_j Y_j(t(x_0, f(x_0, s)), s)\mathfrak{v}^*_{x_0}(\psi).$$

By (5),

(9) $$F_j(x) = \mathfrak{v}_x [Y_j(t(x, f(x, s)), s) Z_j]\, \psi.$$

Hence

(10) $$\sum_{j=1}^{n} \mathfrak{w}_j(Z_j) \circ \psi|_{x=x_0} = \sum_{j=1}^{n} \mathfrak{v}_{x_0}[Y_j(t(x_0, f(x_0, s)), s) Z_j]\, \psi.$$

Similar considerations yield the corresponding assertions about the existence of local values (in the sense of Łojasiewicz) at points y_0.

3. We are going to apply the above results to boundary value problems.

Boundary value problem A

Given the real numbers $x_1, x_2, \ldots, x_n$ $(x_i \neq x_j$ for $i \neq j)$ and the distributions $F_1, F_2, \ldots, F_n$ $(F_i \in \mathcal{D}'(R^1)$ for $i = 1, 2, \ldots, n)$ find a solution U of the differential equation (1) satisfying the conditions

(11) $$U|_{x=x_i} = F_i \qquad (i = 1, 2, \ldots, n).$$

A solution U of the form (2) of equation (1) satisfies the boundary conditions (11) provided

$$U|_{x=x_i} \cdot \psi = U \circ \psi|_{x=x_i} =$$

(12) $$= \sum_{j=1}^{n} \mathfrak{v}_{x_i}[Y_j(t(x_i, f(x_i; s)), s) Z_j] \cdot \psi = F_i \cdot \psi$$

$$(i = 1, 2, \ldots, n).$$

This is equivalent to

$$\sum_{j=1}^{n} Y_j(t(x_i, f(x_i, s)), s) Z_j \psi = \mathfrak{v}_{x_i}^{-1}(F_i) \psi$$

or

$$\sum_{j=1}^{n} Y_j(t(x_i, f(x_i, s)), s) Z_j = \mathfrak{v}_{x_i}^{-1}(F_i) \qquad (i = 1, 2, \ldots, n). \tag{13}$$

Hence the existence problem for the boundary value problem A is reduced to the determination of distributions Z_j satisfying the system of equations (13). We write down the determinant of the system:

$$\begin{vmatrix} Y_1(t(x_1, f(x_1; s)), s) & Y_2(t(x_1, f(x_1; s)), s) & \ldots & Y_n(t(x_1, f(x_1; s)), s) \\ \vdots & \vdots & & \vdots \\ Y_1(t(x_n, f(x_n; s)), s) & Y_2(t(x_n, f(x_n; s)), s) & \ldots & Y_n(t(x_n, f(x_n; s)), s) \end{vmatrix} \tag{14}$$

Since the functions $Y_j(t(x, f(x, s)), s)$ $(j = 1, 2, \ldots, n)$ are linearly independent for fixed s, the determinant (14) is non-zero (see for example [1], p. 201). Thus the distributions Z_j can be uniquely determined from the system (13).

It may happen that the distributions determined in this way are only defined on a subset of $\mathscr{D}(R^1)$.

Anyway, the generalized Hahn – Banach theorem yields extensions to the whole space $\mathscr{D}(R^1)$. Though, in general, there exists an infinite number of different extensions, they all coincide on a certain subset of R^2.

Substituting the distributions Z_j $(j = 1, 2, \ldots, n)$ evaluated from (13), into (2) we obtain the desired solution of the boundary value problem A. As a result, we have the following

Theorem. *For arbitrary distributions* F_i $(i = 1, 2, \ldots, n)$, *boundary value problem* A *admits at least a solution.*

It can be shown that if the function

$$y = f(x, s)$$

is defined for all real values of x and s then the above solution is unique.

If the function $f(x, s)$ is defined on a domain $J_s \times R^1$, where J_s is a proper subdomain of R^1, then boundary value problem A admits infinitely many solutions, which however, coincide on a certain subdomain of R^2.

A similar result holds for

Boundary value problem B

Given the real numbers $y_1, y_2, \ldots, y_n$ $(y_i \neq y_j$ for $i \neq j)$ and the distributions $H_1, H_2, \ldots, H_n$ $(H_i \in \mathscr{D}'(R^1)$ for $i = 1, 2, \ldots, n)$, find a solution U of the differential equation (1) satisfying the conditions

$$(15) \qquad U|_{y=y_i} = H_i \qquad (i = 1, 2, \ldots, n).$$

We mention the following special case: if the coefficient a occurring in equation (1) is independent of x and y, then the conditions for the unique solvability of boundary value problems A and B are satisfied. This yields a former result of I. Fenyő [3].

REFERENCES

[1] J. Aczél, *Lectures on Functional Equation and their Applications.* New York – London, 1966.

[2] I. Bartsch, Über die Differentialgleichung $\sum_{k=0}^{n} c_k(D_1 + aD_2)^k U = 0$. *Wissensch. Zeitschrift der Universität Rostock* XXIII (1974), 9.

[3] I. Fenyő, On a Differential Equation of two Variables. *Demonstratio Math.*, 2 (1973), 6.

[4] I. Fenyő. Sur les équations distributionelles, in: *Functional Equations and Inequalities,* CIME, 3. Ciclo, *La Mendola* 1970; *Roma* 1971.

[5] I. Fenyő – T. Frey, *Modern Mathematical Methods in Technology I,* Amsterdam – London 1969.

[6] L. Hörmander, *Linear Partial Differential Operators.* Berlin – Göttingen – Heidelberg 1963.

[7] S. Łojasiewicz, Sur la valeur et la limite d'une distribution en un point. *Studia Math.*, 16 (1958), 1-36.

I. Bartsch

Institut für Lehrerbildung, Rostock-Lichtenhagen, GDR.

COLLOQUIA MATHEMATICA SOCIETATIS JÁNOS BOLYAI
15. DIFFERENTIAL EQUATIONS, KESZTHELY (HUNGARY), 1975.

THE DIFFERENTIAL EQUATIONS OF QUANTUM MECHANICS

E.M. BRUINS

§1. INTRODUCTION

W.R. Hamilton (1805-1865), after years of intensive research, suddenly discovered – in 1843, October the 16-th – the quaternions

$$\alpha = a_0 + a_1 i + a_2 j + a_3 k,$$

$$a_0, a_1, a_2, a_3 \quad \text{real numbers,}$$

$$i^2 = j^2 = k^2 = -1, \qquad ij = k, \quad jk = i, \quad ki = j,$$

quantities with a non commutative multiplication. This new mathematical tool was immediately exploited by theoretical physicists, who – indeed – succeeded then in writing their formulas very concisely. The reason for this success is easily understood in our times, as considering $i, j, k,$ as symbols for three mutually prependicular unit vectors, $e_1, e_2, e_3,$ with $a_0 = b_0 = 0$ we have

$$-\frac{1}{2}(\alpha\beta + \beta\alpha) = ab, \qquad \text{the scalar vector product,}$$

$$\frac{1}{2}(\alpha\beta - \beta\alpha) = [ab], \quad \text{the vectorial vector product,}$$

and these quantities were badly needed by J.C. Maxwell (1831-1879) and H.A. Lorentz (1853-1928) in their new theories of the electromagnetic field and for the Lorentz-force.

Around the beginning of the XX-th century J.W. Gibbs – presumedly not seeing good reasons for always taking $a_0 = b_0 = 0$ and always determining half of the sum and the difference of two products – isolated from the quaternions the scalar and the vector product . . . and still in 1913 L. Silberstein indicated in his famous "Vectorial Mechanics" initially: "the main object of this little volume is to present the chief principles and theorems of theoretical mechanics in the language of vectors, and thereby to contribute to the diffusion of the use of vectorial methods". Not until 1926 a second edition was needed!

Meanwhile in mechanical theories tendencies were developing to derive everything from the "Hamiltonian" $H(q_k, p_k)$ of systems by canonical equations, describing the changes of place- and momentum-coordinates q_k, p_k by

$$\frac{dq_k}{dt} = \frac{\partial H}{\partial p_k}, \qquad \frac{dp_k}{dt} = -\frac{\partial H}{\partial q_k}$$

corresponding to an infinitesimal contact transformation, developing in time.

During the very same period the great work of H. Grassmann (1809-1877) appeared in 1844, practically simultaneously with Hamilton's discovery of the quaternions. He emphasised the importance of his quantities, determining linear spaces, for theoretical physics.

The line coordinates of a line in three-dimensional space passing through the points (x_1, x_2, x_3, x_4) and (y_1, y_2, y_3, y_4), a one-dimensional space passing through (x) and (y), are the minors

$$p_{ik} = \begin{vmatrix} x_i & x_k \\ y_i & y_k \end{vmatrix}$$

of the matrix

$$\left\| \begin{matrix} x_1 & x_2 & x_3 & x_4 \\ y_1 & y_2 & y_3 & y_4 \end{matrix} \right\|$$

and it is evident that the cartesian coordinates $(x_4 = y_4 = 1)$ provide in the line coordinates the components of the difference vector $x - y$ and of the vector product $[xy]$.

These line coordinates must satisfy a quadratic p-relation

$$p_{12}p_{34} + p_{13}p_{42} + p_{14}p_{23} = 0,$$

which relation, using complex numbers in

$$p_{12} = u_1 + iu_2, \quad p_{34} = u_1 - iu_2, \quad \text{etc.}$$

gave rise to F. Klein's "quadratic surface in five dimensional space", used for the study of line geometry in three dimensional space.

For lines in higher dimensions the set of quadratic p-relations consists of an overdetermined system of equations for p_{ik} and e.g. in four-dimensional space these interdependent relations are five in number of the form

$$p_{ik}p_{lm} + p_{il}p_{mk} + p_{im}p_{kl} = \pi_n = 0$$

which satisfy the identities

$$\sum_{m=1}^{5} p_{km}\pi_m = 0, \ \{k\}.$$

Leaving out these relations R.W. Weitzenböck considered general quantities $p_{ik} = - p_{ki}$ as the coordinates of line-complexes and his work – 1908 – "Komplexsymbolik" introduced the complexsymbols p_i – which should not be confused with complex numbers! –

$$p_i p_k = - p_k p_i = p_{ik}.$$

In 1923 appeared his definitive symbolic method in his "Invariantentheorie".

About that same time quantum mechanics was developed and the new principle stated that an observation causes in general a disturbance of the physical system and thus the order of the observations is not irrelevant, as it was considered to be in classical physics. A non commutative operator became fundamental!

W. Heisenberg took refuge in the non commutative matrix multiplication whereas E. Schrödinger used the non commutativity of the operators x and $\frac{\partial}{\partial x}$. Replacing the component of the momentum in the classical Hamiltonian by

$$\frac{h}{2\pi i}\frac{\partial}{\partial x_k}$$

he derived the "Schrödinger-equation" for the physical problem.

P.A.M. Dirac decomposed the Maxwell operator

$$\square = \frac{\partial^2}{\partial x^2} + \frac{\partial^2}{\partial y^2} + \frac{\partial^2}{\partial x^2} - \frac{1}{c^2}\frac{\partial^2}{\partial t^2},$$

into linear operators. Then he had to introduce his non commutative, alternating in multiplication, Dirac-matrices, whereas his "undor" consists of four functions (f_1, f_2, f_3, f_4). The wave function, however, is a simple means of describing infinite sets of coordinates, determining the physical state, its introduction showing that the finite $(6n + 1)$-space for the n-particles problem in classical physics is not sufficient any more even for one particle in quantum mechanics. Indeed, considering the set of coordinates

$$(a_0, a_1, a_2, \ldots, a_n, \ldots)$$

as coefficients in an arbitrarily chosen set of orthogonal functions, e.g. as Fourier-coefficients, this set is stored in one function $\psi(x)$. It is then clear that four functions are exactly as general as one single: the sets can be stored in one set

$$(a_0, b_0, c_0, d_0, a_1, b_1, c_1, d_1, \ldots)$$

and thus be represented by one function only.

Only the algebraical structure of the physical theory requires, for reasons of simplification, such choices of four special functions. In 1949 (Proc. Acad. Amsterdam, 52, 1135 seq.) I indicated that the Dirac equations could be simply deduced from line geometrical computational schemes, making, moreover, Dirac's postulates for the non free particles superfluous and that Møller's meson equations were just as well covered by this procedure.

Recently the question arose whether other fundamental equations should not be preferred and sought for. For these reasons I develop the method with complexsymbols and classify some theories according to the theory of invariants.

§2. THE EQUATION FOR FREE PARTICLES

The condition that a point x is incident on the line p^2 is expressed by

$$(p^2xy) \equiv 0 \quad \{y\}$$

or written in full

$$\text{(I)} \quad \begin{aligned}
(p^2x)_{234} &= \phantom{p_{34}x_1 + {}} p_{34}x_2 + p_{42}x_3 + p_{23}x_4 = 0 \\
(p^2x)_{134} &= p_{34}x_1 \phantom{{} + p_{41}x_2} + p_{41}x_3 + p_{13}x_4 = 0 \\
(p^2x)_{124} &= p_{24}x_1 + p_{41}x_2 \phantom{{} + p_{41}x_3} + p_{12}x_4 = 0 \\
(p^2x)_{123} &= p_{23}x_1 + p_{31}x_2 + p_{12}x_3 \phantom{{} + p_{12}x_4} = 0.
\end{aligned}$$

This system has the same algebraic structure as the Dirac equations for the free electron, in which – however – the linear differential operators bearing one index are replaced by two index alternating symbols. This leads to the introduction of complexsymbols representing an alternating differential operator of order one half ∂_i

$$\partial_i\partial_k = - \partial_k\partial_i = \partial_{ik}.$$

If we take

$$\partial_{12} = \frac{\partial}{\partial x_5} + i\frac{\partial}{\partial x_6}, \quad \partial_{13} = \frac{\partial}{\partial x_1} - i\frac{\partial}{\partial x_2}, \quad \partial_{14} = \frac{\partial}{\partial x_3} - i\frac{\partial}{\partial x_4},$$

$$\partial_{34} = \frac{\partial}{\partial x_5} - i\frac{\partial}{\partial x_6}, \quad \partial_{42} = \frac{\partial}{\partial x_1} + i\frac{\partial}{\partial x_2}, \quad \partial_{23} = \frac{\partial}{\partial x_3} + i\frac{\partial}{\partial x_4},$$

the relation

$$(\partial^2 fg) \equiv 0 \quad \{g\}$$

yields the system (I) in which p_{ik} is replaced by ∂_{ik}. In matrix notation this gives straightforward

$$\text{(II)} \qquad \Big(\sum_{k=1}^{6} \gamma_k \frac{\partial}{\partial x_k}\Big) \| f \| = 0.$$

The conjugate complex equations read then (f' denoting the conjugated complex value of f)

$$\partial_k(\partial_1 f_1' + \partial_2 f_2' + \partial_3 f_3' + \partial_4 f_4') = 0 \qquad (k = 1, 2, 3, 4).$$

The γ_k are matrices corresponding to permutations of the four functions, which we shall consider in §3.

The main difference with Dirac's equations for the free electron is that two more variables do occur. If one simply puts

$$\frac{\partial f_k}{\partial x_6} = m f_k, \quad \frac{\partial f_k}{\partial x_5} \equiv 0,$$

Dirac's equations are obtained as a special case.

If, however, one expresses one or both of these derivatives by linear combinations of the f_i a generalization of Moller's meson equations arises

$$\Big(\sum \gamma_k \Big(\frac{\partial}{\partial x_k} + A_k\Big) + \gamma_\lambda \gamma_\mu A_{\lambda\mu} + M \| 1 \|\Big) f,$$

$$A_{\lambda\mu} = - A_{\mu\lambda}, \qquad (\lambda, \mu = 1, 2, 3, 4, 5).$$

The variables can in this way be used to express structural properties of the particle.

Without having a mass zero, e.g. a particle can behave as having a mass zero, in the free state. In external fields this mass would become evident. If we take for this situation

$$\frac{\partial f_1}{\partial x_5} + i\frac{\partial f_3}{\partial x_4} = mf_1, \qquad \frac{\partial f_2}{\partial x_5} + i\frac{\partial f_4}{\partial x_1} = mf_2,$$

$$\frac{\partial f_3}{\partial x_5} + i\frac{\partial f_1}{\partial x_4} = -mf_3, \qquad \frac{\partial f_4}{\partial x_5} + i\frac{\partial f_2}{\partial x_1} = -mf_4,$$

properly choosing the constant m, the mass terms in Dirac's equation are cancelled. By differentiation and elimination the equation for each of the functions become

$$\frac{\partial^2 f_k}{\partial x_4^2} + \frac{\partial^2 f_k}{\partial x_5^2} = m^2 f_k, \qquad (k = 1, 2, 3, 4)$$

Finally we remark that, guided by the quadratic p-relation, multiplying each of the equations of the system (I) into the appropriate symbolical differential operators in such a way that one of the functions obtains as its "coefficient" the left hand side of the quadratic p-relation, all other terms cancel and we are left for every function with the integrability condition

$$(\partial^2 \partial_1^2)f = \Big(\sum_{k=1}^{6} \frac{\partial^2}{\partial x_k^2}\Big)f = \bigcirc f = 0.$$

§3. SOME RELATIONS WITH γ-MATRICES

The matrices γ cause a simple permutation and changes of sign of the set (x_1, x_2, x_3, x_4) and, writing for simplicity the columns normally used in a horizontal line, these are

$$\gamma_1 = [x_4, x_3, x_2, x_1], \qquad \gamma_2 = i[x_4, -x_3, x_2, -x_1],$$

$$\gamma_3 = [x_3, -x_4, x_1, -x_2], \qquad \gamma_4 = i[-x_3, -x_4, x_1, x_2],$$

$$\gamma_5 = [x_1, x_2, -x_3, -x_4], \qquad \gamma_6 = -i[x_1, x_2, x_3, x_4],$$

$$\gamma_1\gamma_2 = i[x_1, -x_2, x_3, -x_4], \quad \gamma_1\gamma_3 = [x_2, -x_1, x_4, -x_3],$$

$$\gamma_1\gamma_4 = i[-x_2, -x_1, x_4, x_3], \quad \gamma_1\gamma_5 = [x_4, x_3, -x_2, x_1],$$

$$\gamma_2\gamma_3 = i[x_2, x_1, x_4, x_3], \quad \gamma_2\gamma_4 = [x_2, -x_1, -x_4, x_3],$$

$$\gamma_2\gamma_5 = i[x_4, -x_3, -x_2, x_1], \quad \gamma_3\gamma_4 = i[-x_1, x_2, x_3, -x_4],$$

$$\gamma_3\gamma_5 = [x_3, -x_4, x_1, x_2], \quad \gamma_4\gamma_5 = i[-x_3, -x_4, x_1, x_2].$$

They satisfy

$$\gamma_1^2 = \gamma_2^2 = \gamma_3^2 = \gamma_4^2 = \gamma_5^2 = -\gamma_6^2 = \| 1 \|$$

$$\gamma_i\gamma_k = -\gamma_k\gamma_i \qquad (i, k = 1, 2, 3, 4, 5)$$

$$\gamma_1\gamma_2\gamma_3\gamma_4 = \gamma_5.$$

Taking the conjugate complex values as the coordinates of a contragradient quantity we have the covariant elements

$$(f'\gamma_k f) = P_k, \qquad (f'\gamma_i\gamma_k f) = p_{ik}$$

$$(f'\gamma_i\gamma_k\gamma_l f) = q_{ikl} = q'_{mn}, \quad (f'\gamma_i\gamma_k\gamma_l\gamma_m f) = a_{iklm} = a'_n.$$

Here

$$\gamma_i\gamma_k\gamma_l = -\operatorname{sign}(iklmn)\gamma_m\gamma_n$$

$$\gamma_i\gamma_k\gamma_l\gamma_m = \operatorname{sign}(iklmn)\gamma_n$$

and therefore

$$P_k = a'_k, \quad f_{ik} = -q'_{ik}.$$

We have then invariants and covariants of the type

$$(f'f) = \rho, \ (Pp^2\pi^2), \ (p_1^2p_2^2x), \ldots$$

$$(Pp_1^2p_2^2), \ (p^2q^3), \ldots .$$

One easily verifies that

$$(f'\gamma_i\gamma_k f)(f'\gamma_l f) + (f'\gamma_k\gamma_l f)(f'\gamma_i f) + (f'\gamma_l\gamma_i f)(f'\gamma_k f) \equiv$$

$$\equiv \rho(f'\gamma_i\gamma_k\gamma_l f)$$

or

$$(p^2 P\pi^2) \equiv \rho(q^3\pi^2), \ \{\pi^2\}.$$

Again

$$(f'\gamma_i\gamma_k f)(f'\gamma_l\gamma_m f) + (f'\gamma_i\gamma_l f)(f'\gamma_m\gamma_k f) + (f'\gamma_i\gamma_m f)(f'\gamma_k\gamma_l f) \equiv$$
$$\equiv \rho(f'\gamma_i\gamma_k\gamma_l\gamma_m f)$$

or

$$(p_1^2 p_2^2 x) \equiv \rho(a'x), \ \{x\}.$$

Putting

$$s_{iklm} = (p_1^2 p_2^2)_{iklm} = \rho P_n = \rho t_{n6}$$
$$s_{ikl6} = p_{ik}p_{l6} + p_{i6}p_{kl} + p_{il}p_{6k} =$$
$$= p_{ik}P_l + p_{li}P_k + p_{kl}P_i = \rho q_{ikl} = \rho t_{mn}$$

we see that the algebraic minor of a minor in the sixth order matrix

$$\| t_{ik} \|$$

is proportional to the minor itself. This means, apart from a constant, that the matrix is *orthogonal.* The quantities

$$(f'\gamma_k f) = P_k, \quad (f'\gamma_i\gamma_k f) = p_{ik}$$

are proportional to the elements of the orthogonal matrix $\| t_{ik} \|$. Moreover we have directly

$$\sum_{k=1}^{6} (f'\gamma_k f)^2 = 0.$$

A further algebraic property may be indicated. Considering the sets of coordinates of γ_i and their products as sets of point-coordinates sixteen sets of six points are coplanar. The coordinates of the planes are the same set of sixteen as those of the points. They form together a Kummer configuration. Instead of the big tables formerly in use we have simply for the description of all incidencies the following rule:

In the plane with coordinate set γ_6 the points

$$\gamma_3\gamma_5, \gamma_4, \gamma_1\gamma_5, \gamma_2, \gamma_1\gamma_3, \gamma_2\gamma_4$$

are situated. The plane with coordinate set γ_i or $\gamma_i\gamma_k$ contains the points indicated by multiplying to the left the set in γ_6 by this γ_i or $\gamma_i\gamma_k$!

In variables X the planes of the Kummer configuration are:

$$(X\gamma_k x) = 0, \quad (X\gamma_i\gamma_k x) = 0.$$

Every quadratic matrix of the fourth degree can be represented by

$$\| S \| = \sum (s_k \gamma_k + i s_{k\lambda} \gamma_k \gamma_\lambda)$$

over all indices and all combinations $k\lambda$. We indicate for future reference

$$S_{11} = s_5 - is_6 - s_{12} + s_{34}, \qquad S_{21} = s_{14} - s_{23} + is_{13} + is_{24},$$

$$S_{12} = s_{14} - s_{23} - is_{13} - is_{24}, \qquad S_{22} = s_5 - is_6 + s_{12} - s_{34},$$

$$S_{13} = s_3 + is_4 + s_{45} - is_{35}, \qquad S_{23} = s_1 + is_2 + s_{25} - is_{15},$$

$$S_{14} = s_1 - is_2 - s_{25} - is_{15}, \qquad S_{24} = -s_3 + is_4 + s_{45} + is_{35},$$

$$S_{31} = s_3 - is_4 + s_{45} + is_{35}, \qquad S_{41} = s_1 + is_2 - s_{25} + is_{15},$$

$$S_{32} = s_1 - is_2 + s_{25} + is_{15}, \qquad S_{42} = -s_3 - is_4 + s_{45} - is_{35},$$

$$S_{33} = -s_5 - is_6 - s_{12} - s_{34}, \qquad S_{43} = -s_{14} - s_{23} + is_{13} - is_{24},$$

$$S_{34} = -s_{14} - s_{23} - is_{13} + is_{24}, \qquad S_{44} = -s_5 - is_6 + s_{12} + s_{34},$$

from which follows that if $s_6 = i\sigma_6$, $s_k, s_{k\lambda}, \sigma_6$ real-valued the matrix S is *hermitian.*

We add one result: corresponding to the "equation of continuity of electrical charge" in Maxwell's theory, we have six such equations for the free particle equations (I):

$$\sum_k \frac{\partial}{\partial x_k} (f'\gamma_k f) = 2i \frac{\partial \rho}{\partial x_6}, \qquad \sum_k \frac{\partial}{\partial x_k} (f'\gamma_1\gamma_k f) = -2 \frac{\partial \rho}{\partial x_1}$$

$$\sum_k \frac{\partial}{\partial x_k} (f'\gamma_2\gamma_k f) = 2i \frac{\partial \rho}{\partial x_2}, \qquad \sum_k \frac{\partial}{\partial x_k} (f'\gamma_3\gamma_k f) = -2 \frac{\partial \rho}{\partial x_3}$$

$$\sum_k \frac{\partial}{\partial x_k} (f'\gamma_4\gamma_k f) = 2i \frac{\partial \rho}{\partial x_4}, \qquad \sum_k \frac{\partial}{\partial x_k} (f'\gamma_5\gamma_k f) = -2 \frac{\partial \rho}{\partial x_5}.$$

§4. PARTICLES IN EXTERNAL FIELDS

The first fundamental theorem of invariants learns that the *only* generalisation *possible* is contained in

$$(\partial^2 fg) + (c^2 fg) \equiv 0 \ \{g\}.$$

with $\partial'_{ik} = \partial_{ik} + c_{ik}$ the equations read the same as the system (I) of §2 replacing p_{ik} by ∂'_{ik}. Splitting c_{ik} in the same way as was done with ∂_{ik} into c_t's we have

$$\sum_{k=1}^{6} \gamma_k \left(\frac{\partial}{\partial x_k} + c_k \right) \| f \| = 0.$$

This shows the Dirac postulate for particles in external fields as a consequence of projective invariance. Moreover it follows that an external field can only give rise to terms $\gamma_k c_k$ and never to terms with $\gamma_\lambda \gamma_k$. This means that a particle due to its internal structure can, being free, behave as an other particle in the field A_λ, but that the coefficients $A_{\lambda\mu}$ of terms $\gamma_\lambda \gamma_\mu$ are due to the intrinsic structure of the particle and cannot be influenced by. external fields! Generating for each of the functions the analogue of the quadratic-relation as its coefficient one finds the system

$$(\partial'_{14}\partial'_{23} + \partial'_{31}\partial'_{24} + \partial'_{12}\partial'_{34})f_1 + (\partial'_{14}\partial'_{31} + \partial'_{31}\partial'_{41})f_2 +$$
$$+ (\partial'_{14}\partial'_{12} + \partial'_{12}\partial'_{41})f_3 + (\partial'_{31}\partial'_{12} + \partial'_{12}\partial'_{13})f_4 = 0$$

$$(\partial'_{24}\partial'_{31} + \partial'_{32}\partial'_{41} + \partial'_{12}\partial'_{34})f_2 + (\partial'_{24}\partial'_{23} + \partial'_{32}\partial'_{24})f_1 +$$
$$+ (\partial'_{24}\partial'_{12} + \partial'_{12}\partial'_{42})f_3 + (\partial'_{32}\partial'_{12} + \partial'_{12}\partial'_{23})f_4 = 0$$

$$(\partial'_{34}\partial'_{12} + \partial'_{32}\partial'_{41} + \partial'_{13}\partial'_{42})f_3 + (\partial'_{34}\partial'_{23} + \partial'_{32}\partial'_{34})f_1 +$$
$$+ (\partial'_{34}\partial'_{31} + \partial'_{13}\partial'_{34})f_2 + (\partial'_{32}\partial'_{13} + \partial'_{13}\partial'_{23})f_4 = 0$$

$$(\partial'_{34}\partial'_{12} + \partial'_{42}\partial'_{13} + \partial'_{14}\partial'_{23})f_4 + (\partial'_{34}\partial'_{24} + \partial'_{42}\partial'_{34})f_1 +$$

$$+ (\partial'_{34}\partial'_{41} + \partial'_{14}\partial'_{34})f_2 + (\partial'_{42}\partial'_{41} + \partial'_{14}\partial'_{42})f_3 = 0.$$

Evaluating the operators and introducing

$$R_{\lambda\mu} = \frac{\partial c_\lambda}{\partial x_\mu} - \frac{\partial c_\mu}{\partial x_\lambda}$$

one has

$$(\Omega + K)\| f \| = 0$$

where

$$\Omega = \{\bigcirc + |c|^2 + \mathrm{Div}\, c + 2(c \cdot \mathrm{grad})\}\| 1 \|,$$

$$K_{11} = i(-R_{12} + R_{34} + R_{56}),$$

$$K_{22} = i(R_{12} - R_{34} + R_{56}),$$

$$K_{33} = i(-R_{12} - R_{34} - R_{56}),$$

$$K_{44} = i(R_{12} + R_{34} - R_{56}),$$

$$K_{12} = (R_{13} + R_{24}) + i(R_{14} - R_{23}),$$

$$K_{34} = (R_{13} - R_{24}) - i(R_{14} + R_{23}),$$

$$K_{13} = (R_{35} - R_{56}) + i(R_{45} + R_{36}),$$

$$K_{24} = (-R_{35} - R_{46}) + i(R_{45} - R_{36}),$$

$$K_{14} = (R_{15} + R_{26}) + i(R_{16} - R_{25}),$$

$$K_{23} = (R_{15} - R_{26}) + i(R_{25} + R_{16}).$$

Comparing this with the formula for the matrix $\| S \|$ in § 3 the result is

(III) $$\left\{\Omega \| 1 \| + \sum_{k=1}^{5} iR_{k6}\gamma_k - \sum_{(\lambda\mu)} R_{\lambda\mu}\gamma_\lambda\gamma_\mu\right\}\| f \| = 0.$$

This result can be obtained multiplying the operator

$$[\sum_{k=1}^{5} \gamma_k\left(\frac{\partial}{\partial x_k} + c_k\right) + \gamma_6\left(\frac{\partial}{\partial x_6} + c_6\right)]$$

into

$$[\sum_{k=1}^{5} \gamma_k\left(\frac{\partial}{\partial x_k} + c_k\right) - \gamma_6\left(\frac{\partial}{\partial x_6} + c_6\right)].$$

We indicated that the structure of the particle can be expressed in certain cases by replacing one of the derivatives by a linear combination of the functions, which gives rise to terms with $\gamma_\lambda\gamma_\mu$, which an external field cannot cause. For this reason we compute the product of the operator

$$\sum_{k=1}^{5} \gamma_k\left(\frac{\partial}{\partial x_k} + c_k\right) - \gamma_6\left(\frac{\partial}{\partial x_6} + c_6\right) + \sum c_{\lambda\mu}\gamma_\lambda\gamma_\mu$$

into the same operator with the sign of γ_6 changed into plus. The result is

$$\begin{aligned}[&(\Omega - \sum c_{\lambda\mu}^2)\,\|1\| + \sum_k \left(2\pi_k + \sum_\lambda \frac{\partial c_{k\lambda}}{\partial x_\lambda} + iR_{k6}\right)\gamma_k + \\ &+\left(\sum_{\circlearrowleft} \frac{\partial c_{\lambda\mu}}{\partial x_k} + 2\sum_{\circlearrowleft} c_k c_{\lambda\mu} + 2\sum_{\circlearrowleft} c_{\lambda\mu}\frac{\partial}{\partial x_k}\right)\gamma_\lambda\gamma_\mu\gamma_k + \\ &+\left(i\frac{\partial c_{\lambda\mu}}{\partial x_6} + \sum_{k=1}^{5} c_{\lambda k}c_{k\mu} - R_{\lambda\mu}\right)\gamma_\lambda\gamma_\mu]\,\|f\| = 0.\end{aligned}$$

The terms causing complications in this system are

$$\sum_{\circlearrowleft} c_{\lambda\mu}\frac{\partial}{\partial x_k}.$$

If we wish them to vanish simple algebraic reduction of the ten equation yields

$$\pi_k \frac{\partial f}{\partial x_i} = 0 \quad \{i, k\}.$$

A. In the most general case, $\frac{\partial f}{\partial x_i} \neq 0 \ \{i\}$ we must have

$$\pi_k = 0 \ \{k\}$$

and this means that the quantities $c_{k\lambda}$ are line coordinates.

$$c_{k\lambda} = [AB]_{k\lambda}$$

B. If one derivative for all functions vanishes, e.g. $\frac{\partial f}{\partial x_5} \equiv 0$ we are left with

$$\pi_1 = 0, \quad \pi_2 = 0, \quad \pi_3 = 0, \quad \pi_4 = 0$$

and due to the relation

$$\sum_\lambda c_{k\lambda} \pi_\lambda = 0$$

this yields

$$c_{k5} \cdot \pi_5 = 0$$

and we have

either $\pi_5 = 0$, coming back to case A,

or $c_{k5} = 0 \ \{k\}$,

or both.

If in addition to this $c_{\lambda 6} = 0 \ \{\lambda\}$ we have a physical system related with the skew symmetric matrix

$$\mathcal{M} = \| c_{ik} \| \qquad (i, k = 1, 2, 3, 4)$$

of four rows and columns, all other elements vanishing. The determinant of $\mathcal{M}$ is the square of the "Pfaffian", which is π_5. Putting

$$c_{23} = H_1, \quad c_{31} = H_2, \quad c_{12} = H_3, \quad c_{14} = iE_1,$$

$$c_{24} = iE_2, \quad c_{14} = iE_3$$

we have

$$\det \mathcal{M} = -(\mathfrak{E}\,\mathfrak{H})$$

and the "Maxwell tensor" is then obtained in

$$\mathscr{T} = \mathscr{M}^2 + \frac{1}{2}(\mathfrak{H}^2 - \mathfrak{E}^2)\| 1 \|,$$

having in its last row and column the components of the Poynting vector

$$\mathfrak{S} = [\mathfrak{E}\,\mathfrak{H}]$$

C. $\dfrac{\partial f}{\partial x_4} \equiv 0, \quad \dfrac{\partial f}{\partial x_5} \equiv 0$ leads to

$$\pi_1 = \pi_2 = \pi_3 = 0$$

$$c_{k4}\pi_4 + c_{k5}\pi_5 = 0 \ \{k\}.$$

Here we have

either $\pi_4 = 0, \ \pi_5 = 0,$ in a special case of A,

or $\pi_4 \neq 0, \ \pi_5 = 0, \ c_{k4} = 0,$ analogue of B,

or $\pi_5 \neq 0, \ \pi_4 \neq 0, \ c_{45} = 0, \ \lambda c_{k4} = c_{k5}$

and in fact the problem corresponds to a matrix of five rows and columns.

As for physical systems at least four variables should be involved a further analysis of more degenerated cases seems not to be of interest.

E.M. Bruins

Amsterdam – Z1 Joh. Verhulstrat 185, The Netherlands.

COLLOQUIA MATHEMATICA SOCIETATIS JÁNOS BOLYAI
15. DIFFERENTIAL EQUATIONS, KESZTHELY (HUNGARY), 1975.

ON SINGULAR SOLUTIONS OF NONLINEAR SYSTEMS OF ORDINARY DIFFERENTIAL EQUATIONS

T.A. ČANTURIA

Consider the system of differential equations

$$\frac{dx_i}{dt} = f_i(t, x_1, \ldots, x_n) \qquad (i = 1, 2, \ldots, n) \tag{1}$$

where the functions $f_i\colon [0, +\infty) \times R^n \to R$ $(i = 1, \ldots, n)$ satisfy the local Carathéodory-conditions, i.e. they are summable on any finite subinterval of $[0, +\infty)$ for arbitrary fixed $x_i \in R$ $(i = 1, \ldots, n)$ and depend continuously on $x_1, \ldots, x_n$ in R^n for almost all $t \in [0, +\infty)$ moreover, for arbitrary $\rho \in (0, +\infty)$

$$\sup \Big\{ \sum_{i=1}^{n} |f_i(t, x_1, \ldots, x_n)| \colon \sum_{i=1}^{n} |x_i| \leqslant \rho \Big\} \in L(0, \rho).$$

A solution $x_1(t), \ldots, x_n(t)$ of the system (1) is said to be (right) non-continuable if its domain of definition is of the form $[t_0, \infty)$ or, its domain of definition $[t_0, t_1)$ is finite and

$$\lim_{t \to t_1} \sum_{i=1}^{n} |x_i(t)| = +\infty.$$

A solution $x_1(t), \ldots, x_n(t)$ of system (1) defined on $[t_0, +\infty)$ is said to be proper, if for arbitrary $\tau \in [t_0, +\infty)$

$$\sup\left\{\sum_{i=1}^{n} |x_i(t)|:\ \tau \leqslant t < +\infty\right\} > 0$$

holds.

A nontrivial solution $x_1(t), \ldots, x_n(t)$ of system (1), defined on $[t_0, +\infty)$ is said to be singular of the first kind if there exists a $t_1 \in \in (t_0, +\infty)$ such that

$$\sum_{i=1}^{n} |x_i(t)| \equiv 0 \qquad \text{for} \qquad t \in [t_1, +\infty).$$

A non-continuable solution $x_1(t), \ldots, x_n(t)$ of system (1) is said to be singular of the second kind if its domain of definition is finite interval $[t_0, t_1)$.

In this paper we shall investigate the problem of existence of singular solutions of the system (1). Our theorems generalize certain results of [3] for equation of order n to the case of systems of equations. We mention some earlier papers devoted to analogous questions. In [1] the behaviour of solutions singular of the second kind in the case of systems, consisting of two equations is investigated. In [2], [4], and [5] sufficient conditions are given for the existence of solutions of second-order differential equations, singular of the second kind. In [6], conditions for the existence of solutions of second order differential equations, singular of the first kind are investigated.

Theorem 1. *Suppose that for* $t \geqslant 0$, $0 \leqslant (-1)^{\nu_k} x_1 \leqslant x_0$ $(k = 1, \ldots, n)$ *the following conditions hold:*

$$f_i(t, 0, 0, \ldots, 0) \equiv 0 \qquad (i = 1, 2, \ldots, n)$$

(2) $$(-1)^{\nu_i} f_i(t, x_1, \ldots, x_n) \leqslant -a_i(t) |x_{i+1}|^{\lambda_i} \quad (i = 1, 2, \ldots, n)$$

where $x_0 > 0$, $x_{n+1} = x_1$, $\nu_i = 0$ *or* 1, $\lambda_i > 0$ $(i = 1, 2, \ldots, n)$, $\prod_{i=1}^{n} \lambda_i < 1$, *the functions* $a_i(t)$ $(i = 1, 2, \ldots, n)$ *are continuous and pos-*

itive on $[0, +\infty)$. *Then for* $t_1 \in (0, +\infty)$ *arbitrary, the system* (1) *posses a solution* $x_1(t), \ldots, x_n(t)$, *singular of the first kind, defined on a certain interval* $[t_0, +\infty)$, $t_0 < t_1$ *such that*

$$(3) \qquad \begin{aligned} &(-1)^{\nu_i} x_i(t) > 0 \quad &&\text{for} \quad t \in [t_0, t_1), \\ &x_i(t) \equiv 0 \quad &&\text{for} \quad t \in [t_1, +\infty), \qquad (i = 1, 2, \ldots, n). \end{aligned}$$

For the proof we shall need the following lemmas.

Lemma 1. *Let the conditions of Theorem* 1 *be satisfied and let* $x_1(t), \ldots, x_n(t)$ *be a solution of system* (1), $y_1(t), \ldots, y_n(t)$ *a solution of the system*

$$\frac{dy_i}{dt} = -a_i(t) y_{i+1}^{\lambda_i} \qquad (i = 1, 2, \ldots, n;\ y_{n+1} = y_1)$$

defined on an interval $[t_0, t_1]$ *and suppose that* $0 < (-1)^{\nu_i} x_i(t) \leqslant x_0$, $y_i(t) \geqslant 0$ *hold for* $t \in [t_0, t_1]$, $(i = 1, 2, \ldots, n)$. *If*

$$(-1)^{\nu_i} x_i(t_1) > y_i(t_1) \qquad (i = 1, 2, \ldots, n)$$

then

$$(-1)^{\nu_i} x_i(t) > y_i(t) \qquad \text{for} \qquad t \in [t_0, t_1], \qquad (i = 1, 2, \ldots, n).$$

Proof. Suppose that the converse holds. Then there exist an index $k \in \{1, 2, \ldots, n\}$ and a $t_2 \in [t_0, t_1)$ such that

$$(-1)^{\nu_i} x_i(t) > y_i(t) \qquad \text{for} \qquad t \in (t_2, t_1] \qquad (i = 1, 2, \ldots, n)$$

and

$$(-1)^{\nu_k} x_k(t_2) = y_k(t_2).$$

Thus by (2) we obtain the contradiction

$$\begin{aligned} 0 &= (-1)^{\nu_k} x_k(t_2) - y_k(t_2) = \\ &= (-1)^{\nu_k} x_k(t_1) - y_k(t_1) - \\ &- \int_{t_2}^{t_1} \{(-1)^{\nu_k} f_k(\tau, x_1(\tau), x_2(\tau), \ldots, x_n(\tau)) + \end{aligned}$$

$$+ a_k(\tau)[y_{k+1}(\tau)]^{\lambda_k}\} d\tau >$$

$$> \int_{t_2}^{t_1} a_k(\tau)(|x_{k+1}(\tau)|^{\lambda_k} - [y_{k+1}(\tau)]^{\lambda_k})\, d\tau > 0.$$

The Lemma is proved.

Lemma 2. *Let* $\delta > 0$, $\epsilon > 0$, $\lambda_i > 0$ $(i = 1, 2, \ldots, n)$ *and* $y_1(t), y_2(t), \ldots, y_n(t)$ *a solution of the problem*

(4) $$\frac{dy_i}{dy} = -\delta y_{i+1}^{\lambda_i} \qquad (i = 1, \ldots, n;\ y_{n+1} = y_1)$$

(5) $$y_1(t_1) = \epsilon, \quad y_2(t_1) = \ldots = y_n(t_1) = 0$$

defined on an interval $[t_0, t_1]$. *Then*

(6) $$y_n(t) \leqslant c_0[y_1(t)]^{\alpha_1 - \frac{\alpha_{n-1}}{\beta_{n-1}}}[y_2(t)]^{\frac{1}{\beta_{n-1}}} \qquad \text{for} \qquad t \in [t_0, t_1]$$

where

$$\alpha_i = \prod_{k=n-i+1}^{n} \lambda_k \qquad (i = 1, \ldots, n),$$

$$\beta_i = 1 + \lambda_{n-i+1}\beta_{i-1} \qquad (i = 2, \ldots, n),\ \beta_1 = 1,$$

(7) $$c_0 = \prod_{i=1}^{n-2} c_i^{\frac{1}{\beta_i}},$$

$$c_i = \left(\frac{\beta_{i+1}}{\beta_i} c_{i-1}^{\lambda_{n-i-1}}\right)^{\frac{\beta_i}{\beta_{i+1}}} \qquad (i = 2, \ldots, n-2),\ c_1 = \beta_2^{\frac{1}{\beta_2}}.$$

Proof. We prove first the inequality

(8) $$y_{n-i+1}(t) \leqslant c_i[y_1(t)]^{\frac{\alpha_i}{\beta_{i+1}}}[y_{n-i}(t)]^{\frac{\beta_i}{\beta_{i+1}}}$$
$$\text{for} \quad t \in [t_0, t_1] \qquad (i = 1, \ldots, n-2).$$

Multiplying the n-th equation of the system (4) by $[y_n(t)]^{\lambda_{n-1}}$ and integrating on the interval (t, t_1) we get

$$y_n(t) \leqslant (\lambda_{n-1} + 1)^{\frac{1}{\lambda_{n-1}+1}} [y_1(t)]^{\frac{\lambda_n}{\lambda_{n-1}+1}} [y_{n-1}(t)]^{\frac{1}{\lambda_{n-1}+1}},$$

that is (8) holds for $i = 1$. Suppose the assertion holds for an $i \in \{1, \ldots \ldots, n-3\}$. Taking λ_{n-i}-th power on both sides of inequality (8), multiplicating by $[y_{n-i}(t)]^{\frac{\beta_{i+2}}{\beta_{i+1}} - 1}$ and integrating on the interval $[t, t_1]$, by $\frac{\lambda_{n-i}\beta_i + \beta_{i+2}}{\beta_{i+1}} - 1 = \lambda_{n-i-1}$, (4) and (5) yield

$$y_{n-i}(t) \leqslant \left(\frac{\beta_{i+2}}{\beta_{i+1}} c_i^{\lambda_{n-i}}\right)^{\frac{\beta_{i+1}}{\beta_{i+2}}} [y_1(t)]^{\frac{\alpha_{i+1}}{\beta_{i+2}}} [y_{n-i-1}(t)]^{\frac{\beta_{i+1}}{\beta_{i+2}}}.$$

Hence (8) is proved. From (8) it follows easily that

$$y_n(t) \leqslant c_1 c_2^{\frac{1}{\beta_2}} \ldots c_{n-2}^{\frac{1}{\beta_{n-2}}} [y_1(t)]^{\sum_{i=1}^{n-2} \frac{\alpha_i}{\beta_i \beta_{i+1}}} [y_2(t)]^{\frac{1}{\beta_{n-1}}}.$$

Since $\frac{\alpha_i}{\beta_i \beta_{i+1}} = \frac{\alpha_i}{\beta_i} - \frac{\alpha_{i+1}}{\beta_{i+1}}$, this is equivalent to (6). The Lemma is proved.

Proof of Theorem 1. Let $t_1 \in (0, +\infty)$ be arbitrary, $\epsilon_0 > 0$ and $t_0 \in [0, t_1)$ such that any solution $x_1(t), \ldots, x_n(t)$ of the system (1) satisfying the condition

(9) $\qquad (-1)^{\nu_i} x_i(t_1) = 2\epsilon \qquad (i = 1, \ldots, n)$

for $\epsilon \in (0, \epsilon_0]$ is defined on the interval $[t_0, t_1]$ and also satisfies the inequalities

$$0 < (-1)^{\nu_i} x_i(t) \leqslant x_0 \quad \text{for} \quad t \in [t_0, t_1] \quad (i = 1, 2, \ldots, n).$$

Define δ by

$$\delta = \min \{a_i(t)\colon t \in [t_0, t_1] \ (i = 1, \ldots, n)\}.$$

Let $x_{1\epsilon}(t), \ldots, x_{n\epsilon}(t)$ be a solution of the system (1) satisfying condition

(9), $y_{1\epsilon}(t), \ldots, y_{n\epsilon}(t)$ a solution of the system (4) satisfying condition (5). Lemma 1 implies

(10) $$(-1)^{\nu_i} x_{i\epsilon}(t) > y_{i\epsilon}(t) \quad \text{for} \quad t \in [t_0, t_1] \qquad (i = 1, \ldots, n).$$

First we prove

(11) $$y_{1\epsilon}(t) \geqslant c\Big[\frac{(1-\alpha_n)\delta}{\beta_n c}(t_1 - t)\Big]^{\frac{\beta_n}{1-\alpha_n}} \quad \text{for} \quad t \in [t_0, t_1],$$

where $c = ((\lambda_n + 1)c_0)^{\frac{\beta_{n-1}}{\alpha_{n-1}+\beta_{n-1}}}$ and c_0, α_i, β_i are defined by (7).

It follows from (4), that

(12) $$y_{1\epsilon}(t) = \epsilon + \delta \int_t^{t_1} [y_{2\epsilon}(\tau)]^{\lambda_1}\, d\tau \quad \text{for} \quad t \in [t_0, t_1].$$

On the other hand, by (6):

$$[y_{1\epsilon}(t)]^{\lambda_n+1} = \epsilon^{\lambda_n+1} - (\lambda_n + 1)\int_t^{t_1} [y_{2\epsilon}(\tau)]^{\lambda_1} dy_{n\epsilon}(\tau) \leqslant$$

$$\leqslant \epsilon^{\lambda_n+1} + (\lambda_n + 1)[y_{2\epsilon}(t)]^{\lambda_1} y_{n\epsilon}(t) \leqslant$$

$$\leqslant \Big[\epsilon^{1+\frac{\alpha_{n-1}}{\beta_{n-1}}} + c^{1+\frac{\alpha_{n-1}}{\beta_{n-1}}}[y_{2\epsilon}(t)]^{\lambda_1+\frac{1}{\beta_{n-1}}}\Big] \times$$

$$\times [y_{1\epsilon}(t)]^{\lambda_n - \frac{\alpha_{n-1}}{\beta_{n-1}}} \quad \text{for} \quad t \in [t_0, t_1].$$

Therefore

$$y_{1\epsilon}(t) \leqslant \epsilon + c[y_{2\epsilon}(t)]^{\frac{\beta_n}{\beta_{n-1}+\alpha_{n-1}}} \quad \text{for} \quad t \in [t_0, t_1].$$

Combining this inequality by (12) we obtain

$$[y_{2\epsilon}(t)]^{\frac{\beta_n}{\beta_{n-1}+\alpha_{n-1}}} \geqslant \frac{\delta}{c}\int_t^{t_1} [y_{2\epsilon}(\tau)]^{\lambda_1}\, d\tau \quad \text{for} \quad t \in [t_0, t_1]$$

i.e.

$$[y_{2\epsilon}(t)]^{\lambda_1}\Big(\int_t^{t_1}[y_{2\epsilon}(\tau)]^{\lambda_1}\,d\tau\Big)^{-1+\frac{1-\alpha_n}{\beta_n}} \geqslant \Big(\frac{\delta}{c}\Big)^{1-\frac{1-\alpha_n}{\beta_n}}$$

for $t \in [t_0, t_1]$,

or integrating the latter inequality between t and t_1:

$$\Big(\int_t^{t_1}[y_{2\epsilon}(\tau)]^{\lambda_1}\,d\tau\Big)^{\frac{1-\alpha_n}{\beta_n}} \geqslant \frac{1-\alpha_n}{\beta_1}\Big(\frac{\delta}{c}\Big)^{1-\frac{1-\alpha_n}{\beta_n}}(t_1 - t)$$

for $t \in [t_0, t_1]$.

Hence (11) follows easily from (12).

Now it is easy to prove the existence of a sequence $\{\epsilon_m\}_{m=1}^{+\infty}$ of positive numbers, converging to 0 for $m \to \infty$ such that the sequences $\{x_{i\epsilon_m}(t)\}_{m=1}^{+\infty}$ $(i = 1, \ldots, n)$ converge uniformly on $[t_0, t_1]$. Let

$$x_{i0}(t) = \lim_{m\to\infty} x_{i\epsilon_m}(t) \qquad (i = 1, 2, \ldots, n).$$

Then $x_{10}(t), \ldots, x_{n0}(t)$ is a solution of the system (1) satisfying the initial conditions

$$x_{i0}(t_1) = 0 \qquad (i = 1, 2, \ldots, n).$$

It follows from (10) and (11)

$$(-1)^{\nu_1}x_{10}(t) \geqslant c\Big[\frac{(1-\alpha_n)\delta}{\beta_n c}(t_1 - t)\Big]^{\frac{\beta_n}{1-\alpha_n}} > 0$$

for $t \in [t_0, t_1)$.

Applying also (4), (5) we obtain

$$(-1)^{\nu_i}x_{i0}(t) > 0 \quad \text{for} \quad t \in [t_0, t_1) \qquad (i = 2, \ldots, n).$$

Hence $x_1(t), \ldots, x_n(t)$ defined by

$$x_i(t) = \begin{cases} x_{i0}(t) & \text{for} \quad t \in [t_0, t_1] \\ 0 & \text{for} \quad t \in (t_1, +\infty), \end{cases}$$

is the singular solution of the first kind of the system (1), satisfying condition (3). The theorem is proved.

It is easily seen that Theorem 1 still holds true if the inequalities (2) are not satisfied for all $i \in \{1, \ldots, n\}$, but only for a subset $\{i_1, \ldots, i_m\} \subset \subset \{1, \ldots, n\}$.

Actually the following theorem is valid.

Theorem 1'. *Let*

$$f_i(t, 0, 0, \ldots, 0) \equiv 0 \quad \text{for} \quad t \geqslant 0, \qquad (i = 1, 2, \ldots, n) \tag{13}$$

and suppose that for some $m \in \{1, \ldots, n\}$, *when* $t \geqslant 0$, $0 \leqslant (-1)^{\nu_k} x_{i_k} \leqslant \leqslant x_0$ $(k = 1, \ldots, m)$, $|x_i| < +\infty$ *for the remaining* i,

$$(-1)^{\nu_k} f_{i_k}(t, x_1, x_2, \ldots, x_n) \leqslant -a_k(t) |x_{i_{k+1}}|^{\lambda_k}$$

$$(k = 1, \ldots, m)$$

where $x_0 > 0$, $x_{i_{m+1}} = x_{i_1}$, $i_k \in \{1, \ldots, n\}$ $\nu_k = 0$ *or* 1, $\lambda_k > 0$ $(k = 1, \ldots, m)$, $\prod_{k=1}^{m} \lambda_k < 1$ *the functions* $a_k(t)$ $(k = 1, \ldots, m)$ *are positive and continuous on* $[0, +\infty)$.

Then for arbitrary $t_1 \in (0, +\infty)$, *the system* (1) *has a solution* $x_1(t), \ldots, x_n(t)$ *singular of the first kind, defined on* $[t_0, +\infty)$, $t_0 < t_1$, *and such that*

$$(-1)^{\nu_k} x_{i_k}(t) > 0 \quad \text{for} \quad t \in [t_0, t_1) \quad (k = 1, \ldots, m)$$

$$x_i(t) \equiv 0 \quad \text{for} \quad t \in [t_1, +\infty) \quad (i = 1, \ldots, n).$$

Theorem 2. *Suppose that conditions* (13) *are satisfied and for* $t \geqslant a$, $0 \leqslant (-1)^{\nu_k} x_k \leqslant x_0$ $(k = 1, 2, \ldots, n)$ *the inequalities*

$$(14) \quad \begin{aligned} &(-1)^{\nu_i} f_i(t, x_1, \ldots, x_n) \leqslant -\delta t^{\mu_i} |x_{i+1}|^{\lambda_i} \qquad (i = 1, \ldots, n) \\ &(-1)^{\nu_l} f_l(t, x_1, \ldots, x_n) \geqslant -\delta_l t^{\mu_l} |x_{l+1}|^{\lambda_l} \end{aligned}$$

hold. Here x_0, a, δ *and* δ_l *are positive numbers,* $l \in \{1, \dots, n\}$, $\delta_l \geq \delta$, $\nu_i = 0$ *or* 1, $\lambda_i > 0$ $(i = 1, \dots, n)$.

$$\prod_{i=1}^{n} \lambda_i < 1, \quad x_{n+1} = x_1$$

and

$$(15) \qquad 1 + \mu_l + \sum_{j=1}^{n-1} (1 + \mu_{j+l}) \prod_{k=l}^{j+l-1} \lambda_k \geq 0$$

(where $\mu_{n+i} = \mu_i$, $\lambda_{n+i} = \lambda_i$ $(i = 1, \dots, n-1)$*). Then if* $t_0 \in (0, +\infty)$ *is arbitrary, any nontrivial solution* $x_1(t), \dots, x_n(t)$ *of the system* (1), *defined on the interval* $[t_0, +\infty)$, *and satisfying the conditions*

$$(16) \qquad 0 \leq (-1)^{\nu_i} x_i(t) \leq x_0 \quad \text{for} \quad t \in [t_0, +\infty) \qquad (i = 1, \dots, n)$$

is singular of the first kind.

Proof. Suppose the contrary. Let $x_1(t), \dots, x_n(t)$ be a proper solution of the system (1), defined on an interval $[t_0, +\infty)$ and satisfying the inequalities (16).

Without loss of generality $l = 1$ can be assumed. Then by (14)

$$(17) \qquad \begin{aligned} &-|x_i(t)|' \geq \delta t^{\mu_i} |x_{i+1}(t)|^{\lambda_i} \\ &\text{for} \quad t \in [t_0, +\infty) \qquad (i = 1, \dots, n), \\ &-|x_1(t)|' \leq \delta_1 t^{\mu_1} |x_2(t)|^{\lambda_1} \quad \text{for} \quad t \in [t_0, +\infty). \end{aligned}$$

From (17) it follows that

$$|x_i(t)| > 0 \quad \text{for} \quad t \in [t_0, +\infty) \qquad (i = 1, \dots, n).$$

Let $s \in [t_0, +\infty)$. Then by (17) there obviously exist $\delta_0 > 0$ and $\delta_{10} > 0$ such that

$$(18) \qquad \begin{aligned} &-|x_i(t)|' \geq \delta_0 s^{\mu_i} |x_{i+1}(t)|^{\lambda_i} \\ &\text{for} \quad t \in [s, 2s] \qquad (i = 1, \dots, n), \\ &-|x_1(t)|' \leq \delta_{10} s^{\mu_1} |x_2(t)|^{\lambda_1} \quad \text{for} \quad t \in [s, 2s]. \end{aligned}$$

We shall prove that

$$s^{\mu_1\beta_i}|x_2(t)|^{\lambda_1\beta_i}|x_{n-i+1}(t)| \geqslant b_i s^{\gamma_i}(|x_1(t)|-|x_1(2s)|)^{\alpha_i+\beta_i} \tag{19}$$
$$\text{for} \quad t\in[s,2s] \qquad (i=1,\dots,n-1)$$

where α_i and β_i are defined by the equalities (7), $\gamma_1=\mu_n$, $\gamma_i = \mu_{n-i+1}+\lambda_{n-i+1}\gamma_{i-1}$ $(i=2,\dots,n)$

$$b_1=\frac{\delta_0}{(\lambda_n+1)\delta_{10}},\quad b_i=\frac{\delta_0 b_{i-1}^{\lambda_{n-i+1}}}{\delta_{10}(\alpha_i+\beta_i)} \qquad (i=2,\dots,n-1).$$

By (18) we have

$$s^{\mu_1}|x_2(t)|^{\lambda_1}|x_n(t)| > -s^{\mu_1}\int_t^{2s}|x_2(\tau)|^{\lambda_1}|x_n(\tau)|'\,d\tau \geqslant$$
$$\geqslant -\frac{\delta_0}{\delta_{10}}\int_t^{2s} s^{\mu_n}|x_1(\tau)|^{\lambda_n}|x_1(\tau)|'\,d\tau =$$
$$=\frac{\delta_0}{\delta_{10}(\lambda_n+1)}s^{\mu_n}(|x_1(t)|^{\lambda_n+1}-|x_1(2s)|^{\lambda_n+1})\geqslant$$
$$\geqslant\frac{\delta_0}{\delta_{10}(\lambda_n+1)}s^{\mu_n}(|x_1(t)|-|x_1(2s)|)^{\lambda_n+1} \quad \text{for} \quad t\in[s,2s]$$

i.e. (19) holds true for $i=1$. Suppose, that (19) holds for an $i\in\{1,\dots,n-2\}$. We shall prove, that it holds for $i+1$ also. Taking λ_{n-i}-th powers on both sides of (19), multiplying by $s^{\mu_{n-i}+\mu_1}|x_2(t)|^{\lambda_1}$ and applying the inequality (18) we obtain

$$-s^{\mu_1\beta_{i+1}}|x_2(t)|^{\lambda_1\beta_{i+1}}|x_{n-i}(t)|'\geqslant$$
$$\geqslant -\frac{\delta_0}{\delta_{10}}b_i^{\lambda_{n-i}}s^{\gamma_{i+1}}|x_1(t)|'(|x_1(t)|-|x_1(2s)|)^{\alpha_{i+1}+\beta_{i+1}-1}$$
$$\text{for} \quad t\in[s,2s].$$

Integrating the latter inequality on the interval $[t,2s]$ we obtain (19) for $i+1$, hence (19) is proved.

Applying (19) for $i=n-1$ we obtain

$$s^{\mu_1}|x_2(t)|^{\lambda_1}(|x_1(t)|-|x_1(2s)|)^{-\frac{\alpha_n+\beta_n-1}{\beta_n}} \geqslant b_{n-1}^{\frac{\lambda_1}{\beta_n}} s^{\frac{\gamma_n}{\beta_n}}$$

for $t \in [s, 2s)$,

whence according to (18) we have

$$-|x_1(t)|'(|x_1(t)|-|x_1(2s)|)^{-\frac{\alpha_n+\beta_n-1}{\beta_n}} \geqslant \delta_0 b_{n-1}^{\frac{\lambda_1}{\beta_n}} s^{\frac{\gamma_n}{\beta_n}}$$

for $t \in [s, 2s)$.

Integrating this inequality from s to $2s$ we obtain

(20)
$$(|x_1(s)|-|x_1(2s)|)^{1-\alpha_n} \geqslant \left(\frac{1-\alpha_n}{\beta_n}\delta_0\right)^{\beta_n} b_{n-1}^{\lambda_1} s^{\gamma_n+\beta_n}$$

for $s \in [t_0, +\infty)$.

By the conditions of the theorem

$$\alpha_n = \prod_{k=1}^{n} \lambda_k < 1,$$

$$\gamma_n + \beta_n = 1 + \mu_1 + \sum_{j=1}^{n-1} (1 + \mu_{j+1}) \prod_{k=1}^{j} \lambda_k \geqslant 0,$$

hence letting s to $+\infty$ in (20) we obtain a contradiction which proves the theorem.

We present two theorems for the existence of solutions, singular of the second kind.

Theorem 3. *Suppose that for* $t \geqslant 0$, $(-1)^{\nu_k} x_k \geqslant x_0$ $(k = 1, \ldots, n)$ *the inequalities*

$$(-1)^{\nu_i} f_i(t, x_1, \ldots, x_n) \geqslant a_i(t)|x_{i+1}|^{\lambda_i} \qquad (i = 1, \ldots, n)$$

hold. Here $x_0 \geqslant 0$, $x_{n+1} = x_1$, $\nu_i = 0$ *or* 1, $\lambda_i > 0$ $(i = 1, \ldots, n)$, $\prod_{i=1}^{n} \lambda_i > 1$, *the functions* $a_i(t)$ $(i = 1, \ldots, n)$ *are continuous and positive on the interval* $[0, +\infty)$. *Then for arbitrary* $t_0, t_1 \in [0, +\infty)$, $t_0 < t_1$, *the system* (1) *possesses a solution* $x_1(t), \dots, x_n(t)$, *singular of the second kind, defined on a subinterval* $[t_0, t^*) \subset [t_0, t_1)$, *such that*

$$(-1)^{\nu_i} x_i(t) > 0 \quad \textit{for} \quad t \in [t_0, t^*) \qquad (i = 1, \dots, n).$$

Theorem 4. *Suppose that for* $t \geqslant a$, $(-1)^{\nu_k} x_k \geqslant x_0$ $(k = 1, \dots, n)$ *the inequalities*

$$(-1)^{\nu_i} f_i(t, x_1, \dots, x_n) \geqslant \delta t^{\mu_i} |x_{i+1}|^{\lambda_i} \qquad (i = 1, \dots, n)$$

hold, where x_0, a, δ *are positive numbers,* $\nu_i = 0$ *or* 1, $\lambda_i > 0$ $(i = 1, \dots, n)$, $x_{n+1} = x_1$, $\prod \lambda_i > 1$ *and for some* $l \in \{1, \dots, n\}$

$$1 + \mu_l + \sum_{j=1}^{n-1} (1 + \mu_{j+l}) \prod_{k=l}^{j+l-1} \lambda_k > 0$$

(where $\mu_{n+i} = \mu_i$, $\lambda_{n+i} = \lambda_i$, $(i = 1, \dots, n-1)$*). Then for arbitrary* $t_0 \in [a, +\infty)$, *any non-continuable solution* $x_1(t), \dots, x_n(t)$ *of the system* (1), *satisfying the conditions*

$$(-1)^{\nu_i} x_i(t_0) > x_0 \qquad (i = 1, \dots, n)$$

is singular of the second kind.

Note that Theorems 2-4 also have extensions, in analogy with Theorem 1′.

REFERENCES

[1] N.P. Erugin, *A textbook for the common course on differential equations.* Nauka i Technika, Minsk, 1972 (in Russian).

[2] I.T. Kiguradze, Asymptotic properties of solutions of an Emden – Fowler-type non-linear differential equation. *Izv. A.N. SSSR, ser. mat.*, 29, 5 (1965), 965-986 (in Russian).

[3] I.T. Kiguradze, *Some singular boundary value problem for ordinary differential equations,* Tbilisi Univ. Press, Tbilis, 1975 (in Russian).

[4] A.V. Kostin, On the existence and behaviour of solutions having asymptotas in the case of second order nonlinear equations. *Diff. Urav.*, 6, 12 (1970), 2182-2192 (in Russian).

[5] Ts.V. Tabukashvili – T.A. Čanturia, On some boundary value problem for strongly non-linear second order ordinary differential equations. *Diff. Urav.*, 10, 5 (1974), 961-964 (in Russian).

[6] T.A. Čanturia, On the asymptotical representation of solutions of the equation $u'' = a(t) \mid u \mid^n \operatorname{sign} u$. *Diff. Urav.*, 8, 7 (1972), 1195-1206 (in Russian).

T.A. Čanturia

Inst. Prykl. Mat., Tbilisskovo Gos. Univ., 38 0043 Tbilisi 43, Universitetskaja 2, USSR.

COLLOQUIA MATHEMATICA SOCIETATIS JÁNOS BOLYAI
15. DIFFERENTIAL EQUATIONS, KESZTHELY (HUNGARY), 1975.

THE BOUNDEDNESS OF SOLUTIONS OF SYSTEMS OF DIFFERENTIAL EQUATIONS

M.L. CARTWRIGHT – H.P.F. SWINNERTON-DYER

1. The Liénard equation

$$\ddot{x} + f(x)\dot{x} + g(x) = p(t) \tag{1.1}$$

has been studied by many authors by means of the equivalent system

$$\begin{cases} \dot{x} = y - F(x) + P(t), \quad F(x) = \int\limits_0^x f(\xi)\, d\xi, P(t) = \int\limits_p^t p(\tau)\, d\tau \\ \\ \dot{y} = -\, g(x), \quad G(x) = \int\limits_0^x g(\xi)\, d\xi). \end{cases} \tag{1.2}$$

In particular Graef has given a survey of the literature and proved under very general conditions that the solutions of (1.1) and their derivatives are bounded by constants independent of their initial conditions. In an earlier paper we proved that under similar conditions any solution $x(t), y(t)$ of the system

$$\begin{cases} \dot{x} = y - kF(x, t), \quad 0 < k < \infty \\ \\ \dot{y} = g(x), \end{cases} \tag{1.3}$$

satisfies

$$|x(t)| < B, \quad |y(t)| < B(k+1), \qquad t > t_1$$

where B is independent of the parameter k, for t_1 sufficiently large.

The more general second order equation

$$\ddot{x} + f(x, \dot{x})\dot{x} + g(x) = p(t) \tag{1.4}$$

has also been studied by many authors, Levinson and Smith, Reuter, Levinson and Langenhop, Opial and others*. These authors all used the system

$$\begin{cases} \dot{x} = y \\ \dot{y} = -g(x) - f(x, y)y + p(t). \end{cases} \tag{1.5}$$

We also obtained some results for (1.4) by comparing a specially constructed system with a system of the form (1.2), but our method could not be used to establish the results of earlier authors.

The motivation of most of the work on (1.1) was the importance of the behaviour of solutions of van der Pol's equation, especially with a forcing term, viz.,

$$\ddot{x} + k(x^2 - 1)\dot{x} + x = bk \cos \lambda t, \qquad 0 < k < \infty, \ b > 0, \tag{1.6}$$

and other similar equations. If $b = 0$ and (1.6) is integrated, we obtain Rayleigh's equation

$$\ddot{\xi} + k\left(\frac{\dot{\xi}^3}{3} - \dot{\xi}\right) + \xi = 0, \qquad \xi = \int_0^t x(\tau)\, d\tau. \tag{1.7}$$

This suggested considering the slightly more general equation

$$\ddot{x} + kF(\dot{x}) + g(x) = kp(t), \tag{1.8}$$

and this was done by Reuter for $0 < k \leqslant 1$ by means of the system

*See Reissig, Sansone and Conti.

$$(1.9) \qquad \begin{aligned} \dot{x} &= y, \\ \dot{y} &= -g(x) - \epsilon(F(y) - p(t)), \qquad 0 < \epsilon \leqslant 1. \end{aligned}$$

Now putting $x = y'$, $y = -x'$, $F(y) - p(t) = -F(-x') - p(t) = F(x', t)$ in (1.9), we obtain the system

$$(1.10) \qquad \begin{aligned} \dot{x}' &= g(y') - \epsilon F(x', t), \\ \dot{y}' &= -x', \qquad 0 < \epsilon \leqslant 1. \end{aligned}$$

Here we have a system very similar to (1.3) for $0 < k \leqslant 1$, but y is replaced by the general function $g(y')$ and $g(x)$ by the special function x'. By a similar substitution the equivalent system (1.5) of the more general equation (1.4) can be included in the system

$$(1.11) \qquad \begin{aligned} \dot{x}' &= g(y') - \epsilon F(x', y', t), \\ \dot{y}' &= -x', \\ & 0 < \epsilon \leqslant 1, \end{aligned}$$

where $F(x', y', t) = f(y', -x')x' + p(t)$.

2. A comparison of (1.3) with (1.10) and (1.11) suggests the consideration of the general system

$$(2.1) \qquad \begin{cases} \dot{x} = g_1(y) - \epsilon F(x, y, t), \qquad 0 < \epsilon \leqslant 1, \\ \dot{y} = -g_2(x). \end{cases}$$

In fact van der Pol's equation and many other second-order equations were originally derived from systems more or less like (2.1) and some years ago Prof. Courtney Coleman suggested to me that all results about the boundedness of solutions of second order equations might be included in a single result for a system. However, as I said then, the way in which the special properties of y, or x as the case may be, are used in the proofs seems to make this impossible. Weaker hypotheses in one respect seem to necessitate stronger hypotheses in some other respect. For instance in Theorem of Graef's paper the hypothesis

$$|F(x)| + G(x) \to \infty \quad \text{as} \quad |x| \to \infty$$

is used, and this means *either* $|F(x)| \to \infty$ *or* $G(x) \to \infty$. It is however possible to generalize previous results and prove at least three distinct types of result depending on how closely $g_1(y)$ and $g_2(x)$ approximate to y and x respectively in the behaviour imposed on them by the hypotheses.

In what follows we suppose that all functions are continuous for the values for which they are considered, so that for every three real numbers x_0, y_0, t_0 the system (2.1) has a solution

$$x(t) = x(t; x_0, y_0), \quad y(t) = y(t; x_0, y_0), \quad x(t_0) = x_0, \quad y(t_0) = y_0$$

for $t \geqslant t_0$. Since the equations (1.6) and (1.7) figure in the motivation, there is usually a critical strip, or strips, $|x| \leqslant a'$, $|y| \leqslant b'$ in which the hypotheses on $F(x, t)$ or $F(x, y, t)$ are much weaker. In particular if $b = 0$ the equation (1.6) can be considered *either* as a system of the form (1.3) with $F(x) \operatorname{sign} x < 0$ for $|x| < 3^{-\frac{1}{2}}$ *or* as a system of the form (1.11) with $F(x', y', t) = F(x', y') = ((y')^2 - 1)x'$ in which $F(x', y') \operatorname{sign} x < 0$ for $|y'| < 1$ and large for $|x'|$ large unless y' is near 1. These critical strips can be standardized to $|x| \leqslant 1$, $|y| \leqslant 1$ by putting $x = a'x''$, $y = b'y''$, but this affects the constants used in the hypotheses for (1.8) and (1.4) by some authors in a rather complicated way. However in the theorems which we have obtained for the system (2.1) this standardization does not affect the relationship between the constants.

The following statements hold if $g_1(y) = g_1(y, \epsilon)$, $F(x, y, t) = F(x, y, t, \epsilon)$, $g_2(x) = g_2(x, \epsilon)$, $0 < \epsilon \leqslant 1$, provided that the hypotheses are satisfied uniformly in ϵ. In particular when $G_1(y) = \int_0^y g_1(\eta)\, d\eta$ or $G_2(x) = \int_0^x g_2(\xi)\, d\xi$ tends to infinity as $|x| \to \infty$ this must be interpreted as not tending to infinity too rapidly for small ϵ as well as not too slowly, as the expression tending to infinity uniformly in ϵ usually implies. All constants are independent of ϵ unless otherwise stated.

3. The first theorem is a straightforward generalization of Theorems 1 and 8 of our earlier paper and Reuter's result for (1.8) can easily be deduced from it.

Theorem 1. *Suppose that*

(i) *there is a constant* $c_1 > 0$ *such that* $|F(x, y, t)| \leqslant c_1$ *for* $|x| \leqslant 1$,

(ii) $F(x, y, t) \operatorname{sign} x \geqslant 0$ *for* $|x| \geqslant 1,\ |y| \geqslant 1$,

(iii) *there is a constant* $c_2 > 0$ *such that* $F(x, y, t) \operatorname{sign} x > -c_2$ *for* $|y| \leqslant 1$,

(iv) *there is a constant* $b_1 > 0$ *such that* $F(x, y, t) \operatorname{sign} x \geqslant b_1$ *in one of the four domains defined by* $|x| > 1,\ |y| > 1$,

(v) *there is a constant* $c_3 > 0$ *such that* $|g_1(y)| \leqslant c_3$ *for* $|y| \leqslant 1$,

(vi) $g_1(y) \operatorname{sign} y > 0$ *for* $|y| \geqslant 1$,

(vii) $G_1(y) = \int_0^y g_1(\eta)\, d\eta \to \infty$ *as* $|y| \to \infty$,

(viii) *there are constants* $b_2 > 0,\ B_0 > 1$ *such that* $g_1(y) \operatorname{sign} y > b_2 + \epsilon c_1$ *for* $|y| \geqslant B_0$,

(ix) *there is a constant* $c_4 > 0$ *such that* $|g_2(x)| < c_4$ *for* $|x| \leqslant 1$,

(x) $g_2(x) \operatorname{sign} x > 0$ *for* $|x| \geqslant 1$,

(xi) $G_2(x) = \int_0^x g(\xi)\, d\xi \to \infty$ *as* $|x| \to \infty$.

Then there are constants B_1, B_2 *independent of* ϵ, x_0, y_0, t_0, *but depending on* $c_1, c_2, c_3, c_4, b_1, b_2, B_0$ *and the functions* $g_1(y),\ g_2(x)$, *such that*

$$|x(t)| < B_1, \quad |y(t)| < B_2 \tag{3.1}$$

for all sufficiently large t.

Notice that (v) and (viii) follow from continuity if the functions are independent of ϵ and involve only one variable. Further, if as in (1.2) and (1.10), $F(x, y, t) = F(x) - P(t)$ or $F(x, y, t) = F(x) - p(t)$, where $P(t)$ and $p(t)$ are bounded, (i) follows from the continuity of $F(x)$ and then (iii) from (ii). If $F(x, y, t) = F(x, t)$ is independent of y the inequalities in (iii) and (iv) are satisfied in a half plane $x \geqslant 1$ or $x \leqslant -1$ or not at all, and this was a defect of the comparison method which we used earlier.

Condition (viii) is the only one involving a relationship between constants and it is equivalent to that used by Opial. It can be improved slightly, but the improvement appears at first to be more complicated.

4. It should be noticed that condition (iii) is one-sided, whereas (i) is a condition on $|F|$. The fact that (1.6) with $b = 0$ gives rise to a system with $F(x, y) = (y^2 - 1)x$ (as we observed above), so that $F\left(\frac{1}{2}, y\right) = = \frac{1}{2}(y^2 - 1)$ is large and positive, makes it desirable to have a result with a one-sided condition in place of (i), but this can only be done by strengthening the condition on $g_1(x)$ for $|x| < 1$. A minor adjustment leads to

Theorem 2. *Suppose that the hypotheses of Theorem* 1 *hold, except that in place of* (i) *we have*

(i)′ *there is a constant* $c_1 > 0$ *such that* $F(x, y, t) \operatorname{sign} x > -c_1$ *for* $|x| \leqslant 1$

and $\limsup_{|x| \to \infty} |F(x, y, t)| \leqslant c_1$ *uniformly in* t, *and instead of* (x) *we have*

(x)′ $g_2(x) \operatorname{sign} x > 0$ *for all* $x \neq 0$.

Then the conclusion of Theorem 1 *holds.*

Unfortunately the result is insufficient to cover the case in which $F(x, y) = (y^2 - 1)x$ because condition (iii) of Theorem 1 is far too severe, and if (iii) is replaced by

(iii)′ *there is a constant* c_2' *such that* $F(x, y, t) \operatorname{sign} x > -c_2'x$ *for* $|y| \leqslant 1$,

then (iv) must be replaced by something much stronger, and an additional condition somewhat similar to (viii) imposed on $g_2(x)$. What is more, if (iv) is merely replaced by a stronger inequality in *one* of the four domains, an additional condition must be imposed on $g_2(x)$ for large $|x|$ to ensure that the behaviour of $G_2(x)$ as $x \to \pm\infty$ is not too different. Thus as conditions on $F(x, y, t)$ are relaxed in the critical strips, stricter conditions on $g_1(y)$, or $g_2(x)$ as the case may be, are needed to make them behave more like y or x respectively, in addition to a stronger condition in some cases on $F(x, y, t)$ in at least one domain outside the critical strips.

I shall not attempt to state the various results based on (iii), in plane of (iii), but try to illustrate the principles involved by considering some parts of the proof of Theorem 1.

5. As in most of the earlier work the proof depends very heavily on the use of the function

$$V(x, y) \equiv G_2(x) + G_1(y), \tag{5.1}$$

and the construction of a sequence J_n, $n = 0, 1, 2, \ldots$ of Jordan curves with interior domains D_n, $n = 0, 1, 2, \ldots$. This as usual is done in such a way that the union of the D_n is the whole plane, and all solutions which meet any particular J_n cross it inwards, and in a finite time depending on n enter D_0, and remain in it. Starting from a point $P_1^{(n)} = (-1, y_1^{(n)})$, $y_1^{(n)} > 1$, the curve J_n leaves and enters the critical strips at points $P_2^{(n)}, P_3^{(n)}, \ldots, P_9^{(n)} = (-1, y_9^{(n)})$ as shown in the figure, and is closed by the segment of $x = -1$ from $P_9^{(n)}$ to $P_1^{(n)}$. By (viii) it is essential to have $B_0 < y_9^{(n)} < y_1^{(n)}$, for $\dot{x} > 0$ on $x = -1$, and we make $y_9^{(n+1)} = y_1^{(n)}$, $n = 1, 2, \ldots$. Apart from the segment $P_9^{(n)}P_1^{(n)}$ J_n is made up of arcs of $V(x, y)$, or $V(x, y)$ with the addition of some simple function, equal to a constant, and in some cases a segment of a straight line. The arcs and segments have all been used by previous authors, in particular we make considerable use of Reuter's techniques. I shall therefore only make a few observations.

6. Conditions (vi), (vii), (x), and (xi) ensure that the curve

$$V(x, y) = V_0, \quad \text{a constant}, \tag{6.1}$$

is a Jordan curve, and that any x_0, y_0 belongs to its interior domain for V_0 sufficiently large. Putting $x = x(t)$, $y = y(t)$ in $V(x, y)$ and differentiating, we have

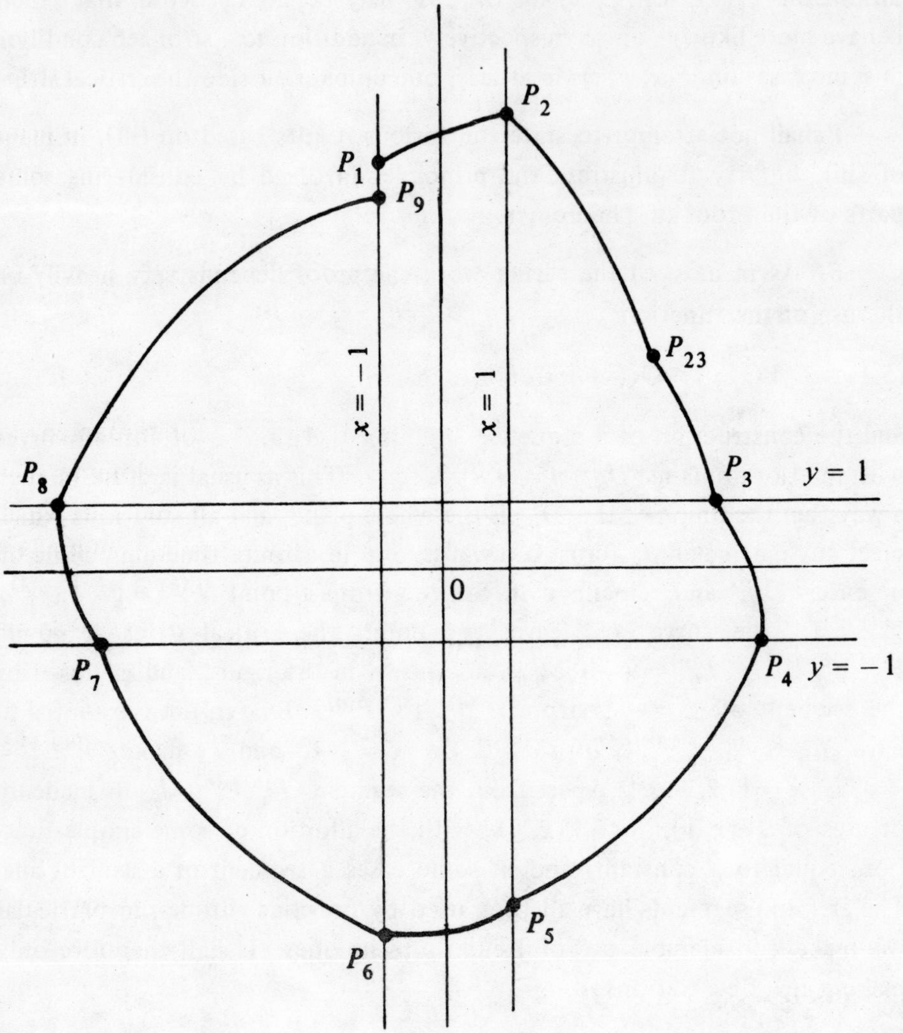
P_2
P_1
P_9
P_{23}
$x = -1$
$x = 1$
P_8
P_3
$y = 1$
0
P_7
P_4
$y = -1$
P_5
P_6

(6.2) $\dot{V}(t) = -\epsilon g_2(x)F(x, y, t),$

and so it follows from (ii) and (x) that arcs of (6.1) can be used outside the critical strips. For $\dot{V} < 0$ there.

For the strip $|y| \leqslant 1$, we use

(6.3) $V(x, y) + \epsilon c_2 y = V(P_3) + \epsilon c_2$ for $x > 1$.

For by (iii)

$$\dot{V} + \epsilon c_2 \dot{y} = -\epsilon g_2(x)(c_2 + F) < 0,$$

so that solutions cross (6.3) inwards and, writing $P_1 = P_1^{(n)}$, $P_2 = P_2^{(n)}$ and so on we have $V(P_4) - V(P_3) = 2\epsilon c_2$.

This increase in $V(x, y)$ and similar ones from the arc P_7P_8 and from the arcs P_1P_2, P_5P_6 have to be compensated by using (iv) in the domain in which it holds which we may assume to be $x > 1$, $y > 1$. However this is only easy for $y > B_0$. We put $\delta = \frac{1}{2}\min(b_1, b_2)$

(6.4) $V(x, y) - \epsilon\delta y = V(P_2) - \epsilon\delta y_2,\quad y_2 \geqslant y \geqslant B_0,$

for the arc from P_2 to $P_{23} = (x_{23}, B_0)$ and use an arc of the form (6.1) for the arc $P_{23}P_3$. By (viii).

(6.5) $\frac{d}{dy}(G_1(y) - \epsilon\delta y) = g_1(y) - \epsilon\delta > \frac{1}{2}b_2 > 0$ for $y \geqslant B_0$

so that $G_1(y) - \epsilon\delta y$ decreases as y decreases. On (6.4) therefore $G_2(x)$, and so by (x) x, increases from P_2 to P_{23}. At P_{23} by (ix) and (6.5)

$$G_2(x_{23}) = G_2(1) + G_1(y_2) - \epsilon\delta(y_2 - B_0) - G_1(B_0) >$$
$$> \frac{1}{2}b_2(y_2 - B_0) - c_4 > \frac{1}{4}b_2(y_2 - B_0),$$

provided that $y_2 > B_0 + \frac{4c_4}{b_2}$. Hence $V(P_{23})$ is large if y_2 is large. It is easy to verify from (iv) that solutions cross the arc P_2P_{23} inwards, and obviously

$$V(P_{23}) - V(P_2) = -\epsilon\delta(y_2 - B_0),$$

so that by making y_2 large we can reduce V by an arbitrarily large multiple of ϵ. Further since $G_2(x_{23})$ and therefore $V(P_{23})$ is large there is no danger that the point P_3 lies within the critical strip.

If (i) holds, we can use

$$V(x, y) - \epsilon\delta x = V(P_1) + \epsilon\delta$$

with suitable δ for the arc P_1P_2, and a corresponding expression for thee arc P_5P_6, but modifications such as those used by Reuter or Opial have to be made for Theorem 2.

REFERENCES

[1] M.L. Cartwright – H.P.F. Swinnerton-Dyer, Boundedness theorems for some second order differential equations I. *Annales Polonici Mathematici,* 29 (1974), 229-254.

[2] J.R. Graef, On the generalized Liénard equation with negative damping, *Journal of Differential Equations,* 12 (1972), 34-62.

[3] R. Reissig – G. Sansone – R. Conti, *Qualitative Theorie nichtlinearer Differentialgleichungen,* Edizioni Cremonese, Roma (1963).

[4] G.E.H. Reuter, Boundedness theorems for non-linear differential equations of the second order (II), *Journal of the London Mathematical Society,* 27 (1952), 48-58.

[5] G. Villari, Criteri di esistenza di soluzioni periodiche per una classe di equazioni differenziali del secondo ordine non lineari, *Annali di Matematica Pure ed Applicata,* IV, 65 (1964), 153-166.

M.L. Cartwright – H.P.F. Swinnerton-Dyer
38 Sherlock Close, Cambridge, England.

COLLOQUIA MATHEMATICA SOCIETATIS JÁNOS BOLYAI
15. DIFFERENTIAL EQUATIONS, KESZTHELY (HUNGARY), 1975.

ON EXISTENCE AND UNIQUENESS FOR CAUCHY'S PROBLEM IN INFINITE DIMENSIONAL BANACH SPACES

K. DEIMLING

Given a Banach space X with dim $X = \infty$, $D \subset X$, $J = [0, a) \subset R^1$, f: $J \times D \to X$ and $x_0 \in D$, we look for local/global solutions of Cauchy's problem

(1) $$x' = f(t, x), \quad x(0) = x_0.$$

This problem appears in many situations. For instance, in the application of Fourier's method and the method of partial discretization to initial/boundary value problems for parabolic or hyperbolic equations, in the theory of Markov processes with discrete states in continuous time and in game theory. In most situations X is a suitable sequence space; see [1, Chap. 15] for references. Since we always assume that f is at least continuous and D is at least closed, the theorems presented here can be applied to evolution equations only indirectly.

Contrary to Peano's existence theorem in case dim $X < \infty$, the continuity of f is not sufficient for local existence. Apparently, the first simple counterexample has been given in [11]; see also [12, Chap. 5] and [2].

The reason for this sad fact is that bounded sets are not necessarily relatively compact if $\dim X = \infty$. Therefore, we have to impose further conditions on f in order to obtain existence theorems.

1. CONDITIONS OF LIPSCHITZEAN TYPE

Almost all results in finite dimensional spaces carry over with essentially the same proofs if f satisfies a local generalized Lipschitz condition

$$(2) \qquad |f(t, x) - f(t, y)| \leqslant \omega(t, |x - y|)$$

and D is an open set. In this case (1) has a unique local solution. It may be obtained by successive approximation in case $\omega(t, \rho)$ is monotone increasing in ρ.

However, the classical theorem on continuation of solutions up to the boundary of $J \times D$ is wrong for $\dim X = \infty$. It may happen that a solution exists only in $[0, \delta)$, for some $\delta < a$, and remains bounded there, a situation impossible for $\dim X < \infty$. The first counterexample was given again in [11]; see also [12], [2]. The reason for this is that even a locally Lipschitzean mapping needs not map bounded sets into bounded sets. Global existence is guaranteed if f satisfies in addition $|f(t, x)| \leqslant \varphi(t, |x|)$, where the maximal solution of $\rho' = \varphi(t, \rho)$, $\rho(0) = |x_0|$ exists on J.

The existence theorems involving Lipschitz conditions are also of interest if f is only continuous. It has been proved in [13] that a continuous map f can be approximated, uniformly on open sets, by locally Lipschitzean maps. This fact gives us at least good approximate solutions for (1).

2. CONDITIONS OF ACCRETIVE TYPE

In case X is a Hilbert space with inner product $(\cdot, \cdot)$ it is natural to replace (2) by the weaker condition

$$(f(t, x) - f(t, y), x - y) \leqslant \omega(t, |x - y|)|x - y|.$$

Such a condition can also be formulated in a general Banach space X if we have a concept like an inner product there. Therefore, let X^* denote the dual space, consider the duality map $F\colon X \to 2^{X^*}$, defined by

$$F(x) = \{x^* \in X^*\colon\ x^*(x) = |x|^2 \quad \text{and} \quad |x^*| = |x|\},$$

and define the generalized pairings $(\cdot,\cdot)_{(\pm)}\colon X \times X \to R^1$ by $(x,y)_+ = = \sup\{x^*(x)\colon x^* \in F(y)\}$ and $(x,y)_- = \inf\{x^*(x)\colon x^* \in F(y)\}$. Properties of F may be found for instance in [6] and [18]. In particular, F may be identified with the identity operator I of X if and only if X is a Hilbert space, and in this case the generalized pairings are identical with the inner product. We speak of an accretive type condition if f satisfies

$$(3) \qquad (f(t,x) - f(t,y), x - y)_- \leqslant \omega(t, |x-y|)|x-y|$$

with a uniqueness function ω.

If x_0 is an inner point of D and f satisfies (3) then the approximate solutions mentioned at the end of Section 1 can be shown to converge to a unique local solution of (1) for a large class of uniqueness functions. If, in addition, f maps bounded sets into bounded sets and satisfies an estimation like $(f(t,x),x)_- \leqslant \varphi(t,|x|)|x|$, where the maximal solution of $\rho' = \varphi(t,\rho)$, $\rho(0) = |x_0|$ exists on J, then one can prove global existence. These results may be found in [19] and [8]; see also [5]. Condition (3) is not strong enough to guarantee convergence of successive approximations; see [9] for a counterexample. However, we have shown in [9] that the solution of (1) is the limit of the Galerkin approximations $x_n \in X_n$, defined by $x_n' = P_n f(t, x_n)$, $x_n(0) = P_n x_0$, in case $D = X$, X^* is strictly convex, (X_n) is a sequence of finite dimensional subspaces of X and $P_n\colon X \to X_n$ are projections with $|P_n| = 1$ and $P_n x \to x$ for every $x \in X$. Now, let us assume that D is only closed. Therefore, x_0 may be in the boundary ∂D of D. In order to obtain local existence in this situation, f has to satisfy the boundary condition

$$(4) \qquad \rho(x + \lambda f(t,x), D) = o(\lambda) \quad \text{as} \quad \lambda \to 0+, \qquad \text{for every} \quad x \in \partial D$$

where $\rho(z, D)$ denotes the distance from z to D. It has been proved in [14] that (1) has a unique global solution if f satisfies (4) and

$$(f(t,x) - f(t,y), x-y)_+ \leqslant k|x-y|^2 \qquad (t \in J;\ x, y \in D)$$

for some $k \in R^1$; see also [14a]. In case D is also convex, $(\cdot,\cdot)_+$ may be replaced by $(\cdot,\cdot)_-$. In particular, if D is closed and convex, $f = g - I$ and $g(J \times \partial D) \subset D$ then (4) is satisfied, since $(1-\lambda)x + \lambda g(t,x) \in D$. These assumptions had been considered earlier in [7]. Condition (4) dates back to [15], at least.

The existence theorems of this section can be applied to obtain zeros of accretive operators, i.e. of operators $T\colon D \to X$ such that

$$(Tx - Ty, x - y)_+ \geqslant 0 \qquad \text{for every} \quad x, y \in D.$$

Since $(\cdot,\cdot)_+$ is the inner product in case X is a Hilbert space, these operators are the natural generalization of monotone operators.

Suppose D is closed and convex, $T\colon D \to X$ is accretive and (4) is satisfied for $f(t,x) = -Tx$. Then, the Cauchy problems

(5) $\qquad u' = -Tu, \quad u(0) = x \in D$

have a unique solution $u(t,x)$ on $[0,\infty)$, since $(-Tu + Tv, u - v)_- = -(Tu - Tv, u - v)_+ \leqslant 0$. Hence, T has a zero if and only if (5) has a constant solution. In order to find a constant solution, define $U(t)\colon D \to D$ by $U(t)x = u(t,x)$. The operators $U(t)$ are at least nonexpansive, for every $t \geqslant 0$. To see this, let $z(t) = U(t)x - U(t)y$, $\varphi(t) = |z(t)|$ and

$$D^-\varphi(t) = \limsup_{h \to 0+} h^{-1}\{\varphi(t) - \varphi(t-h)\}.$$

From the definition of $(\cdot,\cdot)_-$ we obtain

$$\varphi(t)D^-\varphi(t) \leqslant (z'(t), z(t))_- = -(TU(t)x - TU(t)y, z(t))_+ \leqslant 0$$

in $t > 0$, and therefore $\varphi(t) \leqslant \varphi(0) = |x - y|$.

Now, assume for instance that X is uniformly convex and D is also bounded. Then the family $\{U(t)\colon t \geqslant 0\}$ has a fixed point x_0 in common, by a well known fixed point theorem of F. Browder. Hence, $u(t,x_0) \equiv x_0$ and therefore $Tx_0 = 0$. In certain situations it turns out that $U(p)$ is actually a k-contraction with $k < 1$, for some $p > 0$. Then, $U(p)$ has a unique fixed point $x_p \in D$ and the corresponding solution $u(t,x_p)$ must be p-periodic, since (5) is uniquely solvable. Therefore,

$$|u(t, x_p) - x_p| = |u(t+p, x_p) - x_p| =$$

$$= |U(p)u(t, x_p) - U(p)x_p| \leqslant k|u(t, x_p) - x_p|.$$

Hence, $u(t, x_p) \equiv x_p$ and therefore $Tx_p = 0$. See [10] for more details.

3. CONDITIONS INVOLVING MEASURES OF NON-COMPACTNESS

In this section we restrict ourselves to the situation which is typical for a local existence theorem in case x_0 is an inner point of D. We consider $J = [0, a] \subset R^1$, $D = \{x \in X: |x - x_0| \leqslant r\}$ for some fixed $r > 0$ and a continuous map $f: J \times D \to X$ which is bounded, say $|f(t, x)| \leqslant c$ on $J \times D$.

Since the possible non-compactness of $f(J \times D)$ is the reason for (1) to have possibly no local solution, it is natural to look for extra conditions on f in terms of measures of non-compactness. Recall that for any bounded set $B \subset X$ the (Kuratowski-) measure of non-compactness $\alpha(B)$ is defined by

$$\alpha(B) = \inf\{d > 0: B \text{ has a finite cover of diameter } \leqslant d\}.$$

Let us mention only some of the properties of α. See [16, Theorem 1.2.3] for proofs.

(i) $\alpha(B) = 0$ if and only if $\bar{B}$, the closure of B, is compact.

(ii) $\alpha(\overline{\text{co}}\, B) = \alpha(B)$, where $\overline{\text{co}}\, B$ is the closed convex hull of B.

(iii) $\alpha(B_1 + B_2) \leqslant \alpha(B_1) + \alpha(B_2)$; $B_1 \subset B_2$ implies $\alpha(B_1) \leqslant \alpha(B_2)$.

(iv) α is continuous with respect to the Hausdorff metric

$$d_H(B_1, B_2) = \max\{\sup_{B_1} \rho(x, B_2), \sup_{B_2} \rho(x, B_1)\}.$$

(v) $\alpha(B_1 \cup B_2) = \max(\alpha(B_1), \alpha(B_2))$; $\alpha(\lambda B) = |\lambda|\alpha(B)$.

Instead of (2), we consider a condition like

$$\text{(6)} \qquad \alpha(f(t, B)) \leqslant \omega(t, \alpha(B)) \qquad t \in (0, a), \; B \subset D.$$

In order to simplify the statement of our first theorem, we introduce the following class U of uniqueness functions ω (class U_2 in [8]) that contains most of the known ones.

The class U. A function ω: $(0, a) \times R^1_+ \to R^1_+$ is said to be of class U if to every $\epsilon > 0$ there exist a constant $\delta > 0$, sequences (t_i) and (δ_i) with $t_i \to 0+$ and $\delta_i > 0$, and a sequence of functions ρ_i, continuous in $[t_i, a)$ and such that $\rho_i(t_i) \geqslant \delta t_i$, $D^- \rho_i(t) \geqslant \omega(t, \rho_i(t)) + \delta_i$ and $\rho_i(t) \leqslant \epsilon$ in $[t_i, a)$.

Theorem 1. *Let* J, D, f *be as above, and* $b < \min\{a, \frac{r}{c}\}$. *Assume in addition that* f *is uniformly continuous on* $J \times D$ *and satisfies* (6) *with* $\omega \in U$. *Then* (1) *has a solution on* $[0, b]$.

Proof.

(i) Since f is continuous and bounded, we may approximate f by f_n such that $|f_n(t, x) - f(t, x)| \leqslant \frac{1}{n}$ and f_n is locally Lipschitzean on $J \times D$ (cf. [13]). Therefore, (1), with f_n instead of f, has exactly one solution x_n on $[0, b]$. Hence,

$$x'_n = f(t, x_n) + y_n(t), \quad x_n(0) = x_0 \quad \text{and} \quad |y_n(t)| \leqslant \frac{1}{n}. \tag{7}$$

Furthermore, it is easy to see that (x_n) has the following property: To every $\eta > 0$ there exist $t_\eta > 0$ such that $|x_n(t) - x_m(t)| \leqslant \left(\frac{1}{n} + \frac{1}{m} + \eta\right) t$ for $0 \leqslant t \leqslant t_\eta$ (cf. the proof of [8, Theorem 2]). Hence, if $B_k(t) = \{x_n(t)\colon n \geqslant k\}$, we have

$$\alpha(B_k(t)) \leqslant 2\left(\frac{2}{k} + \eta\right) t \quad \text{for} \quad t \leqslant t_\eta. \tag{8}$$

(ii) We want to show that $B_1 = \{x_n\colon n \geqslant 1\}$ is relatively compact in the space $C([0, b])$ of all continuous functions from $[0, b]$ into X with the max-norm, since in this case (x_n) has a convergent subsequence whose limit is a solution of (1), as usual. Obviously, B_1 is bounded and equicontinuous. Thus, according to the theorem of Ascoli – Arzela, we only have to prove that $\overline{B_1(t)}$ is compact, for every $t \in [0, b]$. By property (i) of α, we have to show $\alpha(B_1(t)) = 0$ for every t.

(iii) Let $B_k = \{x_n: n \geqslant k\}$, $\varphi_k(t) = \alpha(B_k(t))$ and $B'_k(t) = \{x'_n(t): n \geqslant k\}$. Since B_k is equicontinuous and (iv) holds, φ_k is continuous Furthermore, (iii), (iv) and the uniform continuity of f imply $D^- \varphi_k(t) \leqslant \alpha(B'_k(t))$ in $t > 0$ (cf. the proofs of [16, Theorem 1.4.2] and of Theorem 2 below). Hence, (6), (7) and (iii) imply

$$(9) \qquad D^- \varphi_k(t) \leqslant \omega(t, \varphi_k(t)) + \frac{2}{k} \quad \text{if} \quad t > 0.$$

Now, assume $\alpha(B(t_0)) = 2\epsilon > 0$ for some $t_0 \in (0, b]$. To ϵ we first choose $\delta > 0$ according to $\omega \in U$, then η and k such that $2\left(\frac{2}{k} + \eta\right) < \delta$, and t_η according to (8). Since $\omega \in U$, we find $t_i < t_\eta$, $\delta_i > 0$ and a function ρ_i. Finally, we choose k such that in addition $\frac{2}{k} < \delta_i$. Now, we have $\varphi_k(t) < \delta t_i \leqslant \epsilon$ in $[0, t_i]$ and $D^- \varphi_k(t) < \omega(t, \varphi_k(t)) + \delta_i$ in $(t_i, b]$, by (8) and (9). On the other hand, $\delta t_i \leqslant \rho_i(t_i)$ and $D^- \rho_i(t) \geqslant \omega(t, \rho_i(t)) + \delta_i$. Therefore, $\varphi_k(t) < \rho_i(t) \leqslant \epsilon$ in $[t_i, b]$. In particular, we have $\alpha(B(t_0)) = \alpha(B_k(t_0) \cup \{x_1(t_0), \ldots, x_{k-1}(t_0)\}) = \varphi_k(t_0) \leqslant \epsilon$, a contradiction to $\alpha(B(t_0)) \geqslant 2\epsilon$.

q.e.d.

A similar result with less general ω has been proved in [16, Theorem 3.5.2] by means of a fixed point theorem for condensing maps; see also [14a]. Unfortunately, we had to assume that f is uniformly continuous in order to obtain $D^- \varphi_k(t) \leqslant \alpha(B'_k(t))$. Our next theorem shows that this condition can be removed if we assume

$$(10) \qquad \alpha(f(I \times B)) \leqslant \omega(\alpha(B)) \qquad \text{for} \qquad B \subset D.$$

instead of (6).

Theorem 2. *Let J and D be as in Theorem* 1; *$f: J \times D \to X$ continuous and $|f(t,x)| \leqslant c$ on $J \times D$. Assume in addition f satisfies* (10) *with $\omega: R^1_+ \to R^1_+$ continuous and such that the initial value problem $\rho' = \omega(\rho)$, $\rho(0) = 0$ has only the trivial solution $\rho(t) \equiv 0$ on J. Then* (1) *has a solution on $[0, b]$ for any $b < \min\left\{a, \frac{r}{c}\right\}$.*

Proof. By the assumption about ω, it is well known that to every

$\epsilon > 0$ there exist a constant $\delta_\epsilon > 0$ and a function ρ_ϵ such that $\delta_\epsilon \leqslant \leqslant \rho_\epsilon(t) \leqslant \epsilon$ and $\rho'_\epsilon \geqslant \omega(t, \rho_\epsilon) + \delta_\epsilon$ in J. Therefore, the proof of Theorem 1 shows that the verification of $D^- \varphi_k(t) \leqslant \omega(\varphi_k(t)) + \frac{2}{k}$ is the only difficulty (cf. (9)). Since $D^- \varphi_k(t) = \lim_{\tau \to 0+} \sup_{(0,\tau]} q(h)$ (with $q(h) = h^{-1}(\varphi_k(t) - \varphi_k(t-h))$, and since $B_k(t) \subset \{x_n(t) - x_n(t-h) : n \geqslant k\} + B_k(t-h)$, we obtain by means of (iii) and (v)

$$(11) \qquad \sup_{(0,\tau]} q(h) \leqslant \alpha\Big(\Big\{\frac{x_n(t) - x_n(t-h)}{h} : n \geqslant k,\ 0 < h \leqslant \tau\Big\}\Big).$$

Now, recall that the set on the right side of (11) is a subset of $\overline{\text{co}}\,\{x'_n(s): n \geqslant k,\ t - \tau \leqslant s \leqslant \tau\}$. Therefore, (11) and (ii) yield

$$(12) \qquad D^- \varphi_k(t) \leqslant \lim_{\tau \to 0+} \alpha\Big(\bigcup_{J_\tau} B'_k(s)\Big) \qquad J_\tau = [t - \tau, t].$$

Since $x'_n(s) = f(s, x_n(s)) + y_n(s)$ with $|y_n(s)| \leqslant \frac{1}{n}$, we have

$$\alpha\Big(\bigcup_{J_\tau} B'_k(s)\Big) \leqslant \alpha\Big(\bigcup_{J_\tau} f\Big(s, \bigcup_{J_\tau} B_k(s)\Big)\Big) + \frac{2}{k} \leqslant \alpha\Big(f\Big(J \times \bigcup_{J_\tau} B_k(s)\Big)\Big) + \frac{2}{k}.$$

Hence, (12) and (10) and the continuity of ω imply

$$(13) \qquad D^- \varphi_k(t) \leqslant \omega\Big(\lim_{\tau \to 0+} \alpha\Big(\bigcup_{J_\tau} B_k(s)\Big)\Big) + \frac{2}{k}.$$

Since B_k is equicontinuous, $\bigcup_{J_\tau} B_k(s)$ tends to $B_k(t)$ as $\tau \to 0+$, with respect to the Hausdorff metric. Therefore, (13) and (iv) imply $D^- \varphi_k(t) \leqslant \leqslant \omega(\varphi_k(t)) + \frac{2}{k}$.

q.e.d.

In case f is a compact map we can take $\omega \equiv 0$ in Theorem 2. The case $\omega(\rho) = k\rho$ has been considered in [17]. In [3], a quite different (and longer) proof has been given for another particular case, namely $\omega(\rho) = L(\rho)\rho$ for $\rho > 0$ and $\omega(0) = 0$, where $L(\rho) = \sup\{[\alpha(B)]^{-1}\alpha(f(I \times B)): \alpha(B) \geqslant \rho\}$ and $\int_{0+} \frac{d\rho}{L(\rho)\rho} = \infty$. It is easy to see that this ω is continuous (even increasing). In [4], there is another condition on f, sufficient for local existence in case X^* is uniformly convex: f is continuous and to every $\epsilon > 0$ there exists a finite cover $\{\mathcal{O}_i\}$ of $J \times D$ such that $(f(t_1, x_1) - f(t_2, x_2), x_1 - x_2)_+ \leq \epsilon$ whenever (t_1, x_1) and (t_2, x_2) belong to the same $\mathcal{O}_i$.

Since the conditions in Theorem 1 and Theorem 2 do not imply uniqueness, our next theorem is concerned with the set of all solutions of (1).

Theorem 3. *Let* J, D, f *and* b *be as in Theorem* 1 *or Theorem* 2. *Then the set* S *of all solutions of* (1) *is a continuum, i.e. a nonempty compact and connected subset of* $C([0, b])$. In particular, $S(t) = \{x(t): x \in S\}$ *is a continuum of* X, *for every* $t \in [0, b]$.

Proof.

(i) Obviously, S is not empty. If we take a sequence (x_n) of solutions instead of the approximate solutions in the proofs, we find a convergent subsequence. Since f is continuous, we have S compact.

(ii) Assume that S is not connected. Then, we have $S = S_1 \cup S_2$, $S_i \neq \phi$ and compact and $S_1 \cap S_2 = \phi$. Hence, $\inf\{|x_1 - x_2|: x_1 \in S_1, x_2 \in S_2\} = \beta > 0$. We consider the functional $\varphi: C(J) \to R^1$, defined by $\varphi(x) = \rho(x, S_1) - \rho(x, S_2)$. φ is continuous, we have $\varphi(x) \leq -\beta$ if $x \in S_1$ and $\varphi(x) \geq \beta$ if $x \in S_2$, and we are going to construct the contradiction $\varphi(x) = 0$ for some $x \in S$.

We choose $\epsilon_0 > 0$ such that $b \leq \dfrac{r}{c + 2\epsilon_0}$. Let $\epsilon \in (0, \epsilon_0]$ be fixed; $g: J \times D \to X$ locally Lipschitzean and such that $|g(t, x) - f(t, x)| \leq \epsilon$ on $J \times D$; $x_1 \in S_1$ and $x_2 \in S_2$ be fixed. Then, the mappings f^i defined by

$$f^i(t, x) = g(t, x) + f(t, x_i(t)) - g(t, x_i(t)) \quad (i = 1, 2)$$

are locally Lipschitzean, satisfy $f^i(t, x_i(t)) = f(t, x_i(t))$ and $|f^i(t, x) - f(t, x)| \leq 2\epsilon$ on $J \times D$. The first and the last property are also true for the mappings f_λ defined by

$$f_\lambda(t, x) = f^1(t, x) + \lambda(f^2(t, x) - f^1(t, x)), \qquad \lambda \in [0, 1].$$

Hence, we have a unique solution $x^\lambda \in C([0, b])$ of $x' = f_\lambda(t, x)$ and $x(0) = x_0$. Since f_λ is linear in λ and the f_λ are uniformly bounded and locally Lipschitzean on $J \times D$, it is easy to see that the map $\lambda \to x^\lambda$ is continuous. Hence, $\psi(\lambda) = \varphi(x^\lambda)$ is continuous in $[0, 1]$. Since $f_0(t, x_1(t)) = f^1(t, x_1(t)) = f(t, x_1(t)) = x_1'(t)$, we have $x^0 = x_1$, and $x^1 = x_2$ correspondingly. This implies $\psi(0) \leqslant -\beta$ and $\psi(1) \geqslant \beta$. Therefore, there is a λ_0 such that $\psi(\lambda_0) = 0$.

Let $\epsilon = \frac{1}{n}$ and x_n the corresponding x^{λ_0}. Since

$$x_n' = f(t, x_n) + y_n(t), \quad x_n(0) = x_0$$

and

$$|y_n(t)| = |f_{\lambda_0}(t, x_n(t)) - f(t, x_n(t))| \leqslant \frac{2}{n},$$

(x_n) has a convergent subsequence (x_m). Its limit x^* is in S, contradicting our assumption since $\varphi(x^*) = \lim_{m \to \infty} \varphi(x_m) = 0$. Therefore S is also connected.

q.e.d.

REFERENCES

[1] R. Bellman, *Methods of nonlinear analysis.* Vol. II. Acad. Press, New York 1973.

[2] N. Bourbaki, *Elements de mathématique* XII, Chap. IV. Hermann, Paris 1961.

[3] A. Cellina, On the existence of solutions of ordinary differential equations in Banach spaces, *Funk. Ekvac.*, 14 (1971), 129-136.

[4] A. Cellina, On the local existence of solutions of ordinary differential equations. *Bull. Acad. Pol. Sci.*, 20 (1972), 293-296.

[5] A. Cellina – G. Pianigiani, On the prolongability of solutions of autonomous differential equations. *Boll. Unione Mat. Ital.*, (4) 9 (1974), 824-830.

[6] I. Cioranescu, *Aplicatii de dualitate in analiza functionala neliniara.* Ed. Acad. Rep. Soc. Romania, Bucharest 1974.

[7] M.G. Crandall, Differential equations on convex sets. *J. Math. Soc. Japan,* 22 (1970), 443-455.

[8] K. Deimling, On existence and uniqueness for differential equations in Banach spaces. *Ann. Mat. Pura Appl,* (IV) 106 (1975), 1-12.

[9] K. Deimling, On approximate solutions of differential equations in Banach spaces. *Math. Annalen,* 212 (1974), 79-88.

[10] K. Deimling, Zeros of accretive operators. *Manuscripta Math.,* 13 (1974), 365-374.

[11] J. Dieudonné, Deux examples singuliers d'équations différentielles. *Acta Sci. Math. (Szeged),* 12 B (1950), 38-40.

[12] G.E. Ladas – V. Lakshmikantham, *Differential equations in abstract spaces.* Acad. Press, New York, 1972.

[13] A. Lasota – J.A. Yorke, The generic property of existence of solutions of differential equations in Banach spaces. *J. Diff. Eqs.,* 13 (1973), 1-12.

[14] R.H. Martin, Differential equations on closed subsets of a Banach space. *Trans. Amer. Math. Soc.,* 179 (1973), 399-414.

[14a] R.H. Martin, Approximation and existence of solutions to ordinary differential equations in Banach spaces. *Funk. Ekvac.* 16 (1973), 195-211.

[15] M. Nagumo, Über die Lage der Integralkurven gewöhnlicher Differentialgleichungen. *Proc. Phys.-Math. Soc. Japan,* 24 (1942), 551-559.

[16] B.N. Sadovskii, Limit compact and condensing operators. *Uspehi Mat. Nauk,* 27, (163) (1972), 81-146. Transl. in Russ. Math. Surveys.

[17] S. Szufla, Some remarks on ordinary differential equations in Banach spaces. *Bull. Acad. Pol. Sci.*, 16 (1968), 795-800.

[18] M.M. Vainberg, *Variational method and method of monotone operators in the theory of nonlinear equations.* J. Wiley & Sons, New York 1973, Russian ed. by Izdatel'stvo Nauk, Moscow 1972.

[19] G. Vidossich, *Existence, comparison and asymptotic behaviour of solutions of ordinary differential equations in finite and infinite dimensional Banach spaces.* (preprint).

K. Deimling

D-23 Kiel, Math. Sem. d. Univ. Olshausenstrasse 40-60, GFR.

COLLOQUIA MATHEMATICA SOCIETATIS JÁNOS BOLYAI
15. DIFFERENTIAL EQUATIONS, KESZTHELY (HUNGARY), 1975.

ON STABLE ALGORITHMS CONCERNING NUMERICAL DIFFERENTIATION AND THE NUMERICAL SOLUTION OF SOME INVERSE PROBLEMS FOR EQUATIONS WITH DEVIATING ARGUMENT

I.F. DOROFEEV

Equations of advanced type have been very seldom used to describe concrete processes. In spite of this, it is of relevance to construct stable algorithms for initial value problems of this type, since the inverse problems concerning equations of retarded or neutral type lead to such problems.

Let us consider the inverse problem for equations with retarded argument in the simplest case.

Assume that the equation

(1) $$y'(x) = F(y(x), y(x-\tau), x) \quad (\tau > 0, \ \tau = \text{const.})$$

has a unique solution on the interval $[0, T]$ ($T = m\tau$, m positive integer) corresponding to the continuous initial function φ, φ,

(2) $$y(x) = \varphi(x), \quad x \in [-\tau, 0].$$

The conditions on the function $F(p, q, x)$ are well-known in this case.

Problem A*. *Suppose that we know an approximate solution of problem* (1)-(2) *on the interval* $[T-\tau, T]$, *in other words, let be given a function* $y_\epsilon(x)$ *that differs from the exact solution* $y(x)$ *of problem* (1)-(2) *by some* $\epsilon > 0$, *the difference being measured by the norm of* $C[T-\tau, T]$. *Find a stable algorithm for the approximate construction of the initial function of problem* (1)-(2) *on the segment* $[-\tau, 0]$ *that generates the precise solution* $y(x)$ *on the interval* $[T-\tau, T]$, *by means of the* ϵ*-approximant of* $y(x)$.

Recall that the notion "stable algorithm" assumes that the approximate solution tends to the corresponding precise one, provided that the errors of the data and those of the calculations tend to zero [1].

It is obvious that Problem A* cannot be solved for an arbitrary function $F(p, q, x)$, however, as it was proved in [2], if equation (1) can be uniquely written in the form

$$y(x-\tau) = \bar{f}(y'(x), y(x), x) \tag{3}$$

and the function $\bar{f}(y', y, x)$ is smooth enough with respect to its first two arguments and continuous with respect to the third one, then there exists a stable algorithm for Problem A*. Moreover, as it usually happens in the case of non-correct problems, the stable algorithm mentioned is parametric. More concretely, the parameter appears in the singularly perturbed equation:

$$\mu y'(x-\tau) + \bar{y}(x-\tau) = \bar{f}(\bar{y}'(x), y(x), x), \tag{4}$$

where μ is a small positive parameter.

The presence of singular perturbations brings into the solution of equation (4) summands containing factors of the type $\exp\left(x - \frac{\kappa\tau}{\mu}\right)$. This situation allows us to obtain only asymptotic estimates for the error [2].

The present work has two objectives. In the first section we give a modification of the stable algorithm mentioned which allows us to obtain a uniform estimate of regular type for the error of the approximate solution on the whole interval of length τ. Besides, we study the possibility of minimizing errors in some special cases. Finally, the modifications considered will be illustrated by the example of numerical differentiation.

Section 1.

For the sake of convenience, instead of Problem A* we will study the corresponding initial value problem for the following equation of advanced type

$$
(5)\qquad \begin{aligned} & u(x+\tau) = f(u'(x), u(x), x) \\ & \left.\begin{aligned} u(x) &= u_0(x) \\ u'(x) &= u_0'(x) \end{aligned}\right\} \quad \text{for} \quad x \in [-\tau, 0], \end{aligned}
$$

having in mind that by an appropriate choice of the functions $f(p, q, x)$, $u_0(x)$ and by a change of the independent variable, Problem A* can be transformed into the form (5). We also note that the derivative $u_0'(x)$ occurring in (5) will be considered given, however, in practice it has to be calculated by means of some stable algorithm, in order to solve Problem A*.

Let us introduce the following auxiliary problems that help us obtain the solution of problem (5) on the interval $[0, \tau]$:

$$
(6)\qquad \begin{aligned} & \mu v'(x) + v(x) = f(v'(x-\tau), v(x-\tau), x) \\ & \left.\begin{aligned} v(x) &= u_0(x) \\ v'(x) &= u_0'(x) \end{aligned}\right\} \quad \text{for} \quad x \in [-\tau, 0] \end{aligned}
$$

$$
(7)\qquad \begin{aligned} & \mu_1 z'(x) + z(x) = f(v'(x-\tau), v(x-\tau), x) e^{\frac{x}{\mu_1}} \\ & z(0) = \frac{v(0)}{2}. \end{aligned}
$$

For the given interval of x, by virtue of the assumption on the sufficient smoothness of $f(v', v, x)$ (the precise stipulation on the order of smoothness of this function can be found in Theorem 1) we obtain:

$$
(8)\qquad v(x) = u_0(0) e^{-\frac{x}{\mu}} + \frac{1}{\mu} \int_0^x e^{-\frac{x-s}{\mu}} u(s)\, ds =
$$

$$= u(x) + \sum_{i=1}^{n-1} (-\mu)^i u^{(i)}(x) + (-\mu)^{n-1} \int_0^x e^{-\frac{x-s}{\mu}} u^{(n)}(s)\, ds -$$

$$- \mu e^{-\frac{x}{\mu}} [u'(0) - \mu u''(0) + \ldots + (-\mu)^{n-1} u^{(n-2)}(0)],$$

$$z(x) = 0.5 v(0) e^{-\frac{x}{\mu_1}} + \frac{1}{\mu_1} \int_0^x e^{-\frac{x-2s}{\mu_1}} u(s)\, ds =$$

$$= 0.5 \cdot e^{\frac{x}{\mu_1}} \Big[u(x) - \frac{\mu_1}{2} u'(x) + \frac{\mu_1^2}{4} u''(x) + \ldots$$

$$\ldots + \left(-\frac{\mu_1}{2}\right)^{n-1} u^{(n-1)}(x) - \tag{9}$$

$$- \left(-\frac{\mu_1}{2}\right)^{n-1} \int_0^x e^{-\frac{x-s}{0.5\mu_1}} u^{(n)}(s)\, ds \Big] -$$

$$- \frac{\mu_1}{2} e^{-\frac{x}{\mu_1}} \Big[u'(0) - \frac{\mu_1}{2} u''(0) + \ldots + \left(-\frac{\mu_1}{2}\right)^{n-1} u^{(n-2)}(0) \Big].$$

We remark that in [2] the function $v(x)$ was proposed as the stable solution of problem (5).

Let us set the following notations:

$$A_\mu^0 = u'(0) - \mu u''(0) + \mu^2 u'''(0) + \ldots + (-\mu)^{n-2} u^{(n-1)}(0)$$

$$B_\mu^0 = u(x) - \mu u'(x) + \ldots + (-\mu)^{n-1} \int_0^x u^{(n)}(s)\, ds.$$

Then from (8) and (9), by putting $\mu_1 = 2\mu$ in (7), we obtain:

$$2z(\mu) = e^{0.5} B_\mu^0 - 2\mu e^{-0.5} A_\mu^0 \tag{10}$$

$$v(\mu) = B_\mu^0 - \mu e^{-1} A_\mu^0,$$

whence

$$\mu A_\mu^0 = v(\mu) e^1 - 2z(\mu) e^{0.5}. \tag{11}$$

Let us now consider the following initial value problem on the same interval as before

$$
(12)\quad \begin{aligned} &\mu\bar{v}'(x) + \bar{v}(x) = f(u_0'(x), u_0(x), x) \\ &\bar{v}(0) = u_0(0) + \mu A_\mu^0, \end{aligned}
$$

for which with the above notations $\bar{v}(x) = B_\mu^0(x)$, and, consequently, in virtue of (8), on this interval we have:

$$
(13)\quad \begin{aligned}
&|u(x) - \bar{v}(x)| \leqslant \sum_{i=1}^{n-1} \mu^i |u^{(i)}(x)| + \mu^n \max_{x\in[0,\tau]} |u^{(n)}(x)| \\
&|u'(x) - \bar{v}'(x)| \leqslant \sum_{i=1}^{n-2} \mu_i |u^{(i+1)}(x)| + \mu^{n-1} \max_{x\in[0,\tau]} |u^{(n)}(x)| \\
&\qquad\vdots \\
&|u^{(n-1)}(x) - \bar{v}^{(n-1)}(x)| \leqslant \mu \max_{x\in[0,\tau]} |u^{(n)}(x)|.
\end{aligned}
$$

Moreover, as it is easy to verify, we also have:

$$
(14)\quad |\bar{v}^{(n)}(x) - u^{(n)}(x - \Theta\mu \ln \mu)| \leqslant \mu \max_{x\in[0,\tau]} |u^{(n)}(x)|.
$$

The estimates obtained under (13) show that if we take $\bar{v}(x)$ as the solution of problem (5) on the interval in question, then the dependence of the parameter will correspond to the error of the solution of the non-singularly ("regularly") perturbed equation. We also remark that in virtue of the presence of the summand μA_μ^0 in (12), we have $\bar{v}(0+0) \neq u_0(0-0)$, however, the jump tends to zeros as $\mu \to 0$.

To construct an approximate solution whose error can be estimated on the interval $x \in [\tau, 2\tau]$ in the same way as in (13)-(14), we consider, on this interval, the following problem:

$$
\begin{aligned}
&\mu\bar{v}'(x) + \bar{v}(x) = f(\bar{v}'(x-\tau), \bar{v}(x-\tau), x) \\
&\lim_{x\to\tau+0} \bar{v}(x) = \lim_{x\to\tau-0} \bar{v}(x) + \mu A_\mu^\tau,
\end{aligned}
$$

where

$$A^{\tau}_{\mu} = \left[\frac{d}{dx} f(\bar{v}'(x-\tau), v(x-\tau), x)\right]_{x=\tau+0} -$$

$$-\mu\left[\frac{d^2}{dx^2} f(\bar{v}'(x-\tau), \bar{v}(x-\tau), x)\right]_{x=\tau+0} + \ldots$$

$$\ldots + (-\mu)^{n-2}\left[\frac{d^{n-2}}{dx^{n-2}} f(\bar{v}'(x-\tau), \bar{v}(x-\tau), x)\right]_{x=\tau+0}.$$

To determine A^{τ}_{μ}, one has to take the solution, at the point $x = \tau + \mu$, of the following problem:

$$\text{(16)} \qquad \begin{aligned} &\mu v'(x) + v(x) = f(\bar{v}'(x-\tau), \bar{v}(x-\tau), x) \\ &v(\tau) = \lim_{x\to\tau+0} \bar{v}(x) \end{aligned}$$

$$\text{(17)} \qquad \begin{aligned} &\mu_1 z'(x) + z(x) = f(\bar{v}'(x-\tau), \bar{v}(x-\tau), x) \\ &z(\tau) = \frac{1}{2} \lim_{x\to\tau+0} \bar{v}(x), \end{aligned}$$

and solve a system of equations of the type (10). It is easy to verify that the solution of problem (15) for $x \in [\tau, 2\tau]$ has the following form:

$$\text{(18)} \qquad \begin{aligned} \bar{v}(x) &= f(\bar{v}'(x-\tau), \bar{v}(x-\tau), x) + \\ &+ \sum_{i=1}^{n-3} (-\mu)^i \frac{d^i}{dx^i} f(\bar{v}'(x-\tau), \bar{v}(x-\tau), x) - \\ &- (-\mu)^{n-3} \int_{\epsilon}^{x} e^{-\frac{x-s}{\mu}} \frac{d^{n-2}}{ds^{n-2}} f(\bar{v}'(s-\tau), \bar{v}(s-\tau), s)\, ds. \end{aligned}$$

From this it follows that

$$\text{(19)} \qquad \begin{aligned} |u(x) - \bar{v}(x)| &\leqslant f(\bar{v}'(x-\tau), \bar{v}(x-\tau), x) - \\ &- f(u'(x-\tau), u(x-\tau), x) + \\ &+ \sum_{i=1}^{n-3} \mu^i \frac{d^i}{dx^i} f(\bar{v}'(x-\tau), \bar{v}(x-\tau), x) - \\ &- f(u'(x-\tau), u(x-\tau), x) + \sum_{i=1}^{n-3} \left|\mu^i \frac{d^i}{dx^i} u(x)\right| + \end{aligned}$$

$$+ \mu^{n-3}\left|\int_{\tau}^{x} e^{-\frac{x-s}{\mu}} \frac{d^{n-2}}{ds^{n-2}} (f(\bar{v}'(s-\tau), \bar{v}(s-\tau), s)\right| -$$

$$- f(u'(s-\tau), u(s-\tau), s)\, ds + \mu^{n-3}\left|\int_{\tau}^{x} e^{-\frac{x-s}{\mu}} \frac{d^{n-2}}{ds^{n-2}} u(s)\, ds\right|.$$

Let us denote by L the largest of the Lipschitz constants of the functions $\frac{d^i f}{dx^i}$ $(i = 0, 1, \ldots, n-2)$ with respect to its arguments, excluding the independent variable. Then we obtain from (19) that

$$|u(x) - v(x)| \leqslant L\mu\{(|u'(x-\tau)| + |u(x-\tau)|)\}$$

$$\cdot (1 + \mu + \ldots + \mu^{n-4}) +$$

$$+ \mu u''(x-\tau)(1 + 2\mu + \ldots + \sum_{i=1}^{n-3} \mu^i \left|\frac{d^i}{dx^i} u(x)\right| +$$

$$+ \mu^{n-1} L \max\{|u'(x-\tau)| + |u''(x-\tau)| + \ldots$$

$$+ |u^{(n-3)}(x-\tau)|\} + \mu^{n-2} \max_{x \in [\tau, 2\tau]} |u^{(n-2)}(x)|$$

$$\vdots$$

$$|u^{(n-2)}(x) - v^{(n-2)}(x)| \leqslant$$

$$\leqslant \mu L \max_{x \in [\tau, 2\tau]} \{|u'(x-\tau)| + \ldots + |u^{(n)}(x-\tau)|\},$$

which shows the "regular" character of the error $u(x) - \bar{v}(x)$ on the given interval of the variation of the argument. The same can be said about all derivatives, of order not greater than $n-2$, of this difference.

Continuing our construction, one can establish the above mentioned property of the error of the function $\bar{v}(x)$ and its derivatives up to the orders $n-3, n-4, \ldots$ on the respective intervals $[2\tau, 3\tau], [3\tau, 4\tau], \ldots$. From this we obtain conditions on the smoothness of the functions $F(p, q, x)$ and $f(v', v, x)$ occurring in (1) and (5), respectively; taking into account that the solutions of equations with retarded argument become smoother and smoother as one goes away from the initial domain.

We note that the accurate estimate of the error (of the same type as the one obtained for $x \in [0, \tau]$) of the algorithm considered are very complicated for $x > 2\tau$, as it is seen from (20); however, to elucidate the regular character of the theoretical error of this algorithm it is not necessary to obtain an accurate estimation. The qualitative structure of the error can be seen from the expressions of the errors on the first three segments of τ:

$$x \in [0, \tau]: \ |u(x) - \bar{v}(x)| \leqslant \mu \bar{v}'(x);$$

$$x \in [\tau, 2\tau]: \ |u(x) - \bar{v}(x)| \leqslant L\mu\{|v'(x-\tau)| + |v''(x-\tau)| + \ldots\};$$

(21)

$$x \in [2\tau, 3\tau]: \ |u(x) - \bar{v}(x)| \leqslant L^2\mu\{2|v'(x-2\tau)| + $$

$$+ 2|v''(x-2\tau)| + 2|v'''(x-2\tau)| + \ldots\}.$$

In virtue of (21) we can say that in the estimate of the error derivatives of higher and higher orders appear as one moves away from the interval $[-\tau, 0]$. Hence it is obvious that it is of great relevance to remove from the estimate the summands containing factors of the type $\exp\left(x - \frac{\kappa\tau}{\mu}\right)$ and thus significantly contributing to the errors of the derivatives in the neighborhoods of the points $\kappa\tau$.

Before going on to the formulation of the results obtained we introduce some auxiliary notions.

Definition 1. Let $y(x)$ be the exact solution of problem (1)-(2) on the segment $[0, T]$. By the ϵ-curved tube of $y(x)$ we mean the set of those points in the space (X, Y), the distance of which from $y(x)$ does not exceed ϵ.

Definition 2. Let $u(x)$ be the exact solution of problem (5). By the ϵ-curved tube of $u(x)$ we mean the set of those points in the space (U', U, X), the distance of which from $u(x)$ does not exceed ϵ.

Theorem 1. *Assume that the function* $F(y(x), y(x-\tau), x)$ *has continuous partial derivatives up to order* m *in a domain of its arguments defined by an* ϵ*-curved tube of* $y(x)$, *and, furthermore, assume that so does the function* $f(u'(x), u(x), x)$ *in the space* (U', U, X), *too. Also* *suppose that the initial function of problem* (1)-(2) *is continuous. Then the piecewise continuous function* $\bar{v}(x)$ *constructed by the above method has a uniform character of deviation from the exact solution of problem* (5) *everywhere on the intervals* $[\kappa\tau, (\kappa + 1)\tau]$, *where* $\kappa < m$.

It follows from the above theorem that in order to solve Problem A*, it is necessary to obtain the solution of problem (5) on the interval $[T - \tau, T]$. We remark that in spite of the fact that the initial value function of problem (1)-(2) is continuous, the construction of the function $\bar{v}(x)$ is possible on the interval $[T - \tau, T]$, because of the properties of the solutions of the equation of neutral type that approach the solution of problem (5). The function $\bar{v}(x)$ has error of the following form on this segment:

$$|u(x) - \bar{v}(x + \Theta\mu \ln \mu)| \leqslant C\mu \qquad (0 \leqslant \Theta \leqslant 1).$$

Section 2.

We are now going to study the problem of diminishing the regular theoretical error obtained in the previous section.

Taking into account what was said above about problem (5), besides the function $\bar{v}(x)$ we also introduce the function $w(x)$ which can be defined, on each interval $[\kappa\tau, (\kappa + 1)\tau]$, as the solution of the Cauchy problem

$$(23)\qquad \begin{aligned} &\mu w'(x) + w(x) = \bar{v}(x) \\ &w(\kappa\tau + 0) = \tilde{v}(\kappa\tau - 0) + \mu A_w^{\kappa\tau}, \end{aligned}$$

where in order to determine $A_w^{\kappa\tau}$ on each segment $[\kappa\tau, (\kappa + 1)\tau]$, it is necessary, as in the previous section, to solve an equation of the type (10) by using the solution of problem (23), however, with a different initial value condition: $\bar{w}(\kappa\tau) = \bar{v}(\kappa\tau + 0)$, and by using also the solution of the auxiliary problem:

$$(24)\qquad \begin{aligned} &\mu_1 \bar{z}'(x) + \bar{z}(x) = w(x) e^{-\frac{\kappa\tau + x}{\mu_1}} \\ &\bar{z}(\kappa\tau) = \bar{w}(\kappa\tau + 0) \cdot 0.5, \end{aligned}$$

where $\bar{w}(x)$ denotes the solution of problem (23) with the above mentioned initial value condition.

By introducing the function $\bar{\bar{v}}(x) = 2\bar{v}(x - w(x))$ one can prove the following theorem, similar in type to that proved in section 1 (under the same assumptions): The regular theoretical error of the function $\bar{\bar{v}}(x)$ has order μ^2 for $x < T - 2\tau$ on the intervals $[\kappa\tau, (k+1)\tau]$.

We remark that in the present work we used asymptotic expressions of the error estimates only because the precise expressions are very bulky. In principle it is possible to obtain an accurate expression for the theoretical error. Taking into account this, we remark that by introducing more complicated functions than $\bar{\bar{v}}(x)$, it is possible to obtain an error for the solution of problem (5) having order μ^3, μ^4, etc. on the respective intervals $x < T - 3\tau$, $x < T - 4\tau$, etc.

Section 3.

We are now going to consider a problem of great relevance from the practical point of view. The solution of this problem uses a method similar to the one studied in this paper. In the article [4] a stable algorithm of numerical differentiation was given, using the simplest equation of advanced type. We will give a modification of this algorithm that has a smaller theoretical error.

We remark a special property of this modification. Differently from the algorithm already considered, it is now essential that the auxiliary parameter be finite: if we take it to be equal to zero, then our method is inadequate for obtaining values of the derivative of the function considered. This can be easily seen from the constructions below.

Suppose that the function $f(x)$ is defined and has continuous derivatives up to order n on the interval $[0, a]$.

Let us consider the solution of the following Cauchy problem:

$$(25)\qquad \begin{aligned} &\mu y'(x) + y(x) = f(x) \\ &y(0) = f(0) - \mu A_\mu, \end{aligned}$$

where $A_\mu = f'(0) - \mu f'' + \ldots + (-\mu)^{n-1} f^{(n)}(0)$. In order to determine A_μ we do the same as above, we establish an equation (5) from the solutions of the auxiliary problems of types (6) and (7). Then we obtain:

$$
\begin{aligned}
y(x) &= f(x) - \mu f'(x) + \ldots + (-\mu)^{n-1} f^{(n-1)}(x) - \\
&\quad - (-\mu)^{n-1} \int_0^x e^{-\frac{x-s}{\mu}} f^{(n)}(s)\, ds.
\end{aligned}
\tag{26}
$$

By introducing the solution of the following problem, analogous to (25):

$$-\mu \bar{y}'(x) + \bar{y}(x) = f(x)$$

$$\bar{y}(a) = f(a) - \mu A_{\bar{\mu}},$$

where $A_{\bar{\mu}} = f'(a) + \mu f''(a) + \ldots + \mu^{n-1} f^{(n)}(a)$; we obtain:

$$
\begin{aligned}
\bar{y}(x) &= f(x) + \mu f'(x) + \ldots + \mu^{n-1} f^{(n-1)}(x) + \\
&\quad + \mu^{n-1} \int_a^x e^{\frac{x-s}{\mu}} f^{(n)}(s)\, ds.
\end{aligned}
$$

Consequently, we have

$$
\begin{aligned}
y(x) &= \frac{\bar{y}(x) - y(x)}{2\mu} = f'(x) + \mu^2 f''(x) + \ldots \\
&\quad \ldots + (-\mu)^{n-1} \Big(\int_0^x e^{-\frac{x-s}{\mu}} f(s)\, ds + \int_a^x e^{\frac{x-s}{\mu}} f(s)\, ds \Big).
\end{aligned}
$$

The function $Y(x)$, considered as the derivative of the given function $f(x)$ on the interval $[0, a]$, has a theoretical error of order μ^2, provided that $f(x)$ has a continuous third derivative.

Other stable algorithms for the calculation of derivatives are studied in detail in [1, 5].

Because of lack of space, in this work we did not study the effect of data errors on the approximate solutions. Concerning this question we remark a significant property of our modifications. For the solution of Problem A*, using $v(x)$, the general error has the form

$$|u(x) - v(x)| \leqslant O(\mu) + O\left(\frac{\epsilon}{\mu^m}\right),$$

using $\bar{v}(x)$ we obtain:

$$|u(x) - \bar{v}(x)| \leqslant O(\mu^2) + O\left(\frac{\epsilon}{\mu^m}\right),$$

where ϵ is the error of the data. If we calculate the derivative with the algorithm of [2], then the general error has the form $O(\mu) + O\left(\frac{\epsilon}{\mu}\right)$, and for our modification it is equal to $O(\mu^2) + O\left(\frac{\epsilon}{\mu}\right)$ (provided that $f(x)$ has a continuous third derivative).

REFERENCES

[1] A.N. Tychonov, On non-correctly posed problems, *Vyč. Met. i Programmirovanie* Izd. MGU, Moscow, 1967, 3-33, (in Russian).

[2] I.F. Dorofeev, *On the solution of some non-correctly posed problems,* Dissertation, 1968, (in Russian).

[3] L.E. El'sgol'c, *Introduction to the theory of retarded differential equations,* Izd. Nauka, 1964, (in Russian).

[4] I.F. Dorofeev, On a regular method for evaluating derivatives, *Sb. Naučn. Raboty. Asp. Univ. Družby Nar. im. Patrisa Lumumby,* Moscow, 1968.

[5] T.F. Dolgopalova – V.K. Ivanova, On numerical differentiation, *Žurn. Vyč. Mat. i Mat. Fiz.* 6, 3, (1966).

I.F. Dorofeev

Univ. Druzby Nar. i. Patrisa Lumumby Moscow B-302, Ul. Ordzonikidze 3, USSR.

COLLOQUIA MATHEMATICA SOCIETATIS JÁNOS BOLYAI
15. DIFFERENTIAL EQUATIONS, KESZTHELY (HUNGARY), 1975.

PERIODIC SOLUTIONS OF GENERALIZED LIÉNARD EQUATION WITH FORCING TERM

S. FUČÍK

1. INTRODUCTION

In this paper the equation

$$\begin{aligned} &x^{(2k)}(t) + a_1 x^{(2k-1)}(t) + \ldots + a_{2k-1} x'(t) + g(x(t)) + \\ &+ f(x(t))x'(t) = p(t) \end{aligned} \tag{E}$$

is considered. Under some assumptions on the coefficients $a_1, \ldots, a_{2k-1}$ and the continuous real valued functions f and g necessary and sufficient conditions are stated on the T-periodic function p which ensure the existence of a T-periodic solution of (E).

The proof in the case $f \equiv 0$ is a direct application of the abstract theorems given in [1]. If $f \not\equiv 0$ then it is impossible to use the method from [1], for it does not imply the necessary and sufficient condition on p in this case. The method is modified and the final result is Theorem 1.

Comparison of our results with the corresponding results from this field (see e.g. A.C. Lazer [2], J. Mawhin [3] and N. Vornicescu

[5]) is given in Sections 4 and 3.

The author is very much indebted to Dr. Štefan Schwabik for his advice, comments and the useful discussions.

2. DIFFERENTIAL OPERATORS WITH CONSTANT COEFFICIENTS

If $l \geq 0$ is an integer we shall denote by C_T^l the (Banach) space of real valued functions defined on the real line R which are continuous and T-periodic together with their first l derivatives with the norm

$$\|x\|_l = \max_{0 \leq j \leq l} \sup_{t \in R} |x^{(j)}(t)|$$

$\left(x^{(j)} = \dfrac{d^j x}{dt^j}\right)$.

We shall introduce the projector

$$P\colon C_T^l \to C_T^l, \quad P\colon x \longmapsto T^{-1} \int_0^T x(t)\, dt.$$

It is immediate that

$$\|P(x)\|_l = \|P(x)\|_0 \leq \|x\|_0 \leq \|x\|_l$$

for every $x \in C_T^l$, and that $\operatorname{Im}[P]$ (the range of P) is the subspace of C_T^l of constant functions.

If $k \geq 1$ is an integer, let us now summarize some properties of the differential operator L with constant coefficients defined on C_T^{2k} by

(2.1) $$L(x) = -(-1)^k x^{(2k)} + a_1 x^{(2k-1)} + \ldots + a_{2k-1} x' + a_{2k} x.$$

It is then clear that $\operatorname{Im}[L] \subset C_T^0$. The following lemma will be particularly useful in the sequel.

Lemma 1 (see J. Mawhin [3, p. 593]). *Denote by* $\operatorname{Ker}[L]$ *the null-space (in* C_T^{2k}*) of the operator* L. *Then*

(2.2) $$\operatorname{Ker}[L] = \operatorname{Im}[P]$$

if and only if the following condition (A) *is fulfilled:*

$$a_{2k} = 0$$

and the equation

$$-(-1)^k \lambda^{2k} + a_1 \lambda^{2k-1} + \ldots + a_{2k-1}\lambda = 0$$

has no root λ *of the form* $im\frac{2\pi}{T}$ *with a nonzero integer* m. *In this case,*

(2.3) $\quad \operatorname{Im} L = \{x \in C_T^0;\ P(x) = 0\}.$

Suppose that the condition (A) is fulfilled. Let X be a fixed topological supplement of $\operatorname{Ker}[L]$ in C_T^{2k} consisting of the functions x with $P(x) = 0$. Then the restriction $\widetilde{L}$ of the operator L on X is a one-to-one mapping from X onto $\operatorname{Im}[L]$ and denote by K its inverse (the so-called right inverse of the operator L) which is a bounded linear mapping. Let $\|K\|$ be the norm of K.

3. NONLINEAR EQUATIONS

Let the notations introduced in Section 2 be observed and let us suppose the condition (A) is fulfilled.

Lemma 2. *Let* $f\colon R \to R$ *and* $g\colon R \to R$ *be two continuous functions. Suppose that there exist the finite limits*

$$\lim_{\xi \to -\infty} g(\xi) = g(-\infty), \quad \lim_{\xi \to +\infty} g(\xi) = g(+\infty)$$

and

(3.1) $\quad g(+\infty) < g(s) < g(-\infty)$

for each $s \in R$. *Let* $p \in C_T^0$. *Then a necessary condition to exist* $x \in C_T^{2k}$ *satisfying*

(3.2) $\quad L(x(t)) + g(x(t)) + f(x(t))x'(t) = p(t)$

is

(B) $$g(+\infty) < T^{-1} \int_0^T p(t)\, dt < g(-\infty).$$

Proof. Suppose that $x_0 \in C_T^{2k}$ satisfies (3.2) and let

$$F(t) = \int_0^t f(\xi)\, d\xi.$$

Then $p(t) - g(x_0(t)) - f(x_0(t))x_0'(t) \in \mathrm{Im}\,[L]$ and according to (2.3) we have

$$T^{-1} \int_0^T p(t)\, dt = T^{-1} \int_0^T g(x_0(t))\, dt + T^{-1} \int_0^T f(x_0(t))\, x_0'(t)\, dt =$$

$$= T^{-1} \int_0^T g(x_0(t))\, dt + T^{-1}[F(x_0(T) - F(x_0(0))] =$$

$$= T^{-1} \int_0^T g(x_0(t))\, dt.$$

Now the assertion follows immediately from the previous calculation and from the assumption (3.1).

We shall give some additional conditions on f and g under which the condition (B) is also sufficient for the existence of the T-periodic solution of the equation (3.2) with $p \in C_T^0$. Our plan is as follows:

1. Let $p \in C_T^0$ satisfying (B) be fixed and let $y \in C_T^{2k}$. We shall prove (see Lemma 8) that the equation

$(3.2)_y$ $$L(x(t)) + g(x(t)) = p(t) - f(y(t))y'(t)$$

has a T-periodic solution. Moreover, we shall give the conditions under which for each $y \in C_T^{2k}$ there exists precisely one solution $x_y \in C_T^{2k}$ of $(3.2)_y$.

2. We shall prove that the mapping $\mathcal{T}$: $C_T^{2k} \to C_T^{2k}$ defined by $\mathcal{T}$: $y \to x_y$ is completely continuous (see Lemma 10) and under some assumptions there exists a nonempty bounded closed and convex subset $\mathcal{K}$ of C_T^{2k} such that $\mathcal{T}(\mathcal{K}) \subset \mathcal{K}$ (see Lemma 9). According to Schauder's fixed point theorem (see e.g. [4]) the mapping $\mathcal{T}$ has a fixed point which

is a T-periodic solution of the equation (3.2).

Now we shall realize our plan. Denote

$$N_0\colon C_T^{2k} \to C_T^0, \qquad N_0\colon x(t) \to -g(x(t)),$$

$$N_p\colon C_T^{2k} \to C_T^0, \qquad N_p\colon x(t) \to p(t) - g(x(t)),$$

$$M\colon C_T^{2k} \to C_T^0, \qquad M\colon y(t) \to -f(y(t))y'(t),$$

$$P^0\colon C_T^0 \to C_T^0, \qquad P^0\colon x \to x - P(x).$$

To find all T-periodic solutions of the equation $(3.2)_y$ is equivalent to find all solutions of the operator equation

(3.3) $\qquad L(x) = N_p(x) + M(y).$

Lemma 3. *Let* $p \in C_T^0$ *and* $y \in C_T^{2k}$ *be fixed. Let for arbitrary* $\rho \in R$ *the equation*

(3.4) $\qquad KP^0N_p(\rho + v) + KM(y) = v$

have a solution $v = v_y(\rho, t) \in X$. *Let the equation*

(3.5) $\qquad \int_0^T g(\rho + v_y(\rho, t))\, dt = \int_0^T p(t)\, dt$

have a solution $\rho = \rho_0$.

Then $\rho_0 + v_y(\rho_0, t)$ *is a solution of the equation* (3.3).

Proof. Since $LK(z) = z$ for each $z \in \mathrm{Im}\,[L]$ and $M(y) \in \mathrm{Im}\,[L]$ for each $y \in C_T^{2k}$ we have

$$L(\rho_0 + v_y(\rho_0, t)) - N_p(\rho_0 + v_y(\rho_0, t)) - M(y) =$$

$$= L(v_y(\rho_0, t)) - P^0N_p(\rho_0 + v_y(\rho_0, t)) -$$

$$- PN_p(\rho_0 + v_y(\rho_0, t)) - M(y) =$$

$$= L(v_y(\rho_0, t)) - [P^0N_p(\rho_0 + v_y(\rho_0, t)) + M(y)] =$$

$$= L(v_y(\rho_0, t)) - L[KP^0N_p(\rho_0 + v_y(\rho_0, t)) + KM(y)] =$$

$$= L(v_y(\rho_0, t)) - L(v_y(\rho_0, t)) = 0.$$

Note that the mappings $N_0, N_p: C_T^{2k} \to C_T^0$ are completely continuous.

Lemma 4. *For arbitrary* $\rho \in \boldsymbol{R}$ *and* $y \in C_T^{2k}$ *the equation* (3.4) *has at least one solution* $v = v_y(\rho, t) \in X$.

Proof. It is sufficient to show that the mapping

$$F_\rho: v \to KP^0N_p(\rho + v) + KM(y) \tag{3.6}$$

has a fixed point in X. Obviously, F_ρ is completely continuous for N_p is completely continuous and K and P^0 are continuous. Moreover

$$\| F_\rho(v) \|_{2k} \leqslant 2 \| K \| (\alpha + \| p \|_0) + \| K \| \| M(y) \|_0 ,$$

where $\alpha = \max(|g(+\infty)|, |g(-\infty)|)$.

Put

$$U = \{v \in X; \ \| v \|_{2k} \leqslant \| K \| (\| M(y) \|_0 + 2(\alpha + \| p \|_0))\}. \tag{3.7}$$

Clearly, U is a bounded closed convex and nonempty subset of X and $F_\rho(U) \subset U$. Schauder's fixed point theorem implies our assertion.

Lemma 5. *Suppose that the assumptions of Lemma* 2 *are fulfilled. Moreover, let*

$$a_1 = a_3 = \ldots = a_{2k-3} = a_{2k} = 0, \ ^* \tag{3.8}$$

$$a_{2k-2j}(-1)^j \leqslant 0 \quad \textit{for} \quad j = 1, \ldots, k-1 \tag{3.9}$$

and let the function g *be nonincreasing.*

Then for arbitrary $\rho \in \boldsymbol{R}$ *and* $y \in C_T^{2k}$ *the equation* (3.4) *has exactly one solution* $v = v_y(\rho, t) \in X$.

Proof. The existence of at least one solution is proved in the previous Lemma. Suppose that $v_1, v_2 \in X$ are the solutions of (3.4), i.e.

$$KP^0N_p(\rho + v_1) + KM(y) = v_1,$$

*This assumption may be omitted.

$$KP^0N_p(\rho + v_2) + KM(y) = v_2.$$

Thus

$$KP^0(N_0(\rho + v_1) - N_0(\rho + v_2)) = v_1 - v_2$$

and

$$L(v_1 - v_2) = P^0(N_0(\rho + v_1) - N^0(\rho + v_2)).$$

The last identity may be rewritten as

$$-(-1)^k(v_1 - v_2)^{(2k)} + a_2(v_1 - v_2)^{(2k-2)} + \dots$$

$$\dots + a_{2k-2}(v_1 - v_2)'' + a_{2k-1}(v_1 - v_2)' =$$

$$= g(\rho + v_2(t)) - g(\rho + v_1(t)) -$$

$$- T^{-1}\int_0^T [g(\rho + v_2(t)) - g(\rho + v_1(t))]\,dt.$$

Multiplying by $(v_1 - v_2)$ and integrating over the interval $\langle 0, T\rangle$ we obtain

$$\sum_{j=1}^{k-1} a_{2k-2j}\int_0^T (v_1(t) - v_2(t))^{(2j)}(v_1(t) - v_2(t))\,dt -$$

$$-(-1)^k\int_0^T (v_1(t) - v_2(t))^{(2k)}(v_1(t) - v_2(t))\,dt +$$

$$+ a_{2k-1}\int_0^T (v_1(t) - v_2(t))'(v_1(t) - v_2(t))\,dt =$$

$$= \int_0^T [g(\rho + v_2(t)) - g(\rho + v_1(t))](v_1(t) - v_2(t))\,dt -$$

$$-\left\{T^{-1}\int_0^T [g(\rho + v_2(t)) - g(\rho + v_1(t))]\,dt\right\}\int_0^T (v_1(t) - v_2(t))\,dt.$$

From this we immediately have using integration by parts and (3.9)

$$0 \leqslant \int_0^T [g(\rho + v_2(t)) - g(\rho + v_1(t))](v_1(t) - v_2(t))\,dt =$$

$$= \sum_{j=1}^{k-1} a_{2k-2j}(-1)^j \int_0^T [v_1^{(j)}(t) - v_2^{(j)}(t)]^2 \, dt -$$

$$- \int_0^T [v_1^{(k)}(t) - v_2^{(k)}(t)]^2 \, dt \leqslant 0.$$

Since $v_1, v_2 \in X$ we have $v_1 = v_2$.

Remark. Note that the assumptions (3.8) and (3.9) immediately imply that the condition (A) is fulfilled.

Lemma 6. *Let the assumptions of Lemma 5 be satisfied. Define*

$$\psi\colon R \to X \quad by \quad \psi\colon \rho \to v_y(\rho, t).$$

Then ψ *is continuous.*

Proof. *Let* $\{\rho_n\} \subset R$, $\lim_{n\to\infty} \rho_n = \rho_0$. We shall prove $\lim_{n\to\infty} \| \psi(\rho_n) - \psi(\rho_0)\|_{2k} = 0$. For this it is sufficient to show that if $\{\rho_{n_k}\}$ is any subsequence of $\{\rho_n\}$ then there exists a subsequence $\{\rho_{n_{k_l}}\}$ of $\{\rho_{n_k}\}$ such that $\lim_{l\to\infty} \| \psi(\rho_{n_{k_l}}) - \psi(\rho_0)\|_{2k} = 0$. We have

$$KP^0 N_p(\rho_{n_k} + \psi(\rho_{n_k})) + KM(y) = \psi(\rho_{n_k}).$$

According to the proof of Lemma 4 we have $\psi(\rho_{n_k}) \in U$ (see (3.7)). Since the set U is bounded (and independent of ρ) and $\{\rho_n\}$ is a bounded sequence we have that $\{\rho_{n_k} + \psi(\rho_{n_k})\}$ is a bounded sequence in C_T^{2k}. The mapping N_p is completely continuous and thus there exists a subsequence $\{\rho_{n_{k_l}}\}$ of $\{\rho_{n_k}\}$ and $w \in C_T^0$ such that

$$\lim_{l\to\infty} \| N_p(\rho_{n_k} + \psi(\rho_{n_k})) - w\|_0 = 0.$$

Further,

$$\lim_{l\to\infty} \| KP^0 N_p(\rho_{n_k} + \psi(\rho_{n_k})) + KM(y) - u\|_{2k} = 0$$

and

$$\lim_{l\to\infty} \| \psi(\rho_{n_{k_l}}) - u\|_{2k} = 0,$$

where

$$u = KP^0(w) + KM(y).$$

From this we have $F_{\rho_0}(u) = u$ (F_{ρ_0} is defined by the relation (3.6)) and since the equation (3.4) has exactly one solution (see Lemma 5) we obtain $u = \psi(\rho_0)$. The proof is complete.

Lemma 7. *Let us suppose that the assumptions of Lemma 5 are satisfied. Then for arbitrary* $y \in C_T^{2k}$ *there exists at least one solution* $\rho(y)$ *of the equation* (3.5).

Proof. Put

$$\varphi(\rho) = \int_0^T g(\rho + \psi(\rho))\,dt - \int_0^T p(t)\,dt.$$

Since the mapping ψ is continuous the function φ is also continuous. Moreover, from the proof of Lemma 4 we have

$$\| \psi(\rho)\|_0 \leqslant \| \psi(\rho)\|_{2k} \leqslant \| K\| (\| M(y)\|_0 + 2(\alpha + \| p\|_0)) = \mu(y).$$

Since g is nonincreasing we have

$$g(\rho + \mu(y)) \leqslant g(\rho + \psi(\rho)) \leqslant g(\rho - \mu(y)).$$

Let $\eta \in R$ be such that

$$g(\eta) = T^{-1} \int_0^T p(t)\,dt.$$

Thus

$$T[g(\rho + \mu(y)) - g(\eta)] \leqslant \varphi(\rho) \leqslant T[g(\rho - \mu(y)) - g(\eta)]$$

and

$$\varphi(\eta - \mu(y)) \geqslant 0, \quad \varphi(\eta + \mu(y)) \leqslant 0.$$

The classical theorem implies the existence of

$$\rho_0 \in \langle \eta - \mu(y), \ \eta + \mu(y) \rangle \quad \text{such that} \quad \varphi(\rho_0) = 0.$$

Lemma 8. *Let the assumptions of Lemma 5 be satisfied. Moreover, suppose that the function g is decreasing. Then for arbitrary $y \in C_T^{2k}$ there exists exactly one solution $x_y \in C_T^{2k}$ of the equation (3.3) and the following inequalities hold:*

$$\eta - \mu(y) \leqslant T^{-1} \int_0^T x_y(t)\, dt \leqslant \eta + \mu(y), \tag{3.10}$$

$$\Big\| x_y - T^{-1} \int_0^T x_y(t)\, dt \Big\|_{2k} \leqslant \mu(y). \tag{3.11}$$

Proof. The existence of the solution follows from Lemma 7, the validity of the inequalities is proved in the proof of Lemma 7. Suppose that for $y \in C_T^{2k}$ we have $x_1, x_2 \in C_T^{2k}$, $x_1 \neq x_2$, such that

$$L(x_i) + g(x_i(t)) + f(y(t))\, y'(t) = p(t), \qquad i = 1, 2,$$

and thus

$$0 < \int_0^T (L(x_1) - L(x_2))(x_1(t) - x_2(t))\, dt =$$

$$= - \int_0^T [x_1^{(k)}(t) - x_2^{(k)}(t)]^2\, dt +$$

$$+ \sum_{j=1}^{k-1} (-1)^j a_{2k-2j} \int_0^T [x_1^{(j)}(t) - x_2^{(j)}(t)]^2\, dt \leqslant 0$$

which is a contradiction.

Lemma 9. *Let the assumptions of Lemma 8 be satisfied. Put*

$$\mathscr{F}(\nu) = \sup_{|s| \leqslant \eta + 2(\nu + 4(\alpha + \|p\|_0)) \|K\|} |f(s)|$$

and suppose that there exists $\nu_0 > 0$ such that

$$\mathscr{F}(\nu_0) \|K\| (\nu_0 + 2(\alpha + \|p\|_0)) \leqslant \nu_0. \tag{3.12}$$

Define

$$\mathcal{K}(\nu_0) = \{y \in C_T^{2k};\ \left|T^{-1}\int_0^T y(t)\,dt\right| \leqslant |\eta| +$$

$$+ \|K\|(\nu_0 + 2(\alpha + \|p\|_0)),$$

$$\left\|y - T^{-1}\int_0^T y(t)\,dt\right\|_{2k} \leqslant \|K\|(\nu_0 + 2(\alpha + \|p\|_0)).$$

Then $\mathcal{K}(\nu_0)$ *is bounded closed convex and nonempty subset of* C_T^{2k}. *Moreover, if* $\mathcal{T}\colon C_T^{2k} \to C_T^{2k}$ *is defined by*

(3.13) $\quad \mathcal{T}\colon y \to x_y$

then $\mathcal{T}(\mathcal{K}(\nu_0)) \subset \mathcal{K}(\nu_0)$.

Proof. The assertions about the set $\mathcal{K}(\nu_0)$ are obvious. To prove $\mathcal{T}(\mathcal{K}(\nu_0)) \in \mathcal{K}(\nu_0)$ it is sufficient to show (according to (3.10) and (3.11)) that $\sup\limits_{\nu \in \mathcal{K}(\nu_0)} \|M(y)\|_0 \leqslant \nu_0$. But this follows immediately from the assumption (3.12) and

$$\|M(y)\|_0 = \|f(y(t))y'(t)\|_0 \leqslant$$

$$\leqslant \mathcal{F}(\nu_0)\|K\|(\nu_0 + 2(\alpha + \|p\|_0)) \leqslant \nu_0.$$

Remark. It is easy to see that if

(3.12)′ $\quad \|K\| \sup\limits_{S \in R} |f(s)| < 1$

then (3.12) is automatically fulfilled.

Lemma 10. *Under the assumptions of Lemma 9 the* **mapping** $\mathcal{T}$ *defined by* (3.13) *is completely continuous on* $\mathcal{K}(\nu_0)$.

Proof.

(a) *Continuity.* Let the sequence $\{y_n\}$ converge to y_0 in C_T^{2k}. $y_n \in \mathcal{K}(\nu_0)$. Thus $\{f(y_n(t))y_n'(t)\}$ converges uniformly to $f(y_0(t))y_0'(t)$. Define $x_n \in C_T^{2k}$ which is a solution of

$$L(x_n) + g(x_n(t)) = p(t) - f(y_n(t))y_n'(t).$$

Since $x_n \in \mathscr{K}(\nu_0)$ there exists a subsequence $\{x_{n_k}\}$ of $\{x_n\}$ and $x_0 \in$ $\in C_T^{2k-1}$ such that $\{x_{n_k}^{(j)}\}$ converges uniformly to $x_0^{(j)}$ for $j = 0, 1, \ldots$ $\ldots, 2k-1$. Now it is easy to see that $\{x_{n_k}^{(2k)}\}$ converges uniformly to

$$(-1)^k [a_2 x_0^{(2k-2)}(t) + \ldots + a_{2k-1} x_0'(t) + g(x_0(t)) +$$
$$+ f(y_0(t)) y_0'(t) - p(t)]$$

and

$$L(x_0) + g(x_0(t)) + f(y_0(t)) y_0'(t) = p(t),$$

i.e., $\{\mathscr{T}(y_{n_k})\}$ converges to $\mathscr{T}(y_0)$. From this it follows by repeating the same procedure that $\{\mathscr{T}(y_n)\}$ converges to $\mathscr{T}(y_0)$ and thus $\mathscr{T}$ is continuous on $\mathscr{K}(\nu_0)$.

(b) *Compactness.* Let $\{y_n\} \subset \mathscr{K}(\nu_0)$. According to Arzèla's theorem there exists a subsequence $\{y_{n_k}\}$ converging to y_0 in C_T^{2k-1}. By the same way as in part (a) of this proof we obtain our assertion.

Now we shall formulate the result obtained in this Section.

Theorem 1. *Suppose that* (3.8) *and* (3.9) *hold. Let* g *be a decreasing continuous function satisfying* (3.1) *and let* f *be a continuous function satisfying* (3.12)'. *Let* $p \in C_T^0$.

Then (B) *is the necessary and sufficient condition of the existence of an* $x \in C_T^{2k}$ *satisfying* (3.2).

Remark. N. Vornicescu [5] proved the following result: let $\lim\limits_{|\xi| \to \infty} \dfrac{g(\xi)}{\xi} = 0$, $\xi g(\xi) \geqslant 0$ for $|\xi| \geqslant b$. Let f be bounded on R such that $\inf\limits_{\xi \in R} |f(\xi)| > \dfrac{1}{2} \sup\limits_{\xi \in R} |f(\xi)|$ and $p \in C_T^0$, $T^{-1} \int\limits_0^T p(t)\, dt = 0$. Then there exists a T-periodic solution of $x'' + f(x)x' + g(x) = p$.

4. THE CASE $f \equiv 0$

If $f \equiv 0$ then we can prove the same assertion as in Theorem 1 under more general assumptions on g. Analogously as in Lemma 3 we immediately see that if the system

$$(4.1) \qquad KP^0 N_p(\rho + \nu) = \nu,$$

$$(4.2) \qquad \int_0^T g(\rho + \nu(t))\, dt = \int_0^T p(t)\, dt$$

has a solution $\rho = \rho_0 \in \mathbf{R}$, $\nu = \nu_0 \in X$ then $x(t) = \rho_0 + \nu_0(t) \in C_T^{2k}$ is a solution of

$$(4.3) \qquad L(x) + g(x(t)) = p(t).$$

Theorem 2. *Let g be a continuous function on R with finite limits $g(-\infty)$, $g(+\infty)$ and suppose that* (3.1) *holds. Let $p \in C_T^0$.*

Then (B) *is a necessary and sufficient condition of the existence of an $x \in C_T^{2k}$ satisfying* (4.3).

Proof. The necessity is proved in Lemma 2. For the sufficiency let $\epsilon > 0$ and $G_\epsilon : \mathbf{R} \times X \to \mathbf{R} \times X$ be defined by

$$G_\epsilon : [\rho, \nu] \to \Big[\rho + \epsilon \Big\{ \int_0^T g(\rho + \nu(t))\, dt - \int_0^T p(t)\, dt \Big\}, KP^0 N_p(\rho + \nu)\Big].$$

The mapping G_ϵ is completely continuous. It is easy to see that there exists $\epsilon_0 > 0$ and a bounded convex closed and nonempty set $\mathscr{K} \subset \mathbf{R} \times X$ such that $G_{\epsilon_0}(\mathscr{K}) \subset \mathscr{K}$ (for details see [1, Lemma 2.3.6]). Thus according to Schauder's fixed point theorem the mapping G_{ϵ_0} has a fixed point $[\rho_0, \nu_0]$ and the system (4.1), (4.2) admits a solution.

Remark. Let

$$(3.1)' \qquad g(-\infty) < g(s) < g(+\infty)$$

for each $s \in \boldsymbol{R}$. Then the same assertion as in Theorem 2 is valid if (B) is replaced by

$$\text{(B)}' \qquad g(-\infty) < T^{-1} \int_0^T p(t)\,dt < g(+\infty).$$

The proof is same as that of Theorem 2. Note that this is impossible to do in Section 3.

From the abstract theorem in [1] it is possible immediately to derive

Theorem 3. *Let* g *be an odd continuous monotone function on* $\boldsymbol{R}$ *and* $\lim_{\xi \to \infty} g(\xi) = +\infty$. *Let* $\alpha \geqslant 0$ *and* $\beta > 0$ *be such that*

$$|g(\xi)| \leqslant \alpha + \beta |\xi|$$

for each $\xi \in \boldsymbol{R}$.

Then there exists $\beta_0 > 0$ *with the following property: If* $\beta < \beta_0$ *then for arbitrary* $p \in C_T^0$ *there exists an* $x \in C_T^{2k}$ *satisfying* (4.3).

Comparison of Theorem 2 with the corresponding result of J. Mawhin [3] goes as follows.

(i) Mawhin's paper deals with vector differential equations. It seems that it is possible to extend our method to vector equations only if the i-th component of the vector function g depends only on x_i.

(ii) Mawhin's assumptions on the coefficients $a_1, \ldots, a_{2k}$ are more general. In the case $k = 1$ they are same.

(iii) The assumptions on the function g are different.

(iv) J. Mawhin gives the assertion about the solvability of (4.3) under the assumption $T^{-1} \int_0^T p(t)\,dt = 0$. Our extension is in the formulating of a necessary and sufficient condition.

(Note that paper [3] extends the result of A.C. Lazer [2] and the other papers from the same filed.)

REFERENCES

[1] S. Fučík, Nonlinear equations with noninvertible linear part, *Czech. Math. J.*, 24 (99) (1974), 467-495.

[2] A.C. Lazer, On Schauder's fixed point theorem and forced second-order nonlinear oscillation, *J. Math. Anal. Appl.*, 21 (1968), 421-425.

[3] J. Mawhin, Periodic solutions of some vector retarded functional differential equations, *J. Math. Anal. Appl.*, 45 (1974), 588-603.

[4] J.T. Schwartz, *Nonlinear Functional Analysis,* Gordon and Breach, New York, 1969.

[5] N. Vornicescu, Asupra existenţei unei soluţii periodice pentru ecuaţia oscilaţiilor neliniare fortate, *St. Cerc. Mat.*, 22 (1970), 683-689.

S. Fučík

Dept. of Math. Anal., Fac. of Math. + Ph., Charles U., 83 Sokolovská, 186 00 Prague 8, Czechoslovakia.

COLLOQUIA MATHEMATICA SOCIETATIS JÁNOS BOLYAI
15. DIFFERENTIAL EQUATIONS, KESZTHELY (HUNGARY), 1975.

NUMERICAL SOLUTION OF A NONLINEAR BOUNDARY VALUE PROBLEM

J. GERGELY

ABSTRACT

The paper suggests an iteration method to solve a system of nonlinear equations. The method is applied to the solution of a nonlinear boundary value problem.

1. INTRODUCTION

First we give a description of the iteration method. Let us consider the system of equations

$$f(\boldsymbol{x}) = 0, \quad \boldsymbol{x} \in \boldsymbol{R}^n, \quad f \in \boldsymbol{R}^n. \tag{1}$$

In order to solve it, we suggest the following iteration procedure. Starting from a first approximation $\boldsymbol{x}^0 = x_1^0, \ldots, x_n^0$, let us solve first the equation

$$f_1(x_1^0 + \delta x_1, x_2^0, \ldots, x_n^0) = 0 \tag{2}$$

for δx_1. Let the solution of (2) be δx_1^0 and $x_1^1 = x_1^0 + \delta x_1^0$, then we solve the equation

(3) $$f_2(x_1^1 + c_1^1 \delta x_2, x_2^0 + \delta x_2, x_3^0, \ldots, x_n^0) = 0$$

for δx_2. Furthermore let us require that

$$df_1 = \frac{\partial f_1}{\partial x_1} \delta x_1 + \frac{\partial f_1}{\partial x_2} \delta x_2 = 0$$

from these we derive a connection between δx_1 and δx_2

$$\delta x_1 = - \frac{\dfrac{\partial f_1}{\partial x_2}}{\dfrac{\partial f_1}{\partial x_1}} \delta x_2 = c_1^1 \delta x_2$$

and therefore equation (3) contains only one unknown, δx_2. Let δx_2^1 be the solution of equation (3), $x_1^2 = x_1^1 + c_1^1 \delta x_2^1$ and $x_2^2 = x_2^0 + \delta x_2^1$. Then we solve the equation

(4) $$f_3(x_1^2 + c_1^2 \delta x_3, x_2^2 + c_2^2 \delta x_3, x_3^0 + \delta x_3, x_4^0, \ldots, x_n^0) = 0$$

for δx_3 and let us require that

(5) $$\begin{aligned} df_1 &= \frac{\partial f_1}{\partial x_1} \delta x_1 + \frac{\partial f_1}{\partial x_2} \delta x_2 + \frac{\partial f_1}{\partial x_3} \delta x_3 = 0 \\ df_2 &= \frac{\partial f_2}{\partial x_1} \delta x_1 + \frac{\partial f_2}{\partial x_2} \delta x_2 + \frac{\partial f_2}{\partial x_3} \delta x_3 = 0. \end{aligned}$$

From (5) we can express δx_1 and δx_2 by δx_3

$$\begin{bmatrix} \delta x_1 \\ \delta x_2 \end{bmatrix} = - \begin{bmatrix} \dfrac{\partial f_1}{\partial x_1} & \dfrac{\partial f_1}{\partial x_2} \\ \dfrac{\partial f_2}{\partial x_1} & \dfrac{\partial f_2}{\partial x_2} \end{bmatrix}^{-1} \begin{bmatrix} \dfrac{\partial f_1}{\partial x_3} \\ \dfrac{\partial f_2}{\partial x_3} \end{bmatrix} \cdot \delta x_3 = \begin{bmatrix} c_1^2 \cdot \delta x_3 \\ c_2^2 \cdot \delta x_3 \end{bmatrix}$$

Hence equation (4) contains only one unknown, δx_3 and so on. In general, we solve the equation

(6) $$f_{k+1}(x^k + q^{k+1} \cdot \delta x_{k+1}) = 0$$

for $k = 0, \ldots, n-1$, where $x^k = (x_1^k, \ldots, x_k^k, x_{k+1}^0, \ldots, x_n^0)$,

$$q^{k+1} = \begin{bmatrix} p^k \\ 1 \\ 0 \\ \vdots \\ 0 \end{bmatrix}, \qquad p^k = -A_k^{-1} \frac{\partial f}{\partial x_{k+1}}$$

and A_k is the first $k \cdot k$ part of the Jacobian matrix of the sytem of equations (1). In equation (6) there is only one unknown δx_{k+1}. If we solve equation (6) for $k = 0, \ldots, n-1$, we get from x^0 a new iteration x^n.

2. THE LINEAR CASE

In the linear case $f(x^k) = Ax^n = 0$, that is in one step x^n is the exact solution, and the solving procedure is the same as the socalled order-increasing (or dimension expansion) matrix inversion method. (see [1]).

In this case we partition A_{k+1} as follows:

$$A_{k+1} = \begin{Vmatrix} A_k & w_k \\ v_k^T & a_{k,k} \end{Vmatrix}$$

and we try to find its inverse in the form

$$A_{k+1}^{-1} = \begin{Vmatrix} P_k & r_k \\ q_k^T & \beta_k \end{Vmatrix}$$

where

$$\beta_k = \frac{1}{\alpha_k}, \qquad \alpha_k = a_{k,k} - v_k^T A_k^{-1} w_k,$$

$$r_k = -\beta_k A_k^{-1} w_k, \qquad q_k^T = -\beta_k v_k^T A_k^{-1},$$

$$P_k = A_k^{-1} + \beta_k A_k^{-1} w_k v_k^T A_k^{-1}$$

3. SOLUTION OF A BOUNDARY VALUE PROBLEM

The method can be applied for solving boundary value problems for nonlinear elliptic partial differential equations. Consider for example the equation

$$\text{(7)} \qquad \Delta u = \frac{\partial^2 u}{\partial x^2} + \frac{\partial^2 u}{\partial y^2} = g(x, y, u)$$

in a square domain with some boundary condition. We assume that there is a unique solution of (7). Let us have a square mesh (the division is $s \times s$)

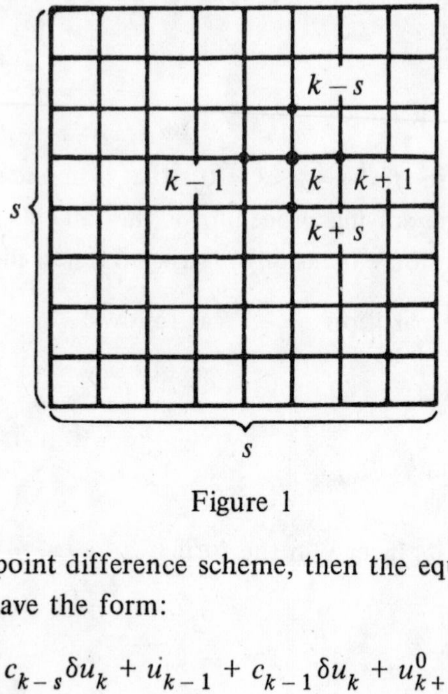

Figure 1

and use the five-point difference scheme, then the equations for the k-th mesh point will have the form:

$$u_{k-s} + c_{k-s}\delta u_k + u_{k-1} + c_{k-1}\delta u_k + u_{k+1}^0 + u_{k+s}^0 -$$
$$- 4(u_k^0 + \delta u_k) = h^2 g(x_k, y_k, u_k + \delta u_k)$$

or in ordered form

$$
(8) \quad \begin{aligned} & h^2 g(x_k, y_k, u_k + \delta u_k) + (4 - c_{k-s} - c_{k-1})\delta u_k - \\ & - (u^0_{k-s} + u^0_{k-1} + u^0_{k+1} + u^0_{k+s} - 4u^0_k) = 0 \end{aligned}
$$

Obviously (8) is a nonlinear equation, with one unknown, δu_k. Solving this equation we obtain

$$u_k = u^0_k + \delta u_k$$

and carry out a linear correction

$$u_i = u^0_i + c_i \delta u_k, \quad \text{for} \quad i < k.$$

We get the coefficients c_i from

$$df_i = 0, \quad i < k.$$

Repeating this computation for $k = 1, \ldots, n = s^2$ we obtain an iteration process.

In our case the Jacobian matrix is

$$
A = \begin{Vmatrix} B_1 & I & & 0 \\ I & B_2 & I & \\ & \cdot & \cdot & \cdot \\ 0 & & & \\ & & I & B_s \end{Vmatrix}
$$

where

$$
B_i = \begin{Vmatrix} d_{(i-1)s+1} & 1 & & 0 \\ 1 & d_{(i-1)s+2} & & \\ & \cdot & \cdot & \cdot \\ 0 & & & \\ & & 1 & d_{i \cdot s} \end{Vmatrix}
$$

$d_k = - [4 + h^2 \frac{\partial}{\partial u} g(x_k, y_k, u^0_k)]$, and I denotes the identity matrix.

Comparing this with the linear case treated above we see that the only deviation occurs in the diagonal of the matrices B_i. These are equal to -4 in the linear case, while here we have entries of the form $-4+$ $+h^2 \frac{\partial}{\partial u} g(x_k, y_k, u_k^0)$. The assumption $\frac{\partial}{\partial u} g(x, y, u) \geqslant 0$ increases the computational stability, hence it also increases the stability of the exact solution.

An iteration step requires the same amount of work as the solution of a linear system of equations.

4. EXAMPLES

We have carried out the computations for the equations $\Delta u = u^2$ and $\Delta u = u^3$ in a unit square with the boundary values given on Figure 2 (with $t = 1, 10, 100, 1000$).

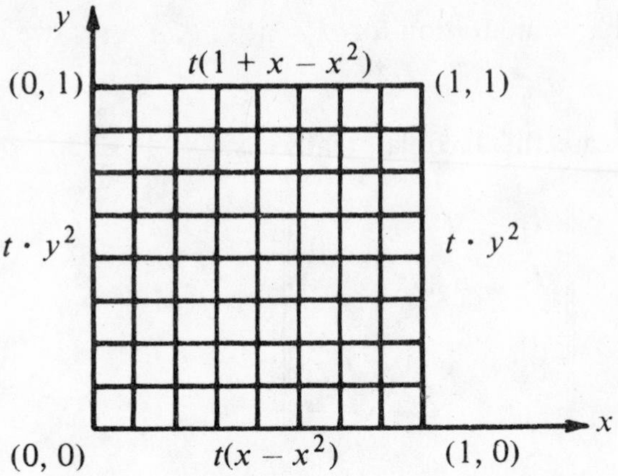

Figure 2

We have used the subdivisions corresponding to $s = 3, 5, 10$ and 20. As starting function we put either $u^0 = 0$ or the solution of equation $\Delta u = 0$. The computations show good convergence and stability. For $t = 100$ and 1000 the non-linearity becomes dominating and influences the speed of the convergence.

The table shows the speed of convergence and the computation time in computer CDC 3300 in case $s = 3$ and with the test:

$$\sum_{i=1}^{n} |u_i^{k-1} - u_i^k| < \epsilon = 10^{-7}.$$

Let k_2 and k_3 denote the number of iterations, T_2 and T_3 denote the computation time in seconds (in the case $s = 3$), then we have the table

t	k_2	T_2	k_3	T_3
1	3	1.8	4	2.1
10	4	2.3	6	3.8
100	5	4.3	8	6.2
1000	8	6.6		

Even for greater values of s the speed of convergence did not become worse.

CONCLUSION

The numerical solution of the boundary value problem is reduced to the successive solution of equations with one unknown. The method gives a quick convergence and stable computation for the solution of the boundary value problem.

REFERENCE

[1] D.K. Faddeev – V.N. Faddeeva, *Computational Methods of Linear Algebra,* Freeman, San Francisco, California, 1963.

J. Gergely
Hungarian Academy of Sciences, Computing Center, H-1111 Budapest, Kende u. 13-17.

COLLOQUIA MATHEMATICA SOCIETATIS JÁNOS BOLYAI
15. DIFFERENTIAL EQUATIONS, KESZTHELY (HUNGARY), 1975.

CONSERVATIVE DISCRETE MODELS WITH COMPUTER EXAMPLES OF NONLINEAR PHENOMENA IN SOLIDS AND FLUIDS

D. GREENSPAN

1. INTRODUCTION

Modern digital computers enable us not only to approximate solutions of differential equations but also to formulate and study some very new types of models of complex physical phenomena [1]. These models are called discrete models because, like modern digital computers, they are entirely arithmetic in nature and finite in structure [2]. Their dynamical equations are not differential equations, but are difference equations which require high speed arithmetic for their solution. It is to these new types of models that we will direct our attention.

2. GRAVITY

In order to know how to begin, it will be convenient to develop first some intuition. To do this, let us consider a simple physical experiment with a force with which we are all familiar, namely, gravity.

Suppose then that a particle P of mass m, situated h feet above ground, is dropped from a position of rest. Suppose also that with a camera whose shutter time is Δt, we take a sequence of pictures at the times $t_k = k\Delta t$, $k = 0, 1, 2, \ldots$. From these pictures, P's height x_k above ground at time t_k can be approximated. Assume, then, that this has been done, say, for $\Delta t = 1$, and that, to the nearest foot, one finds

$$x_0 = 400, \; x_1 = 384, \; x_2 = 336, \; x_3 = 256, \; x_4 = 144, \; x_5 = 0.$$

These data are recorded in column A of Table I. By rewriting x_0, x_1, x_2, x_3, x_4, and x_5 as

$$x_0 = 400 - 0, \qquad x_1 = 400 - 16, \qquad x_2 = 400 - 64,$$

$$x_3 = 400 - 144, \quad x_4 = 400 - 256, \quad x_5 = 400 - 400,$$

Time	A Measured Height	B Velocity by Calculus	C Accelera-tion by Calculus	D Velocity by Arithmetic	E Accelera-tion by Arithmetic
$t_0 = 0$	$x_0 = 400$	$v_0 = 0$	$a_0 = -32$	$v_0 = 0$	$a_0 = -32$
$t_1 = 1$	$x_1 = 384$	$v_1 = -32$	$a_1 = -32$	$v_1 = -32$	$a_1 = -32$
$t_2 = 2$	$x_2 = 336$	$v_2 = -64$	$a_2 = -32$	$v_2 = -64$	$a_2 = -32$
$t_3 = 3$	$x_3 = 256$	$v_3 = -96$	$a_3 = -32$	$v_3 = -96$	$a_3 = -32$
$t_4 = 4$	$x_4 = 144$	$v_4 = -128$	$a_4 = -32$	$v_4 = -128$	$a_4 = -32$
$t_5 = 5$	$x_5 = 0$	$v_5 = -160$	$a_5 = -32$	$v_5 = -160$	

Table I

which express the height above ground as the difference of the initial height and the distance fallen, one readily finds the interesting relationships

$$x_0 = 400 - 16(0)^2, \quad x_1 = 400 - 16(1)^2, \quad x_2 = 400 - 16(2)^2,$$

$$x_3 = 400 - 16(3)^2, \quad x_4 = 400 - 16(4)^2, \quad x_5 = 400 - 16(5)^2,$$

which can be written concisely as

$$(2.1) \qquad x_k = 400 - 16(t_k)^2, \qquad k = 0, 1, 2, 3, 4, 5.$$

The importance of (2.1) is that it describes a relationship between x_k and t_k which, previously, was unknown.

In the traditional manner, one would now interpolate from (2.1) to obtain the continuous formula

$$(2.2) \qquad x = 400 - 16t^2, \qquad 0 \leqslant t \leqslant 5.$$

Defining velocity as the *instantaneous* rate of change of distance with respect to time yields, by differentiation of (2.2),

$$(2.3) \qquad v(t) = x'(t) = -32t, \qquad 0 \leqslant t \leqslant 5,$$

while defining acceleration as the *instantaneous* rate of change of velocity with respect to time yields, by differentiation of (2.3),

$$(2.4) \qquad a(t) = v'(t) = -32, \qquad 0 \leqslant t \leqslant 5.$$

The particle's velocities $v_0 = v(0)$, $v_1 = v(1)$, $v_2 = v(2)$, $v_3 = v(3)$, $v_4 = v(4)$ and $v_5 = v(5)$ at the times when the corresponding heights x_0, x_1, x_2, x_3, x_4 and x_5 have been recorded, are now determined directly from (2.3) and are recorded in column B of Table I. The particle's accelerations at these times are determined from (2.4) and are recorded in column C of Table I.

Note that formulas (2.3) and (2.4), and the interesting conclusion that the acceleration due to gravity is constant, with the value -32, have all been *deduced* from the given distance measurements x_0, x_1, x_2, x_3, x_4 and x_5.

Let us show now that *all* the above conclusions could have been deduced without ever having introduced the limit concept and the methodology of the calculus. To do so, let us define the particle's velocity $v_k = v(t_k)$; $k = 0, 1, 2, 3, 4, 5$, as an *average* (rather than *instantaneous*) rate of change of height with respect to time by the arithmetic formula

$$(2.5) \qquad \frac{v_{k+1} + v_k}{2} = \frac{x_{k+1} - x_k}{\Delta t}; \qquad k = 0, 1, 2, 3, 4.$$

Since averaging procedures are both common and useful in the analysis of experimental data, the left-hand side of (2.5) is perfectly reasonable. Next, for computational convenience, let us rewrite (2.5) in the form

$$(2.6) \qquad v_{k+1} = -v_k + \frac{2(x_{k+1} - x_k)}{\Delta t}; \qquad k = 0, 1, 2, 3, 4.$$

Assuming that $v_0 = 0$ when a particle is dropped from a position of rest, one finds from (2.6) that

$$v_1 = -v_0 + \frac{2(x_1 - x_0)}{\Delta t} = 0 + \frac{2(284 - 400)}{1} = -32$$

$$v_2 = -v_1 + \frac{2(x_2 - x_1)}{\Delta t} = 32 + \frac{2(336 - 384)}{1} = -64$$

$$v_3 = -v_2 + \frac{2(x_3 - x_2)}{\Delta t} = 64 + \frac{2(256 - 336)}{1} = -96$$

$$v_4 = -v_3 + \frac{2(x_4 - x_3)}{\Delta t} = 96 + \frac{2(144 - 256)}{1} = -128$$

$$v_5 = -v_4 + \frac{2(x_5 - x_4)}{\Delta t} = 128 + \frac{2(0 - 144)}{1} = -160,$$

which are *identical* with the results of column B in Table I, and are recorded in column D.

Next, since x_0 and v_0, but not a_0, are known initially, let us define a_k, as the *average* (rather than *instantaneous*) rate of change of velocity with respect to time by the arithmetic formula

$$(2.7) \qquad a_k = \frac{v_{k+1} - v_k}{\Delta t}; \qquad k = 0, 1, 2, 3, 4.$$

From the values v_k just generated, one finds from (2.7) that $a_0 = a_1 = = a_2 = a_3 = a_4 = -32$, which are identical with entries in column C of Table I, and are recorded in column E. Formula (2.7) does not allow a determination of a_5 because this would require knowing v_6. Nevertheless, the entries do indicate quite clearly that the acceleration due to gravity is constant, with the value -32.

Now, just because our arithmetic formulas (2.5) and (2.7) have given the same results as (2.3) and (2.4) does not mean that we have, as yet, a formulation which is of physical significance. Indeed, the physical significance of Newtonian dynamics is characterized by the laws of *conservation* of energy, linear momentum, and angular momentum, and by *symmetry*, that is, by the invariance of its laws of motion under fundamental coordinate transformations [3]. Surprisingly enough, our approach to gravity will also yield conservation and symmetry. We will, however, confine attention here only to the conservation of energy, not only for simplicity, but because of the intimate relationship between energy conservation and computational stability [4].

For completeness, recall now the fundamental Newtonian dynamical equation

(2.8) $F = ma,$

the classical formula for kinetic energy K:

(2.9) $K = \frac{1}{2} mv^2,$

and, for a falling body with $a = -32$, the formula for potential energy V:

(2.10) $V = 32mx.$

The classical energy conservation law then states that if K_0 and V_0 are the kinetic and potential energies, respectively, at time $t_0 = 0$, while K_n and V_n are the kinetic and potential energies, respectively, at time $t_n > t_0$, then

(2.11) $K_n + V_n \equiv K_0 + V_0,$

for all $t_n > t_0$.

It will be instructive, for the discussion later, to recall now the derivation of (2.11). For this purpose, let P be at x_0 when $t = t_0$ and let P be at x_n when $t = t_n$. Then the work W done by gravity in the time interval $0 \leqslant t \leqslant t_n$ is defined by

(2.12) $$W = \int_{x_0}^{x_n} F\,dx.$$

Hence,

$$W = \int_{x_0}^{x_n} ma\,dx = m\int_0^{t_n} av\,dt = m\int_0^{t_n} \frac{d}{dt}\left(\frac{1}{2}v^2\right) dt =$$

$$= \frac{1}{2}mv_n^2 - \frac{1}{2}mv_0^2,$$

so that

(2.13) $$W \equiv K_n - K_0.$$

Note that (2.13) is independent of the actual structure of F. If one then reconsiders (2.12) and uses the knowledge that F is gravity, then

$$W = -32m\int_{x_0}^{x_n} dx = -32mx_n + 32mx_0,$$

so that, from (2.10),

(2.14) $$W \equiv -V_n + V_0.$$

Finally, conservation follows immediately from the elimination of W between (2.13) and (2.14).

Let us now return to our arithmetic formulation. Recall that the experimental data in column A of Table I were obtained from photographs at the distinct times $t_k = k\Delta t$. For this reason, we will concentrate only on these times, so that (2.8)-(2.10) need be considered only as follows:

(2.15) $$F_k = ma_k; \qquad k = 0, 1, 2, \ldots$$

(2.16) $$K_k = \frac{1}{2}m(v_k)^2; \qquad k = 0, 1, 2, \ldots$$

(2.17) $$V_k = 32mx_k; \qquad k = 0, 1, 2, \ldots .$$

In analogy with (2.12), define W_n, $n = 1, 2, 3, \ldots$, by

(2.18) $$W_n = \sum_{i=0}^{n-1} (x_{i+1} - x_i)F_i.$$

Then, by (2.5), (2.7) and (2.15)

$$W_n = m \sum_{i=0}^{n-1} (x_{i+1} - x_i) \left(\frac{v_{i+1} - v_i}{\Delta t}\right) =$$

$$= \frac{m}{2} \sum_{i=0}^{n-1} (v_{i+1} + v_i)(v_{i+1} - v_i) = \frac{m}{2} v_n^2 - \frac{m}{2} v_0^2,$$

so that

(2.19) $\quad W_n \equiv K_n - K_0, \qquad n = 1, 2, 3, \ldots$

which is in complete analogy with (2.13) and is, also, independent of the structure of F. On the other hand, since $a \equiv -32$, one has from (2.15) and (2.18) that

$$W_n = -32m \sum_{i=0}^{n-1} (x_{i+1} - x_i) = -32mx_n + 32mx_0,$$

so that, from (2.17)

(2.20) $\quad W_n = -V_n + V_0, \qquad n = 1, 2, 3, \ldots$

in complete analogy with (2.14).

Finally, elimination of W_n between (2.19) and (2.20) yields

(2.21) $\quad K_n + V_n \equiv K_0 + V_0, \qquad n = 1, 2, 3, \ldots$

in complete analogy with (2.11). Moreover, since K_0 and V_0 are determined from the initial conditions x_0 and v_0, it follows from (2.9), (2.10), (2.16) and (2.17) that $K_0 + V_0$ is the same in both (2.11) and (2.21), so that our strictly arithmetic approach conserves *exactly* the same total energy, independently of Δt, as does classical Newtonian dynamics.

It is also worth noting that in the derivations of (2.19) and (2.20), the telescopic sums

$$\sum_{i=0}^{n-1} (v_{i+1}^2 - v_i^2) \equiv v_n^2 - v_0^2$$

$$\sum_{i=0}^{n-1} (x_{i+1} - x_i) = x_n - x_0$$

play the same roles in the derivations of (2.19) and (2.20) as does integration in the derivations of (2.13) and (2.14)

3. EXTENSIONS

Recently, arithmetic, conservative formulas have been found for forces which are more complex than gravity, and these will now be summarized. In each case the proof of conservation of energy follows in the same spirit as that of Section 2, while the proofs of the other conservation laws and of symmetry are to be found in the references.

Consider first the planar motion of a single particle under the influence of gravitation. For this purpose, if $\Delta t > 0$ and $t_k = k\Delta t$, $k = 0, 1, 2, \ldots$, let particle P of mass m be located at $\vec{r}_k = (x_k, y_k)$, have velocity $\vec{v}_k = (v_{k,x}, v_{k,y})$, and have acceleration $\vec{a}_k = (a_{k,x}, a_{k,y})$ at time t_k. In analogy with (2.5) and (2.7), let

$$(3.1) \qquad \frac{\vec{v}_{k+1} + \vec{v}_k}{2} = \frac{\vec{r}_{k+1} - \vec{r}_k}{\Delta t}, \qquad k = 0, 1, 2, \ldots$$

$$(3.2) \qquad \vec{a}_k = \frac{\vec{v}_{k+1} - \vec{v}_k}{\Delta t}, \qquad k = 0, 1, 2, \ldots .$$

To relate force and acceleration at each time t_k, we assume a discrete Newtonian dynamical equation

$$(3.3) \qquad \vec{F}_k = m\vec{a}_k$$

where

$$(3.4) \qquad \vec{F}_k = (F_{k,x}, F_{k,y}).$$

Suppose now that a massive object, like the sun, whose mass is M, is positioned at the origin of the XY coordinate system and is assumed to have no motion. Then, in analogy with the continuous, conservative gravitional force on P, which has components

$$F_x = -\frac{GMm}{r^2}\frac{x}{r}, \quad F_y = -\frac{GMm}{r^2}\frac{y}{r},$$

where G is a constant, the arithmetic and conservative gravitational force on P is taken to have components [5]:

$$(3.5)\qquad F_{k,x} = -\frac{GMm}{r_k r_{k+1}} \cdot \frac{\frac{x_{k+1}+x_k}{2}}{\frac{r_{k+1}+r_k}{2}} = -\frac{GMm(x_{k+1}+x_k)}{r_k r_{k+1}(r_k+r_{k+1})},$$

$$(3.6)\qquad F_{k,y} = -\frac{GMm(y_{k+1}+y_k)}{r_k r_{k+1}(r_k+r_{k+1})},$$

where

$$(3.7)\qquad r_k^2 = x_k^2 + y_k^2, \qquad k = 0, 1, 2, \ldots .$$

From the computational point of view, the motion of P from prescribed initial conditions $\vec{r}_0$ and $\vec{v}_0$ can be found easily as follows. From (3.2) and (3.3), one has

$$(3.8)\qquad \frac{\vec{F}_k}{m} = \frac{\vec{v}_{k+1} - \vec{v}_k}{\Delta t}, \qquad k = 0, 1, 2, \ldots .$$

Then (3.1) and (3.8) are, by (3.5)-(3.7), four implicit recursion formulas for $x_{k+1}, y_{k+1}, v_{k+1,x}$ and $v_{k+1,y}$ in terms of $x_k, y_k, v_{k,x}$ and $v_{k,y}$. Application of the generalized Newton's method [6] at each time step, beginning with $k = 0$, then yields the motion of P. Of course, the corresponding vector formulation in three dimensions would yield six implicit recursion formulas which can be solved in the same way. The Newtonian iteration converges very rapidly if one chooses for an initial guess $\vec{r}_{k+1}^{(0)} = \vec{r}_k$ and $\vec{v}_{k+1}^{(0)} = \vec{v}_k$.

Now, gravitation is a $\frac{1}{r^2}$ law. Suppose, as in classical molecular mechanics, one would desire an arithmetic and conservative formulation of a $\frac{1}{r^p}$, $p \geqslant 2$, law of attraction. Then, in this case, (3.5) and (3.6) need be modified only as follows [2]:

$$(3.9)\qquad F_{k,x} = -\frac{GMm\left[\sum_{j=0}^{p-2} (r_k^j r_{k+1}^{p-j-2})\right](x_{k+1}+x_k)}{r_k^{p-1} r_{k+1}^{p-1}(r_{k+1}+r_k)}, \qquad G \geqslant 0$$

while $F_{k,y}$ is the same as $F_{k,x}$ except that x and y are exchanged. In the particular case where $p = 2$, (3.9) reduces to (3.5).

In classical molecular mechanics, however, particles attract like $\frac{1}{r^p}$ only when they are relatively far apart. When they are close, they repel like $\frac{1}{r^q}$, $q > p$ [3]. To simulate both these effects, simultaneously, it follows directly from (3.9) that the conservative formulas are

$$(3.10) \qquad F_{k,x} = -\frac{GMm\left[\sum_{j=0}^{p-2} (r_k^j r_{k+1}^{p-j-2})\right](x_{k+1} + x_k)}{r_k^{p-1} r_{k+1}^{p-1}(r_{k+1} + r_k)} +$$

$$+ \frac{HMm\left[\sum_{j=0}^{q-2} (r_k^j r_{k+1}^{q-j-2})\right](x_{k+1} + x_k)}{r_k^{q-1} r_{k+1}^{q-1}(r_{k+1} + r_k)}, \qquad G \geqslant 0, \; H \geqslant 0,$$

while $F_{k,y}$ is the same as $F_{k,x}$ except that x and y are exchanged.

Finally, with regard to the motion of a single particle, it is of interest to note that *all* the arithmetic conservative formulas developed thus far are special cases of the following general formula [7]. For any Newtonian potential $\Phi(r)$, let

$$(3.11) \qquad \vec{F}_k = -\frac{\Phi(r_{k+1}) - \Phi(r_k)}{r_{k+1} - r_k} \cdot \frac{\vec{r}_{k+1} + \vec{r}_k}{r_{k+1} + r_k}.$$

Arithmetic formula (3.11) conserves exactly the same energy, linear momentum and angular momentum as does its continuous, limiting counterpart

$$(3.12) \qquad \vec{F} = -\frac{\partial \Phi}{\partial r}\left(\frac{\vec{r}}{r}\right)$$

Since we have now explored rather completely the motion of a single particle, the next extension is to a system of particles $P_1, P_2, \ldots, P_n$. To do this, let particle P_i of mass m_i be at $\vec{r}_{i,k} = (x_{i,k}, y_{i,k})$, have velocity $\vec{v}_{i,k} = (v_{i,k,x}, v_{i,k,y})$ and have acceleration $\vec{a}_{i,k} = (a_{i,k,x}, a_{i,k,y})$ at time t_k. Position, velocity and acceleration are assumed to be related by

(3.13) $$\frac{\vec{v}_{i,k+1} + \vec{v}_{i,k}}{2} = \frac{\vec{r}_{i,k+1} - \vec{r}_{i,k}}{\Delta t}$$

(3.14) $$\vec{a}_{i,k} = \frac{\vec{v}_{i,k+1} - \vec{v}_{i,k}}{\Delta t}.$$

If $\vec{F}_{i,k} = (F_{i,k,x}, F_{i,k,y})$ is the force acting on P_i at time t_k, then force and acceleration are assumed to be related by

(3.15) $$\vec{F}_{i,k} = m_i \vec{a}_{i,k}.$$

If, in particular, we assume that all particles interact with all other particles with attraction like $\frac{1}{r^p}$ and repulsion like $\frac{1}{r^q}$, then the arithmetic, conservative force on each P_i, $i = 1, 2, \ldots, n$, is given, in analogy with (3.10), by [8]:

(3.16) $$\vec{F}_{i,k} = m_i \sum_{\substack{j=1 \\ j \neq i}}^{n} \left\{ m_j \left(- \frac{G\left[\sum_{\xi=0}^{p-2} (r_{ij,k}^{\xi} r_{ij,k+1}^{p-\xi-2})\right]}{r_{ij,k}^{p-1} r_{ij,k+1}^{p-1} (r_{ij,k} + r_{ij,k+1})} + \right.\right.$$
$$\left. + \frac{H\left[\sum_{\xi=0}^{q-2} (r_{ij,k}^{\xi} r_{ij,k+1}^{q-\xi-2})\right]}{r_{ij,k}^{q-1} r_{ij,k+1}^{q-1} (r_{ij,k} + r_{ij,k+1})} \right) \times$$
$$\left. \times (\vec{r}_{i,k+1} + \vec{r}_{i,k} - \vec{r}_{j,k+1} - \vec{r}_{j,k}) \right\},$$

where $G \geqslant 0$, $H \geqslant 0$, $q > p \geqslant 2$, and $r_{ij,k}$ is the distance between P_i and P_j at t_k.

4. APPLICATIONS

By considering solids and fluids to be composed of particles which interact according to force law (3.16), a variety of conservative, arithmetic n-body models have been developed to study fully nonlinear physical phenomena. For illustrative purposes, we will give two such examples, one for a solid and one for a fluid.

Example 1. *Oscillation of an Elastic Bar.*

The problem is formulated physically as follows. Let the region bounded by rectangle 0ABC, as shown in Figure 4.1, represent a bar which can be deformed, and which, after deformation, tends to return to its original shape. The problem is to describe the motion of such a bar after the external force, which has deformed the bar, is removed. Equivalently, the problem is to describe the motion of an elastic bar after release from a position of tension.

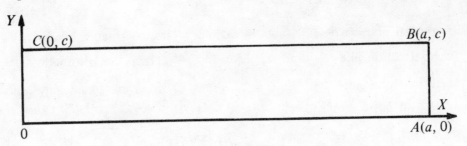

Figure 4.1

Our discrete approach proceeds as follows. The given region is first triangulated, as shown in Figure 4.2. Then, deformation results in the compression of certain particles and the stretching apart of others. Release from a position of deformation, or tension, results in repulsion between each pair of particles which has been compressed and attraction between each pair which has been stretched, the net effect being the motion of the bar.

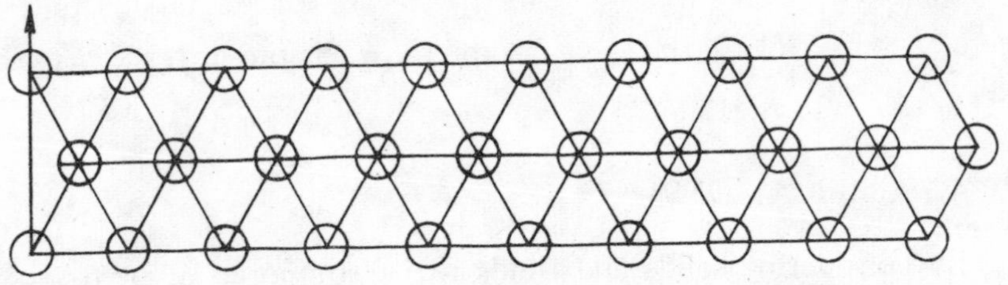

Figure 4.2

As a particular example, let $m_i \equiv 1$, $p = 7$, $q = 10$, $G = 425$, $H = 1000$, and $\Delta t = 0.025$. Let the side of each small triangle in Figure

4.2 have length $r = 1.52254$, to assure a zero force between each pair of particles in any such triangle [9]. The three left most particles are held fixed throughout. In order to obtain an initial position of tension, reset the middle row of particles to (2.28357, 1.29198), (380588, 1.26541), (5.32632, 1.18573), (6.84052, 1.02658), (8.33992, 0.76219), (9.81058, 0.36813), (11.23199, – 0.17750), (12.57631, – 0.89228), and (13.80807, – 1.78721). Any two consecutive particles of this row are now positioned r units apart. The movable points of the uppermost row are now repositioned to form triangular units of sides r with consecutive pairs of points in the middle row, as shown in Figure 4.3a, while the movable points of the bottommost row are also repositioned to form triangular units of side r with consecutive pairs of points of the middle row [9]. In this fashion, consecutive pairs of points in the top row are now separated by distances greater than r, while consecutive pairs of points in the bottom row are now separated by distances less than r. Thus, the top row is in a stretched position while the bottom row is compressed.

From this initial position of tension, the oscillatory motion of the bar is determined from (3.16) with all initial velocities set as $\vec{0}$. The upward swing of the bar was plotted automatically at every twenty time steps and is shown in Figure 4.3a-l from t_0 to t_{220}. It is of interest to note that, as the bar moves, each row of particles exhibits wave oscillation and reflection.

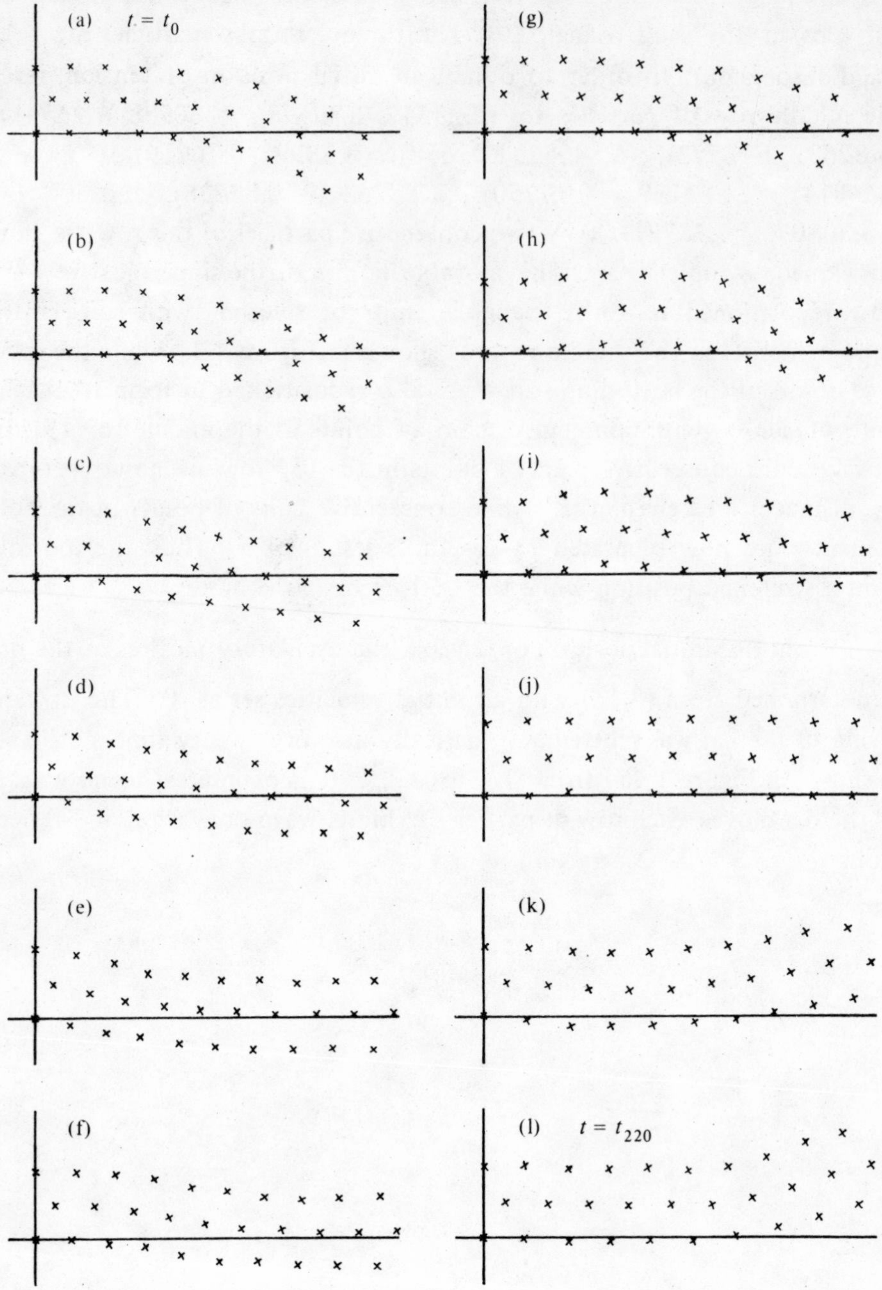

Figure 4.3

Example 2. *Fluid Transition from Laminar to Turbulent Flow*

Consider a small portion of a gas, as shown in Figure 4.4, which is emanating from a jet engine [10]. Take the same parameters as in Example 1 with the exception that $G = H = 1$. This time [2], each small triangle has side $r = \sqrt[3]{1.5}$ in order to assure zero force between each pair of particles of such a triangle.

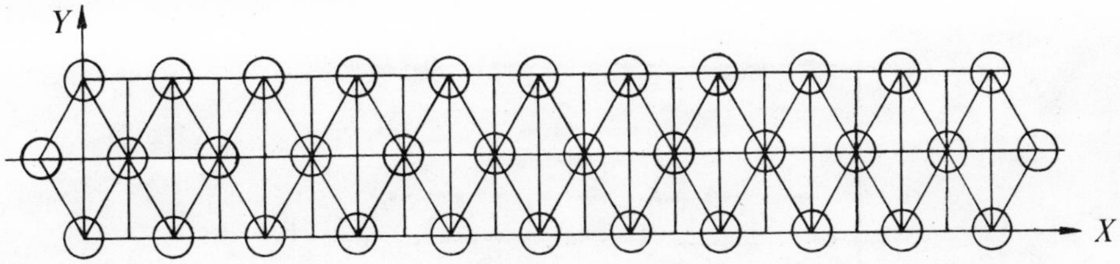

Figure 4.4

To simulate the initial velocities of the particles, set

$$v_{i,0,x} = V + \xi_{i,1}, \quad v_{i,0,y} = \xi_{i,2}$$

where $\xi_{i,1}$ and $\xi_{i,2}$ are random numbers, independent of V, which satisfy

$$|\xi_{i,j}| \leqslant (1\%)\, V.$$

With the initial conditions now fixed, the motion of the particles is completely determined by (3.16). Gravity has been neglected because of the relatively short time interval to be considered. Of course, by the discussion of Section 2, the inclusion of gravity into (3.16) would not disturb the conservation properties of the formulas.

Figure 4.5 shows the particle's motion for $V = 50$ and $\Delta t = 0.02$ at $t = 0.2, 0.4, 0.6, 0.8, 1.0, 1.2,$ and 1.4. A gentle wave motion develops in each row while the rows maintain their relative positions. The flow is essentially of a classical laminar nature.

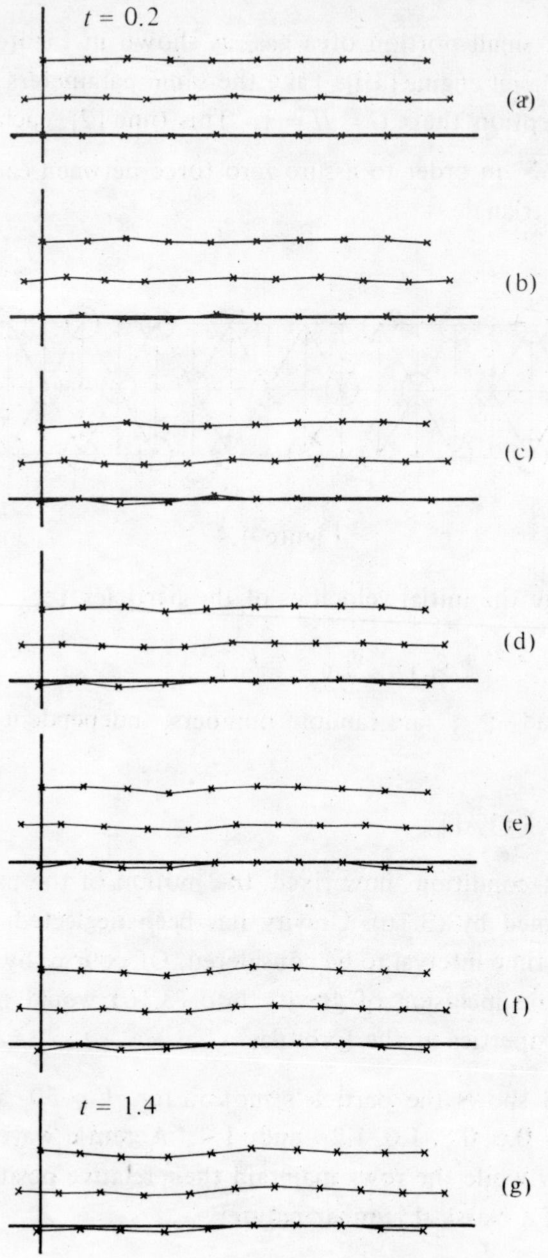

Figure 4.5

Classical laminar flow with $V = 50$.

Figure 4.6 shows the motion for $V = 300$ and $\Delta t = 0.02$ at $t = 0.2, 0.4, 0.6, 0.8$, and 1.0. Repulsion between the particles has assumed a greater significance and, though the rows still maintain their relative positions, the motion is becoming more chaotic. Figure 4.7 shows this same motion from a direction field point of view and exhibits quite clearly the strong effect of repulsion. The rotational effects evident in Figure 4.6 also become more reasonable when viewed from this direction field point of view.

Figure 4.8 shows the motion for $V = 1000$ with $\Delta t = 0.01$ at $t = 0.2, 0.4, 0.6, 0.8$ and 1.0. So much motion results that the choice $\Delta t = 0.01$ was necessary for the convergence of Newton's method. Here, the laminar character of the flow has disappeared in that the rows no longer maintain their relative positions, and the motion becomes relatively chaotic. Thus, with the increase in velocity, particles can come nearer to other particles, which results in increased repulsive forces and more complex motion.

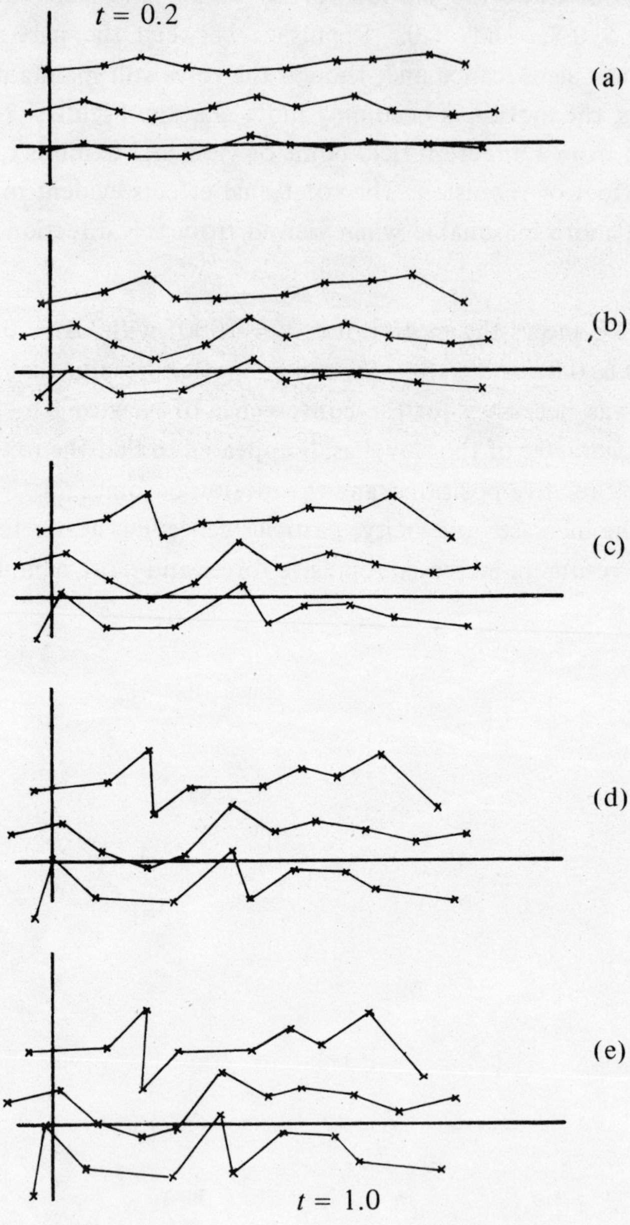

Figure 4.6
Perturbed laminar flow with $V = 300$.

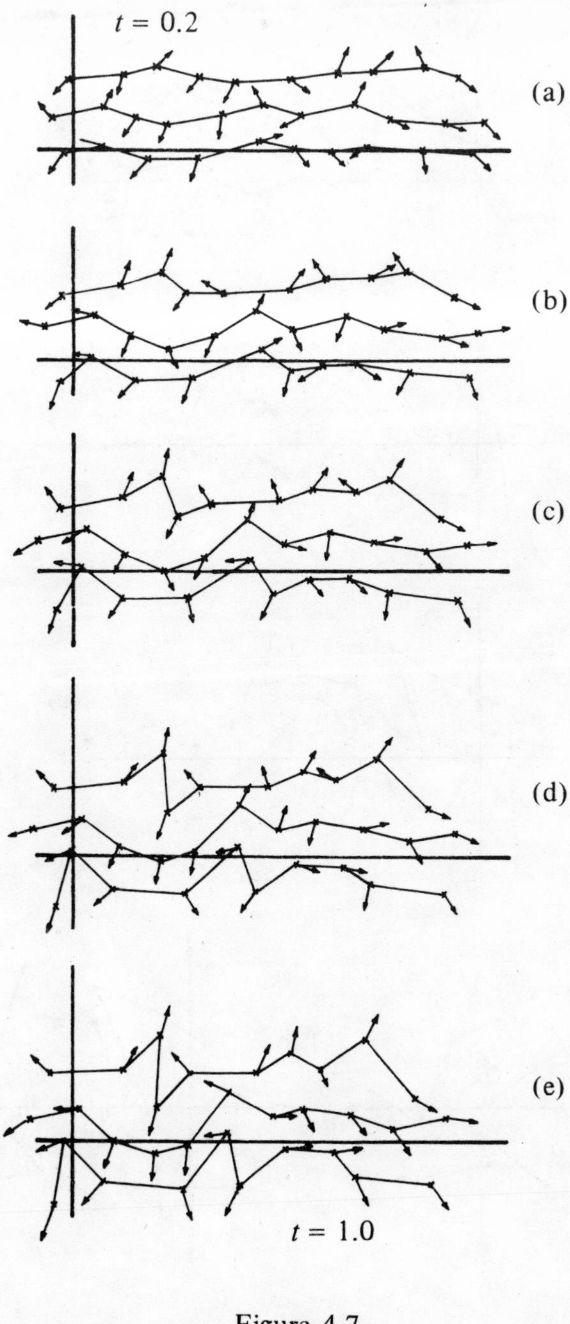

Figure 4.7
Direction field of the flow for $V = 300$.

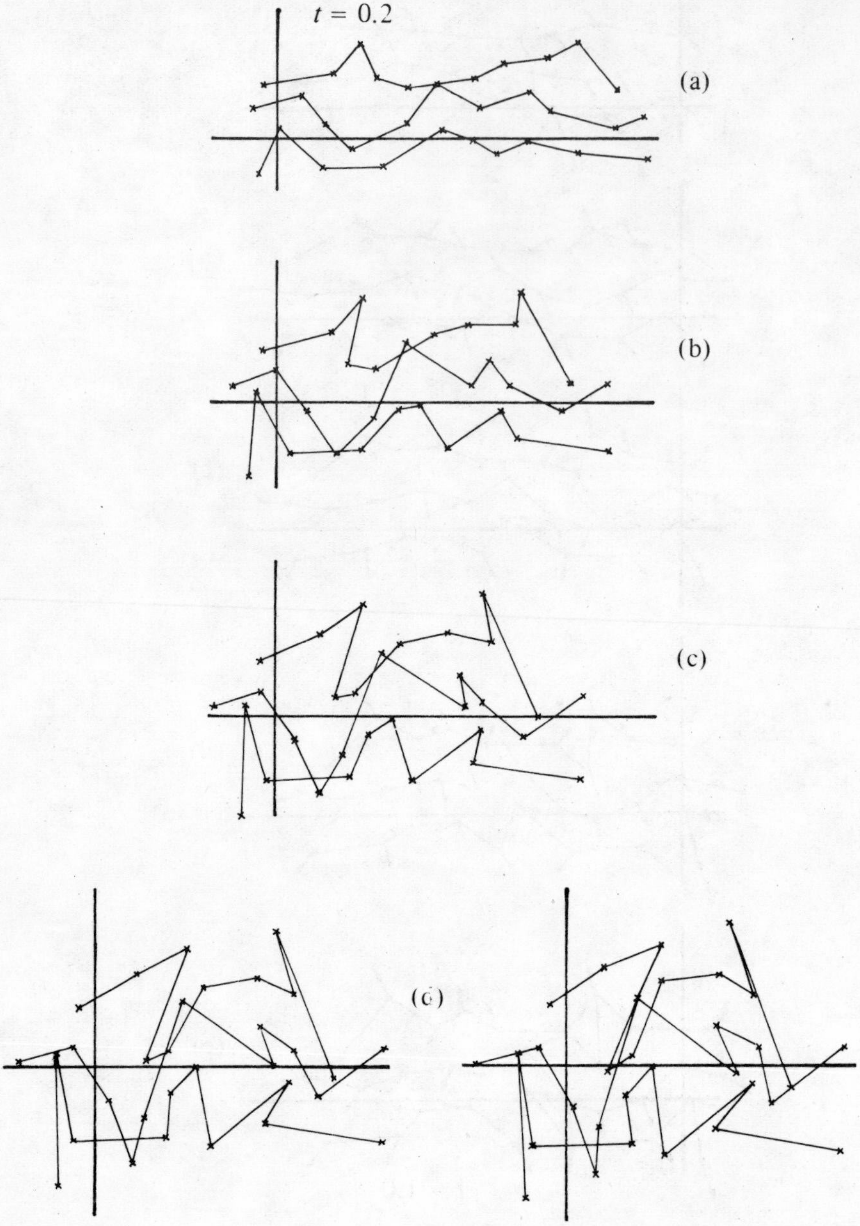

Figure 4.8
Transition from laminar character for $V = 1000$.

REFERENCES

[1] D. Greenspan, Computer power and its impact on applied mathematics, in *Studies in Mathematics,* vol. 7, edited by A.H. Taub, published by the Mathematical Association of America, distributed by Prentice-Hall, Englewood Cliffs, N.J., 1971, pp. 65-89.

[2] D. Greenspan, *Discrete Models,* Addison-Wesley, Reading, Mass., 1973.

[3] R.D. Feynman – R.B. Leighton – M. Sands, *The Feynman Lectures on Physics,* vol. I, Addison-Wesley, Reading, Mass., 1963.

[4] R.D. Richtmyer – K.W. Morton, *Difference Methods for Initial-Value Problems,* 2nd ed., Wiley, N.Y., 1967.

[5] D. Greenspan, Discrete Newtonian gravitation and the three-body problem, *Foundation of Physics,* 4 (1974), 299-310.

[6] D. Greenspan, *Discrete Numerical Methods in Physics and Engineering,* Academic Press, N.Y., 1974.

[7] R.A. La Budde – D. Greenspan, Discrete mechanics – A general treatment, *J. Comp. Physics,* 15 (1974), 134-167.

[8] D. Greenspan, A physically consistent, discrete n-body model, *Bull. Amer. Math. Soc.,* 80 (1974), 553-555.

[9] D. Greenspan, Discrete bars, conductive heat transfer and elasticity, *Computers and Structures,* 4 (1974), 243-251.

[10] D. Greenspan, An arithmetic, particle theory of fluid dynamics, *Computer Methods in Applied Mech. and Engg.,* 3 (1974), 293-303.

D. Greenspan

University of Wisconsin, 1210 W. Dayton St., Madison, Wisconsin 53706, USA.

COLLOQUIA MATHEMATICA SOCIETATIS JÁNOS BOLYAI
15. DIFFERENTIAL EQUATIONS, KESZTHELY (HUNGARY), 1975.

SIMPLIFICATION AND IMPROVEMENT OF A NUMERICAL METHOD FOR NAVIER – STOKES PROBLEMS

D. GREENSPAN – D. SCHULTZ

ABSTRACT

Previously, a viable numerical method for the Navier – Stokes equations was developed and applied to two-dimensional, steady state problems; to three-dimensional, axially symmetric, steady state problems; and to a class of nonsteady problems which had steady state solutions. The method applied for all Reynolds numbers. Among other things, it required the construction of a double sequence of stream and vorticity functions and an appropriate selection of smoothing parameters to assure convergence. Both these complexities are eliminated in the method of this paper. Moreover, illustrative examples show that the new method is faster than the previous one and more accurate for physically sensitive problems.

1. INTRODUCTION

In this paper we will develop a new method for the numerical solution of a class of Navier – Stokes problems. This new method is a significant improvement over the one applied by D. Greenspan to a variety of temperature independent problems for arbitrary Reynolds number [3], [4], [5], [6], [7] and extended by D. Schultz [10] to problems which include temperature dependence. The new method has been found to be approximately ten times faster than the original method. In addition, the new method requires less computer storage, does not require smoothing, and has been found to be more accurate for physically sensitive problems.

2. GENERAL DESCRIPTION

The numerical method to be developed is a finite difference method and is applicable to nonlinear coupled systems of differential equations similar in structure to Navier – Stokes equations which have a "stream-vorticity" formulation.

In the previous method, we would have started with some initial numerical estimate of the solution of the system. Then the algebraic system of equations which approximated only the first differential equation would have been solved completely. Next, this new solution would have been smoothed (averaged) with the initial approximation. Using these results the next differential equation would have been solved and the new solution averaged with its initial approximation. This process would be repeated for all differential equations of the system, in order. After all equations were solved, one would then return to the first equation and repeat the process. This step-by-step iteration for each equation continued until the results converged to within some tolerance.

Analytically, the above numerical method was shown to be convergent for the biharmonic problem (Reynolds number zero) by J. Smith [11].

In the new method we have found that convergence can be increased by a factor of ten (or more) by solving all equations *simultaneously*. In this manner, we will eliminate the need for smoothing and reduce the amount of computer core required. For example, if we have n coupled differential equations to solve, the old method required $2n + 1$ storage arrays while the new method requires only $n + 1$ arrays. Since the step size h is limited by the amount of computer core available we can now work with a much smaller h and thus increase the accuracy of our results.

The general procedure described above will be illustrated next, in detail, by considering several problems of physical interest.

3. THE EDDY PROBLEM IN A RECTANGULAR CAVITY

This problem is defined over a rectangle with interior R and boundary S. The vertices of the rectangle are taken to be $(0, 0), (a, 0), (0, b)$ and (a, b). The equations of motion are

$$(3.1) \qquad \Delta\psi = -\omega$$

$$(3.2) \qquad \Delta\omega + \mathscr{R}\left(\frac{\partial\psi}{\partial x}\frac{\partial\omega}{\partial y} - \frac{\partial\psi}{\partial y}\frac{\partial\omega}{\partial x}\right) = 0,$$

where ψ is the stream function, ω is the vorticity and $\mathscr{R}$ is the Reynolds number. On S the boundary conditions are

$$(3.3) \qquad \psi = 0, \quad \frac{\partial\psi}{\partial x} = 0; \quad \text{on} \quad x = 0$$

$$(3.4) \qquad \psi = 0, \quad \frac{\partial\psi}{\partial y} = 0; \quad \text{on} \quad y = 0$$

$$(3.5) \qquad \psi = 0, \quad \frac{\partial\psi}{\partial x} = 0; \quad \text{on} \quad x = a$$

$$(3.6) \qquad \psi = 0, \quad \frac{\partial\psi}{\partial y} = -1; \quad \text{on} \quad y = b.$$

In general, the boundary value problem (3.1) to (3.6) cannot be solved by existing analytical techniques.

3.1. *Numerical Method for the Eddy Problem*

Consider the particular case $a = b = 1$ (other cases can be treated in a completely analogous fashion). For a fixed positive integer n, set

$h = \frac{1}{n}$, and construct and number the set of interior grid points R_h and the set of boundary grid points S_h in the usual way. Initially, set

$$\psi^{(0)} = C_1 \quad \text{on} \quad R_h$$

$$\omega^{(0)} = C_2 \quad \text{on} \quad R_h + S_h.$$

Then, as in [6], at each point of R_h of the form (h, ih), $i = 2, \ldots, n-2$ approximate (3.3) by

$$\psi(h, ih) = \frac{\psi(2h, ih)}{4}. \tag{3.7}$$

At each point of R_h of the form (ih, h), $i = 1, 2, \ldots, n-1$ approximate (3.4) by

$$\psi(ih, h) = \frac{\psi(ih, 2h)}{4}. \tag{3.8}$$

At each point of R_h of the form $(1-h, ih)$, $i = 2, 3, \ldots, n-2$ approximate (3.5) by

$$\psi(1-h, ih) = \frac{\psi(1-2h, ih)}{4}. \tag{3.9}$$

At each point of R_h of the form $(ih, 1-h)$, $i = 1, 2, \ldots, n-1$ approximate (3.6) by

$$\psi(ih, 1-h) = \frac{h}{2} + \frac{\psi(ih, 1-2h)}{4}. \tag{3.10}$$

At each remaining point of R_h write down the following difference analogue of (3.1):

$$\begin{aligned} &-4\psi(x, y) + \psi(x+h, y) + \psi(x, y+h) + \psi(x-h, y) + \\ &+ \psi(x, y-h) = -h^2\omega(x, y). \end{aligned} \tag{3.11}$$

As in [6], to obtain ω on the boundary S_h, set

$$\omega(ih, 0) = -\frac{2\psi(ih, h)}{h^2} \qquad i = 0, 1, 2, \ldots, n \tag{3.12}$$

$$\omega(0, ih) = -\frac{2\psi(h, ih)}{h^2} \qquad i = 1, 2, \ldots, n-1 \tag{3.13}$$

$$(3.14) \qquad \omega(1, ih) = -\frac{2\psi(1-h, ih)}{h^2} \qquad i = 1, 2, \ldots, n-1$$

$$(3.15) \qquad \omega(ih, 1) = \frac{2}{h} - \frac{2\psi(ih, 1-h)}{h^2} \qquad i = 0, 1, \ldots, n.$$

Finally, to assure diagonal dominance of the difference equation, at each point (x, y) in R_h set

$$\alpha = \psi(x+h, y) - \psi(x-h, y)$$

$$\beta = \psi(x, y+h) - \psi(x, y-h)$$

and approximate (3.2) by

$$\frac{-4\omega(x,y) + \omega(x+h,y) + \omega(x,y+h) + \omega(x-h,y) + \omega(x,y-h)}{h^2} +$$

$$(3.16) \qquad + \mathscr{R}\Big(\frac{\psi(x+h,y) - \psi(x-h,y)}{2h} F -$$

$$- \frac{\psi(x,y+h) - \psi(x,y-h)}{2h} G\Big) = 0$$

where

$$(3.17) \qquad F = \frac{\omega(x,y+h) - \omega(x,y)}{h} \qquad \text{if} \quad \alpha \geqslant 0$$

$$(3.18) \qquad F = \frac{\omega(x,y) - \omega(x,y-h)}{h} \qquad \text{if} \quad \alpha < 0$$

$$(3.19) \qquad G = \frac{\omega(x,y) - \omega(x-h,y)}{h} \qquad \text{if} \quad \beta \geqslant 0$$

$$(3.20) \qquad G = \frac{\omega(x+h,y) - \omega(x,y)}{h} \qquad \text{if} \quad \beta < 0.$$

To start the method applied previously [6], make an initial guess $\psi^{(0)}$ on R_h and $\omega^{(0)}$ on $R_h + S_h$. From these initial guesses two sequences of discrete stream and vorticity functions, called outer iterates, are produced. The iterations performed in getting each one of the outer iterates are called inner iterations. The sequences

$$\begin{array}{ccccc} \psi^{(0)}, & & \psi^{(1)}, & & \psi^{(2)}, \ldots \\ & \nearrow \; \downarrow & & \nearrow \; \downarrow & \nearrow \\ \omega^{(0)}, & & \omega^{(1)}, & & \omega^{(2)}, \ldots \end{array}$$

are calculated in the indicated order until the differences between the k-th and $(k+1)$-st outer iterates agree to within fixed, positive tolerances. Thus, to obtain $\psi^{(1)}$ solve equations (3.7) to (3.11) with $\omega = \omega^{(0)}$ by successive overrelaxation (SOR) and denote this solution by $\bar{\psi}^{(1)}$. Then $\psi^{(1)}$ would be defined by

$$\psi^{(1)} = \rho\psi^{(0)} + (1-\rho)\,\bar{\psi}^{(1)}$$

where

$$0 \leqslant \rho \leqslant 1.$$

Using $\psi = \psi^{(1)}$, one now finds $\bar{\omega}^{(1)}$ on the boundary from equations (3.12) to (3.15) and $\bar{\omega}^{(1)}$ on R_h by solving system (3.16) to (3.20) by SOR. Finally, $\omega^{(1)}$ is defined on $R_h + S_h$ by the smoothing formula

$$\omega^{(1)} = \mu\omega^{(0)} + (1-\mu)\,\bar{\omega}^{(1)}$$

where

$$0 \leqslant \mu \leqslant 1.$$

$\psi^{(2)}$ and then $\omega^{(2)}$ are now calculated in the same fashion as were $\psi^{(1)}$ and $\omega^{(1)}$. This process is continued until the outer iterates have converged, that is, when

$$|\,\psi^{(k)} - \psi^{(k+1)}\,| < \epsilon_\psi \qquad \text{on} \qquad R_h$$

and

$$|\,\omega^{(k)} - \omega^{(k+1)}\,| < \epsilon_\omega \qquad \text{on} \qquad R_h + S_h,$$

for fixed positive tolerances ϵ_ψ and ϵ_ω.

In the new method we solve the system of linear algebraic equations generated by (3.7) to (3.20) *simultaneously* by SOR with overrelaxation factors r_ψ used in the inner ψ iterations and r_ω used for the inner ω iterations. This procedure eliminates the need for smoothing and the necessary search for adequate smoothing parameters, since there are no longer any outer iterates.

3.2. *Results*

For $h = 0.1$, $R = 10$, $\epsilon_\psi = \epsilon_\omega = 0.001$, $r_\omega = 1.0$, $r_\psi = 1.8$, $\rho = 0.03$, $\mu = 0.95$, $\psi^{(0)} = 0$ on R_h, and $\omega^{(0)} = 0$ on $R_h + S_h$ the old method converged in 36 seconds. The new method with the same parameters converged in one second. For $h = 0.05$, $R = 10$, $\epsilon_\psi = \epsilon_\omega = 0.001$, $r_\omega = 1.0$, $r_\psi = 1.8$, $\rho = 0.03$, $\mu = 0.95$, $\psi^{(0)} = 0$ on R_h, and $\omega^{(0)} = 0$ on $R_h + S_h$ the old method did not converge in 2 minutes. The new method converged to $\epsilon_\psi = \epsilon_\omega = 0.0001$ in 6 seconds. For $h = 0.05$, $R = 100000$, $r_\omega = 1.0$, $r_\psi = 1.0$, $\epsilon_\psi = \epsilon_\omega = 0.005$, $\rho = 0.03$, $\mu = 0.70$, $\psi^{(0)} = 0$ on R_h, and $\omega^{(0)} = 0$ on $R_h + S_h$ the old method converged in one minute 30 seconds. The new method converged in 16 seconds. The results obtained by the new method are shown in Figures 1 and 2. Note that the r_ψ and r_ω used above were those which were relatively optimal for the old method, so that it is possible that the new method could have been made to converge even faster if a search had been made for new optimal r_ψ and r_ω.

ψ Contours

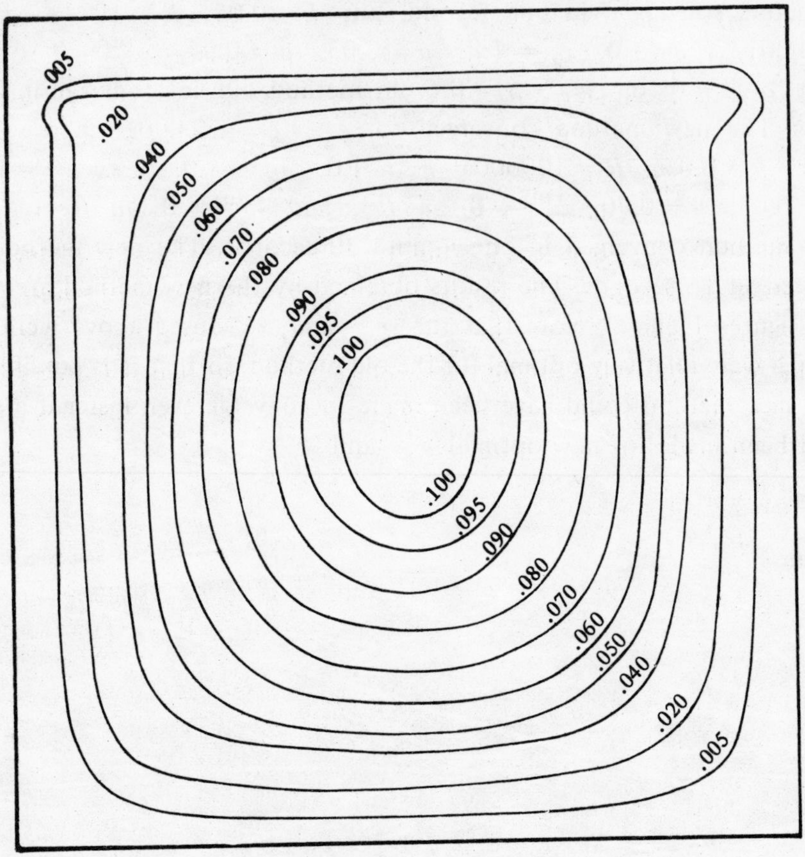

Fig. 1

Eddy Problem, $R = 10^5$

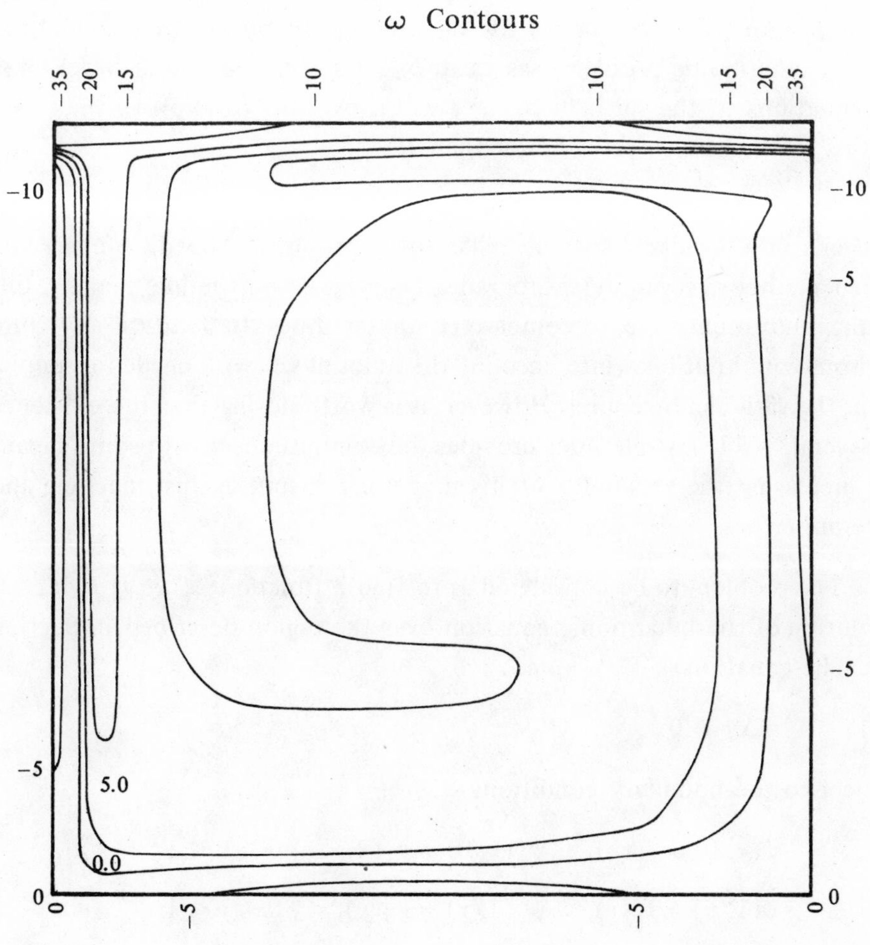

Fig. 2

Eddy Problem, $R = 10^5$

4. BIHARMONIC PROBLEM

In the recent literature [1], [2], [9], [11] there has been a renewed interest in numerical techniques for biharmonic problems. In our method [9], the biharmonic problem was treated as a system of second order elliptic equations in the spirit of [6]. We will show how to improve the speed of this method by a factor better than 10.

We will not make any attempt to rank the different types of fast new methods now available for the biharmonic problem. The differences between methods, computers, programs, languages, and the like make it difficult, if not impossible, to compare computer times. In addition any comparisons could not take into account the amount of work needed to implement the various procedures. However, it is worth noting that the existence of several, viable fast methods provides the computer user with some means for increasing the reliability of his numerical output against machine and program errors.

The problem to be considered is to find a function $\psi(x, y)$ which is a solution of the biharmonic equation over the region described in Section 3.1. The equation is

(4.1) $$\Delta\Delta\psi = 0$$

subject to the boundary conditions

$$\begin{aligned}
&\psi(x, 0) = f_1(x), \quad \psi_y(x, 0) = g_1(x); \quad 0 \leqslant x \leqslant 1\\
&\psi(1, y) = f_2(y), \quad \psi_x(1, y) = g_2(y); \quad 0 \leqslant y \leqslant 1\\
&\psi(x, 1) = f_3(x), \quad \psi_y(x, 1) = g_3(x); \quad 0 \leqslant x \leqslant 1.\\
&\psi(0, y) = f_4(y), \quad \psi_x(0, y) = g_4(y); \quad 0 \leqslant y \leqslant 1.
\end{aligned}$$

We replace equation (4.1), as in [9], by the system of coupled equations.

(4.2) $$\Delta\psi = -\omega$$

(4.3) $$\Delta\omega = 0.$$

Physically, one can interpret ψ and ω from, say, the fluid dynamics point of view, as stream and vorticity functions respectively. Indeed, for $\mathscr{R} = 0$, equations (3.1) and (3.2), reduce to (4.2) and (4.3). Therefore, the new method described in Section 3 can be used for the biharmonic problem also. The only difference will be that the biharmonic problem does not require the special inner boundary equations for the stream function, i.e., (3.7)-(3.10) are not used and (3.11) is applied at all interior grid points.

4.1. *Example*

Using the boundary conditions $f_1 = x^3$, $f_2 = 2y + 1 - 3y^2$, $f_3 = x^3 + 2x - 3$, $f_4 = -3y^2$, $g_1 = 2x$, $g_2 = 2y + 3$, $g_3 = 2x - 6$ and $g_4 = 2y$, the methods of Section 3 were executed with $\mathscr{R} = 0$, $h = 0.05$, $\rho = 0.2$, $\mu = 0.85$, $\epsilon_\psi = 10^{-4}$, $\epsilon_\omega = 10^{-3}$, $r_\psi = 1.8$ and $r_\omega = 1.0$, $\psi^{(0)} = 0$, $\omega^{(0)} = 0$. Using the original method [9], convergence resulted in approximately 6 mintues on the Univac 1108. Using the new method the problem converged in 24 seconds on the same computer. Both methods yielded results which agreed with the exact solution $u = x^3 - 3y^2 + 2xy$ to at least three decimal places.

5. HEATED CAVITY FLOW

Consider now convective flow in a heated cavity [10]. Again, the region of interest will be the square cavity. The equations of motion are

$$\Delta\psi = -\omega \tag{5.1}$$

$$\Delta\theta + \left(\frac{\partial\psi}{\partial x}\frac{\partial\theta}{\partial y} - \frac{\partial\psi}{\partial y}\frac{\partial\theta}{\partial x}\right) = 0 \tag{5.2}$$

$$\Delta\omega + \frac{1}{\sigma}\left(\frac{\partial\psi}{\partial x}\frac{\partial\omega}{\partial y} - \frac{\partial\psi}{\partial y}\frac{\partial\omega}{\partial x}\right) + A\frac{\partial\theta}{\partial y} = 0 \tag{5.3}$$

where ψ is the stream function, ω is the vorticity, θ is a measure of temperature, σ is the Prandtl number, and A is the Rayleigh number. The boundary conditions are

$$\psi = 0, \quad \frac{\partial\psi}{\partial y} = 0, \quad \theta = 0; \qquad \text{on} \quad y = 0 \tag{5.4}$$

$$(5.5) \qquad \psi = 0, \quad \frac{\partial \psi}{\partial x} = 0, \quad \theta = y; \qquad \text{on} \qquad x = 1$$

$$(5.6) \qquad \psi = 0, \quad \frac{\partial \psi}{\partial y} = 0, \quad \theta = 1; \qquad \text{on} \qquad y = 1$$

$$(5.7) \qquad \psi = 0, \quad \frac{\partial \psi}{\partial x} = 0, \quad \theta = y; \qquad \text{on} \qquad x = 0.$$

5.1. *Numerical Method*

To solve the problem numerically approximate (5.1) by (3.11). Equation (5.2) is approximated by

$$(5.8) \qquad \frac{-4\theta_0 + \theta_1 + \theta_2 + \theta_3 + \theta_4}{h^2} + \left(\frac{\psi_1 - \psi_3}{2h} F - \frac{\psi_2 - \psi_4}{2h} G\right) = 0$$

where F and G are defined in (3.17) to (3.20) with ω_n replaced by θ_n. To approximate ω on the boundary use (3.12) to (3.14) with (3.15) replaced by

$$(5.9) \qquad \omega(ih, 1) = \frac{-2\psi(ih, 1-h)}{h^2}.$$

Finally, the approximation to (5.3) is of the form

$$(5.10) \qquad \frac{-4\omega_0 + \omega_1 + \omega_2 + \omega_3 + \omega_4}{h^2} + \\ + \frac{1}{\sigma}\left(\frac{\psi_1 - \psi_3}{2h} F - \frac{\psi_2 - \psi_4}{2h} G\right) + A\,\frac{\theta_2 - \theta_4}{2h} = 0$$

where F and G are defined as in (3.17) to (3.20). The subscripts in (5.8) and (5.10) refer to the same points as in (3.16).

The old numerical method for generating the numerical solution is analogous to the old method described in Section 3.1. To apply the improved method, equations (3.7)-(3.11), (5.8), (3.12) to (3.14), (5.9) and (5.10) are, in the spirit of Section 3.1, simply solved simultaneously for ψ, θ, ω. Note, of course, that in equation (3.10) the term $\frac{h}{2}$ will not appear for this problem.

5.2. *Results*

For $h = 0.1$, $A = 500$, $\sigma = 0.73$, $\epsilon_\psi = \epsilon_\theta = 0.0002$, $\epsilon_\omega = 0.001$, $\psi^{(0)} = 0$, $\theta^{(0)} = 0$, $\omega^{(0)} = 0$, the original method converged in 11 seconds. The new method converged in 1 second. For $h = 0.05$, $A = 500$, $\sigma = 0.73$, $\epsilon_\omega = 0.001$, $\epsilon_\psi = \epsilon_\theta = 0.00001$, $\psi^{(0)} = 0$, $\theta^{(0)} = 0$, $\omega^{(0)} = 0$, the original method converged in 2 minutes. The new method converged in 13 seconds. For $h = 0.025$, $A = 10$, $\sigma = 0.73$, $\epsilon_\omega = 0.002$, $\epsilon_\psi = \epsilon_\theta = 0.00003$, $\psi^{(0)} = 0$, $\theta^{(0)} = 0$, $\omega^{(0)} = 0$, the original method converged in 3 minutes. The new method converged in 43 seconds. For $h = 0.1$, $A = 10000$, $\sigma = 0.73$, $\epsilon_\omega = 0.001$, $\epsilon_\psi = \epsilon_\theta = 0.00002$, $\psi^{(0)} = 0$, $\theta^{(0)} = 0$, $\omega^{(0)} = 0$, the original method converged in 25 seconds. The new method converges in 2.4 seconds. For $h = 0.05$, $A = 10000$, $\epsilon_\omega = 0.001$, $\epsilon_\psi = 0.00002$, $\psi^{(0)} = 0$, $\theta^{(0)} = 0$, $\omega^{(0)} = 0$, the original method converged in 10 minutes. The new method converged in 36 seconds. For $h = 0.1$, $A = 20000$, $\sigma = 0.73$, $\epsilon_\psi = \epsilon_\theta = 0.000002$, $\epsilon_\omega = 0.001$, $\psi^{(0)} = 0$, $\theta^{(0)} = 0$, $\omega^{(0)} = 0$, the original method converged in 25 seconds. The new method converged in 2.4 seconds.

The graphs of the case $A = 10000$ are the same for both the old and the new methods and are shown in Figures 3, 4, 5.

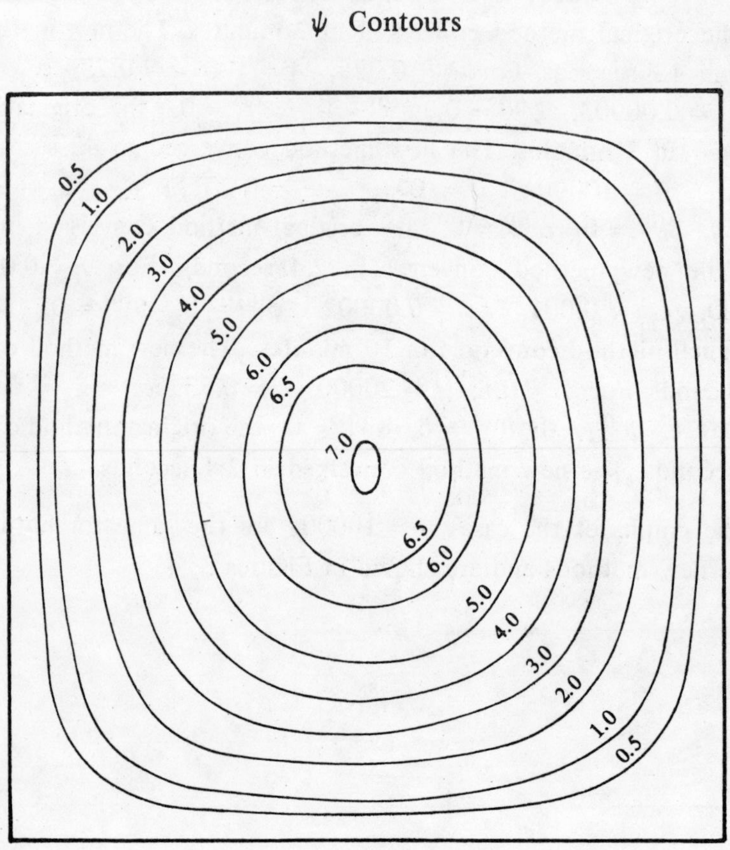

Fig. 3

Heated Cavity, $A = 10^4$

Temperature Contours

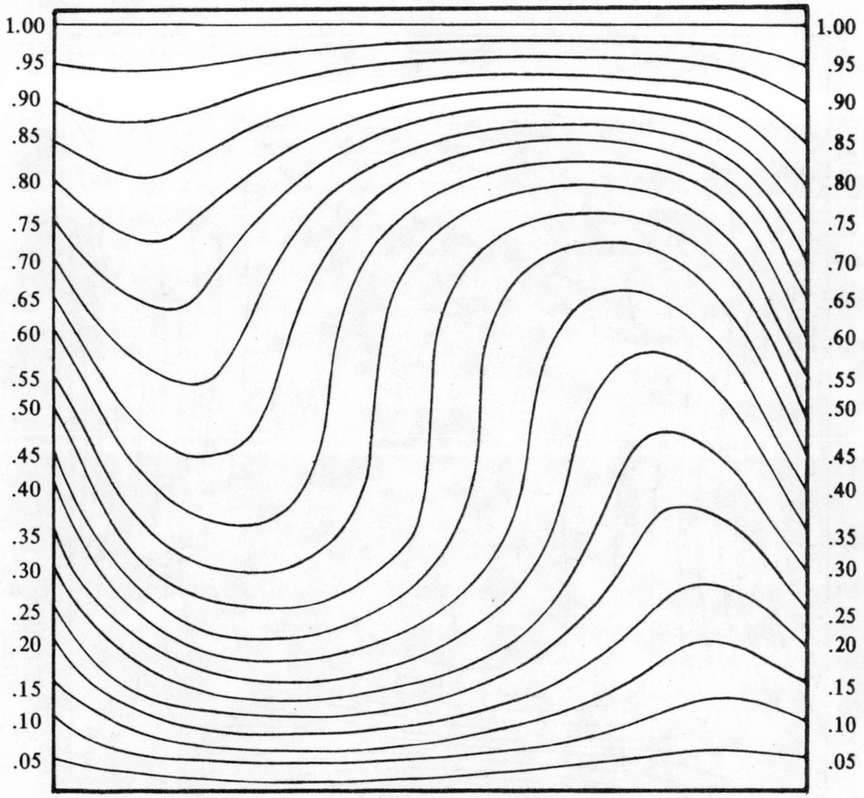

Fig. 4

Heated Cavity, $A = 10^4$

ω Contours

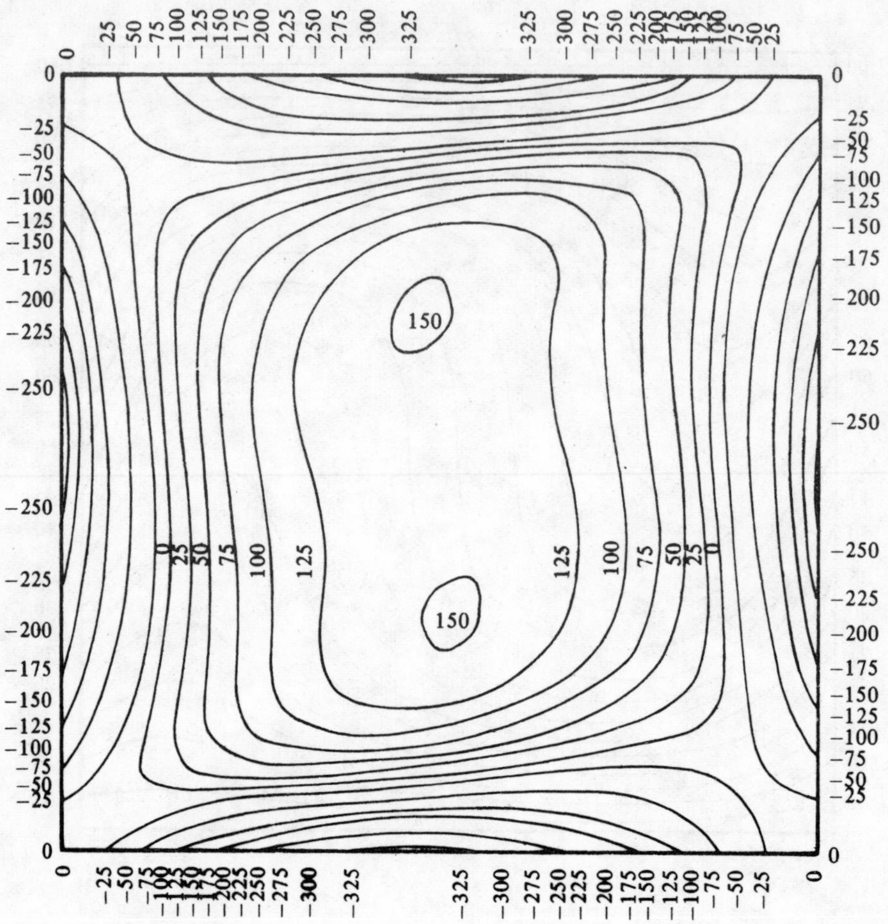

Fig. 5

Heated Cavity, $A = 10^4$

6. ROTATING COAXIAL DISKS

The problem we consider now is the steady motion of a viscous, incompressible fluid between two rotating, infinite coaxial disks [8]. The first disk is, in (x, y, z) space, in the plane $z = 0$ with its center at $(0, 0, 0)$ and has an angular velocity Ω_1. The second disk is in the plane $z = 1$ with its center at $(0, 0, 1)$ and has an angular velocity Ω_2. If the cylinderical coordinates of (x, y, z) are (r, θ, z), and if the fluid at (x, y, z) has velocity components (u, v, w) then the substitutions

$$u = -\frac{1}{2} rH'(z), \qquad v = rG(z), \qquad w = H(z)$$

transform the dimensionless, steady state Navier – Stokes equations to

$$(6.1) \qquad H'' = M, \qquad 0 \leqslant z \leqslant 1$$

$$(6.2) \qquad G'' + R(GH' - G'H) = 0, \qquad 0 \leqslant z \leqslant 1$$

$$(6.3) \qquad M'' - R(HM' + 4GG') = 0, \qquad 0 \leqslant z \leqslant 1,$$

where differentiation is with respect to z. The boundary conditions are

$$(6.4) \qquad G(0) = \Omega_1, \qquad G(1) = \Omega_2,$$

$$(6.5) \qquad H(0) = 0, \qquad H(1) = 0,$$

$$(6.6) \qquad H'(0) = 0, \qquad H'(1) = 0.$$

To obtain the solution we subdivide $0 \leqslant z \leqslant 1$ into n equal parts of length $h = \Delta z = \frac{1}{n}$. Let the points of subdivision be $0 = z_0 < z_1 < < z_2 < \ldots < z_n = 1$. For convenience define $F_i = F(z_i)$. The original method [8] again approximates H, G, M by generating three sequences $H^{(k)}, G^{(k)}$, and $M^{(k)}$ of outer iterates and requires smoothing. The difference equations used were

$$(6.7) \qquad H_{i-1} - 2H_i + H_{i+1} = h^2 M_i^{(k)}, \qquad i = 2, 3, \ldots, n-2$$

$$(6.8) \qquad 4H_1 = H_2$$

$$(6.9) \qquad 4H_{n-1} = H_{n-2}.$$

$$G_{i-1} + [-2 + \mathscr{R}hH_i^{(k+1)}]G_i + [1 - \mathscr{R}hH_i^{(k+1)}]G_{i+1} =$$

$$(6.10) \qquad = -\frac{1}{2}\mathscr{R}hG_i^{(k)}[H_{i+1}^{(k+1)} - H_{i-1}^{(k+1)}]; \quad \text{if} \quad H_i^{(k+1)} < 0,$$

$$i = 1, 2, \ldots, n-1$$

$$[1 + \mathscr{R}hH_i^{(k+1)}]G_{i-1} + [-2 - \mathscr{R}hH_i^{(k+1)}]G_i + G_{i+1} =$$

$$(6.11) \qquad = -\frac{1}{2}\mathscr{R}hG_i^{(k)}[H_{i+1}^{(k+1)} - H_{i-1}^{(k+1)}]; \quad \text{if} \quad H_i^{(k+1)} \geqslant 0,$$

$$i = 1, 2, \ldots, n-1$$

$$(6.12) \qquad M_0^{(k+1)} = \frac{2H_i^{(k+1)}}{h^2}$$

$$(6.13) \qquad M_n^{(k+1)} = \frac{2H_{n-1}^{(k+1)}}{h^2}$$

$$M_{i-1} + [-2 + \mathscr{R}hH_i^{(k+1)}]M_i + [1 - \mathscr{R}hH_i^{(k+1)}]M_{i+1} =$$

$$(6.14) \qquad = 2\mathscr{R}hG_i^{(k+1)}[G_{i+1}^{(k+1)} - G_{i-1}^{(k+1)}], \quad \text{if} \quad H_i^{(k+1)} < 0,$$

$$i = 1, 2, \ldots, n-1$$

$$[1 + \mathscr{R}hH_i^{(k+1)}]M_{i-1} + [-2 - \mathscr{R}hH_i^{(k+1)}]M_i + M_{i+1} =$$

$$(6.15) \qquad = 2\mathscr{R}hG_i^{(k+1)}[G_{i+1}^{(k+1)} - G_{i-1}^{(k+1)}], \quad \text{if} \quad H_i^{(k+1)} \geqslant 0,$$

$$i = 1, 2, \ldots, n-1.$$

For the new method, we simply solve (6.7) and (6.10)-(6.15), simultaneously, with (6.7) extended to the range $i = 1, 2, \ldots, n-1$.

6.1. *Results*

For each of the following cases the original method required approximately 30 seconds of computer time on the Univac 1108. The parameters for the first case are $\mathscr{R} = 10$, $r_H = 1.8$, $r_G = 1.0$, $r_M = 1.0$, $\epsilon_H = \epsilon_G = \epsilon_M = 0.005$, $H^{(0)} = 0$, $G^{(0)} = 0$, $M^{(0)} = 0$. The parameters for the second case are $\mathscr{R} = 100$, $r_H = 1.8$, $r_G = 1.0$, $r_M = 1.5$, $\epsilon_H = \epsilon_G = 0.001$, $\epsilon_M = 0.05$, $H^{(0)} = 0$, $G^{(0)} = 0$, $M^{(0)} = 0$. The parameters for the third case are $\mathscr{R} = 1000$, $r_H = 1.8$, $r_G = 1.1$, $r_M = 1.5$, $\epsilon_H = 0.005$, $\epsilon_G = 0.03$, $\epsilon_M = 0.3$, $H^{(0)} = 0$, $G^{(0)} = 0$, $M^{(0)} = 0$. For all three cases $h = \frac{1}{50}$, $\Omega_1 = 1$, $\Omega_2 = 0$. The new method with $r_G = r_M = 1.0$, $r_H = 1.8$, $\epsilon_G = \epsilon_H = \epsilon_M = 0.001$ converged in a *total* of 7 seconds for all 3 cases.

For $h = \frac{1}{50}$, $\Omega_1 = 1$, $\Omega_2 = -1$ the old method converged in a maximum of 30 seconds and yielded spurious results for each of the cases $\mathscr{R} = 10, 100, 1000$, (see [8] for the parameters). In the new method with $\mathscr{R} = 10$, $h = \frac{1}{50}$, $\Omega_1 = 1$, $\Omega_2 = -1$, $r_G = r_M = 1.5$, $r_H = 1.8$ and $\epsilon_H = \epsilon_G = \epsilon_M = 0.0002$ the problem converged in fewer than 5 seconds. For the case $\mathscr{R} = 100$ with $r_G = r_M = 1.0$, $r_H = 1.8$, $\epsilon = 0.0002$ for all parameters the new method converged in approximately 5 seconds. For the case $\mathscr{R} = 1000$ with $r_G = r_M = 0.8$, $r_H = 1.8$, $\epsilon = 0.001$ for all variables the method converged in 2.7 seconds. In addition the new method gave correct results (see Fig. [6]).

The fact that the computer results by the old method for the sensitive class of problems with $\Omega_1 = -\Omega_2$ were incorrect has now been established analytically (personal communication from S.V. Parter).

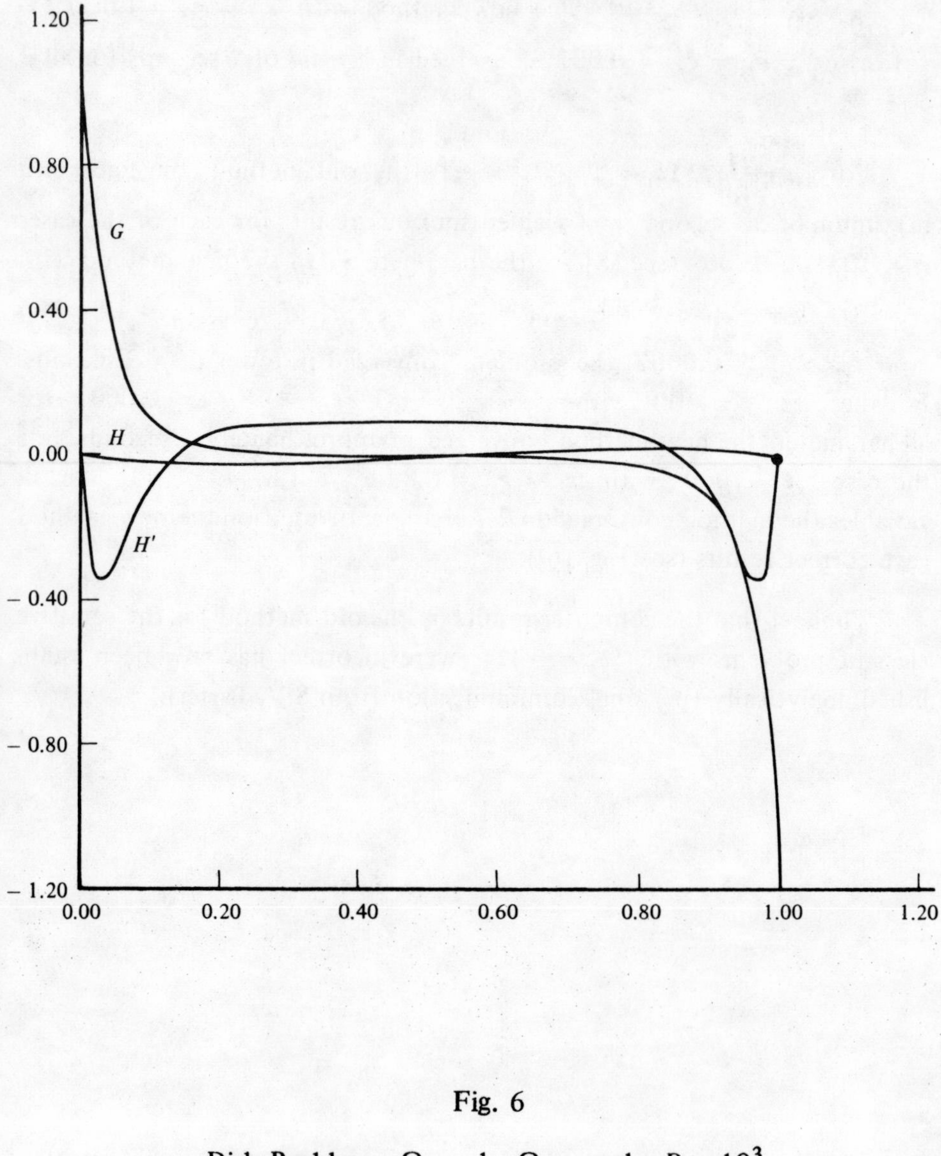

Fig. 6

Disk Problem, $\Omega_1 = 1$, $\Omega_0 = -1$, $R = 10^3$

7. CONCLUSIONS

The new method has been found to work on all problems tested. It has been found to be faster, and thus more economical, to require less storage space and to free the researcher from the problem of trying to find the correct smoothing parameters necessary for convergence. In addition, because it requires less space one can go to much smaller grid sizes and thereby improve the results. Finally, and perhaps most importantly, it has been found to give more accurate results in at least one area of sensitive problems, that is, the case of counter rotating disks with $\Omega_1 = -\Omega_2$.

It is worth noting, in addition, that other computer centers also are experimenting with the new method described in the paper and have noticed the increased speed of convergence (personal communication from M. Friedman), but no published results are as yet available.

REFERENCES

[1] B.L. Buzbee – F.W. Dorr, *The direct solution of the biharmonic equation on rectangular regions and the Poisson equation on irregular regions.* Tech. Report LA-UR-73-636, Los Alamos Scientific Laboratory, 1973.

[2] L.W. Ehrlich, Solving the biharmonic equation in a square: a direct versus a semidirect method, *Communications of the ACM,* Vol. 16, Nov., 1973.

[3] D. Greenspan, *Numerical studies of two dimensional, steady state, Navier – Stokes equations for arbitrary Reynolds number,* Tech. Report No. 9, Dept. of Comp. Sci., University of Wisconsin, Madison, Wisconsin, 1968.

[4] D. Greenspan, *Numerical study of viscous, incompressible flow for arbitrary Reynolds number,* Tech. Report No. 11, Dept. of Comp. Sci, University of Wisconsin, Madison, Wisconsin, 1968.

[5] D. Greenspan, Numerical solution of a class of nonsteady cavity flow problems, *BIT, Vol.* 8, *No.* 4, 1968.

[6] D. Greenspan, Numerical studies of prototype cavity flow problems, *Comp. J.*, Vol. 12, 1969.

[7] D. Greenspan, Numerical studies of steady, viscous, incompressible flow in a channel with a step, *J. Engng. Math.*, Vol. 3, 1969.

[8] D. Greenspan, Numerical studies of flow between rotating coaxial disks, *J.I.M.A.*, Vol. 9, 1972.

[9] D. Greenspan – D.H. Schultz, Fast finite difference solution of biharmonic problems, *Communications of the ACM.*, Vol. 15, May, 1972.

[10] D.H. Schultz, Numerical solution for the flow of a fluid in a heated closed cavity, *Quarterly Journal of Mechanics and Applied Mathematics,* Vol. XXVI, Part 2, May, 1973.

[11] J. Smith, The coupled equation approach to the numerical solution of the biharmonic equation by finite differences, *SIAM Journ. Num. Anal.*, Vol. 5, 1968.

D. Greenspan

Department of Computer Sciences, University of Wisconsin, Madison, Wisconsin 53706, USA.

D. Schultz

Department of Mathematics, University of Wisconsin-Milwaukee, Milwaukee, Wisconsin 53201, USA.

COLLOQUIA MATHEMATICA SOCIETATIS JÁNOS BOLYAI
15. DIFFERENTIAL EQUATIONS, KESZTHELY (HUNGARY), 1975.

EXTENSIONS OF LETTENMEYER'S THEOREM

L.J. GRIMM — L.M. HALL

1. INTRODUCTION

In 1926, F. Lettenmeyer [7] proved that a linear homogeneous ordinary differential system with a (possibly irregular) singular point at $z = z_0$ may have several independent solutions holomorphic at z_0, and estimated the number of such solutions. In this paper we extend Lettenmeyer's result to a class of functional differential equations; our methods are based on the work of W.A. Harris, Jr., Y. Sibuya and L. Weinberg [5] and Ju.F. Korobeĭnik [6].

2. THE HOMOGENEOUS CASE

Consider the neutral functional-differential equation (NFDE)

$$(1) \qquad z^D \frac{dy}{dz} = A(z)y(z) + B(z)y(h(z)) + C(z)y'(h(z)),$$

where A, B, and C are $n \times n$ matrices holomorphic at $z = 0$, $D =$ $= \operatorname{diag}(d_1, \ldots, d_n)$ is an $n \times n$ constant matrix with nonnegative integer entries d_i, $h(z) = \lambda z + \sum_{k=2}^{\infty} h_k z^k$ is holomorphic at zero, with $0 < |\lambda| < 1$, and y is an n-vector. Lettenmeyer's theorem can be extended to this case as a corollary of the following theorem:

Theorem 1. *Let* A, B, C, D, *and* h *be as above. Then for each positive integer* N *sufficiently large, and each polynomial* $\Phi(z)$ with $z^D \Phi(z)$ *of degree* N, *there exists a polynomial* $f(z)$ *(depending on* A, B, C, h, Φ, *and* N*) of degree* $N-1$ *such that the linear neutral differential system*

$$(2) \qquad z^D y'(z) = A(z)y(z) + B(z)y(h(z)) + C(z)y'(h(z)) + f(z)$$

has a solution $y(z)$ *holomorphic at* $z = 0$. *Further,* f *and* y *are linear and homogeneous in* Φ, *and*

$$z^D (y - \Phi) = O(z^{N+1}) \quad \text{as} \quad z \to 0.$$

The proof is similar to the proof of the corresponding result in [2].

The following extensions of Lettenmeyer's theorem are corollaries of this result.

Corollary 1. *Let* $d = \operatorname{trace} D$ *and* $n - d \geqslant 0$. *Then the system* (1) *has at least* $n - d$ *linearly independent solutions holomorphic at* $z = 0$.

Corollary 2. *Let* $h(z) = \lambda z$, *where* $0 < |\lambda| < 1$. *Let* r_k *be the rank of the* $n(k+1) \times n(k+2)$ *matrix*

$$
\begin{bmatrix}
A_0 + B_0 & C_0 & 0 & 0 & \cdots & 0 \\
A_1 + B_1 & A_0 + \lambda B_0 + C_1 & 2\lambda C_0 & 0 & & 0 \\
A_2 + B_2 & A_1 + \lambda B_1 + C_2 - I & A_0 + \lambda^2 B_0 + 2\lambda C_1 & 3\lambda^2 C_0 & & \\
\vdots & \vdots & \vdots & \vdots & & \vdots \\
A_k + B_k & A_{k-1} + \lambda B_{k-1} + C_k & \cdots & U_k & A_0 + \lambda^k B_0 + k\lambda^{k-1} C_1 & (k+1)\lambda^k C_0
\end{bmatrix}
$$

for each $k = 1, 2, \ldots,$ *where* $U_k = A_1 + \lambda^{k-1}B_1 + (k-1)\lambda^{k-2}C_2 - (k-1)I$, *with* I *the identity matrix, and the* $A_i, B_i,$ *and* C_i *the respective coefficients of the power series expansions about* $z = 0$ *of the functions* $A(z), B(z),$ *and* $C(z)$. *Set* $s = \max_k \{nk - r_k\}$. *Then if* $s > 0$, *the NFDE*

$$z^2 \frac{dy}{dz} = A(z)y(z) + B(z)y(\lambda z) + C(z)y'(\lambda z)$$

has at least s *linearly independent solutions holomorphic at* $z = 0$.

Corollary 1 extends the Lettenmeyer theorem to a class of NFDE; Corollary 2 proves the existence of non-trivial holomorphic solutions in cases where the Lettenmeyer theorem is inapplicable.

Both corollaries follow from Theorem 1 by setting $f = 0$ in equation (2) to obtain a system of determining equations for the coefficients of Φ. The conditions of Corollary 1 ensure the existence of at least $n - d$ linearly independent solutions of the determining system, and independent solutions of the NFDE correspond to independent solutions of the determining system. Corollary 2 follows from similar considerations.

3. THE INHOMOGENEOUS CASE

The following question has been raised by H.L. Turrittin [9]: Let $A(t)$ and $g(z)$ be holomorphic at $z = 0$, where A is an $n \times n$ matrix and g is an n-vector, and let $w(z) = \sum_{j=0}^{\infty} z^j w_j$ be a formal power series solution of the system

$$z^2 \frac{dw}{dz} = A(z)w(z) + g(z).$$

What are necessary and sufficient conditions that $w(z)$ converge in a neighborhood of $z = 0$?

In case the formal solution $w(z)$ is uniquely determined, an answer to this question can be obtained as a consequence of the theory of Noetherian operators on the spaces $\mathfrak{A}_k$ of functions holomorphic for

$|z| < 1$ and k times continuously differentiable for $|z| \leqslant 1$, see [1]. The first result in this direction is the following.

Theorem 2. *Let* $n > \operatorname{tr} D$. *Then the differential system*

$$z^D y'(z) = A(z)y(z) + B(z)y(\lambda z) + C(z)y'(\lambda z) + g(z) \tag{3}$$

with A, B, C, D, λ *as in Section* 2, *has a solution holomorphic at* $z = 0$ *for every function* $g(z)$ *holomorphic at* $z = 0$ *if, and only if, the associated homogeneous system*

$$z^D w'(z) = A(z)w(z) + B(z)w(\lambda z) + C(z)w'(\lambda z)$$

has exactly the number of linearly independent solutions holomorphic at $z = 0$ *guaranteed by Lettenmeyer's theorem, viz.,* $n - \operatorname{tr} D$.

In the general case, where the hypothesis of Theorem 2 may not hold, necessary and sufficient conditions for solvability of equation (3) have been derived recently by the authors [3]-[4], who have employed a representation due to A. E. Taylor [8] of the dual space of $\mathfrak{A}_k$ to apply a suitable alternative theorem. The procedure is rather involved; here we will only give an example of the results obtainable.

Example. Consider the 1-dimensional equation

$$z^2 \frac{dy}{dz} + y(z) + y(\lambda z) = h(z) = \sum_{k=0}^{\infty} h_k z^k. \tag{4}$$

An application of our procedure shows that this equation has a solution holomorphic at $z = 0$ if and only if the coefficients of h satisfy

$$\sum_{k=1}^{\infty} \frac{(-1)^{k-1}}{(k-1)!} \Big(\prod_{j=1}^{k-1} (1 + \lambda^j) \Big) h_k = 0.$$

Remarks.

1. The method of Section 2 can be used to prove the existence of holomorphic solutions for certain equations with polynomial inhomogeneous terms, but we have not succeeded in applying it to the general case.

2. The method of Section 2 can also be applied to certain non-linear systems.

Further results and examples will appear elsewhere.

REFERENCES

[1] N. Dunford – J.T. Schwartz, *Linear Operators, Part* I: *General Theory,* New York: Interscience Publishers, 1957, Chapter 4.

[2] L.J. Grimm – L.M. Hall, Holomorphic solutions of functional differential systems near singular points, *Proc. Amer. Math. Soc.,* 42 (1974), 167-170.

[3] L.J. Grimm – L.M. Hall, An alternative theorem for singular differential systems, *J. Differential Equations,* (to appear).

[4] L.M. Hall, *Solvability of differential systems near singular points,* Ph.D. Dissertation, University of Missouri-Rolla, 1974.

[5] W.A. Harris Jr. – Y. Sibuya – L. Weinberg, Holomorphic solutions of linear differential systems at singular points, *Arch. Rational Mech. Anal.,* 35 (1969), 245-248.

[6] Ju.F. Korobeĭnik, Normal solvability of linear differential equations in the complex plane, *Izv. Akad. Nauk SSSR,* 36 (1972), 450-474; *English transl., Math. USSR Izvestija,* 6 (1972), 445-466; *Erratum, Izv. Akad. Nauk SSSR,* 37 (1973), 247.

[7] F. Lettenmeyer, Über die an einer Unbestimmtheitsstelle regulären Lösungen eines systemes homogener linearen Differentialgleichungen, *S.-B. Bayer. Akad. Wiss. München Math.-nat. Abt.,* (1926), 287-307.

[8] A.E. Taylor, Banach spaces of functions analytic in the unit circle, I, II, *Studia Math.,* 11 (1950), 145-170; *Studia Math.,* 12 (1951), 25-50.

[9] H.L. Turrittin, My mathematical expectations, *Symposium on Ordinary Differential Equations,* Lecture Notes in Mathematics #312, 1973, pp. 3-22.

L.J. Grimm

Department of Mathematics, University of Missouri-Rolla, Rolla, Missouri, USA.

L.M. Hall

Department of Mathematics, University of Nebraska, Lincoln, Nebraska, USA.

COLLOQUIA MATHEMATICA SOCIETATIS JÁNOS BOLYAI
15. DIFFERENTIAL EQUATIONS, KESZTHELY (HUNGARY), 1975.

ON THE STABILITY OF SOLUTIONS OF DIFFERENTIAL-DIFFERENCE EQUATIONS IN THE CRITICAL CASE

P.S. GROMOVA

In connection with the recent progress in the technique of big velocities and powers, one has to take more and more into account the effect of the past in the problems of mechanics and techniques. Differential-difference equations describe the effect of the past quite well. Hence it is of great relevance to study the asymptotic behavior of solutions of differential-difference equations. In the theory of ordinary differential equations there exists a rich collection of results in this direction, connected first of all with Liapunov's stability theory. In the theory of differential-difference equations, even in the case of linear equations with constant coefficients and constant deviations of the argument there are many open questions. This can be explained by the fact that the characteristic function of a differential-difference equation is a quasi-polynomial, an integer function with infinitely many zeros, while for an ordinary differential equation it is a polynomial. Consequently, the stability of solutions can be destroyed even in the case when the real parts of all roots of the characteristic equation are negative. This phenomenon occurs in the so-called critical case, when the characteristic function has infinitely many zeros approaching to the imaginary exis, or infinitely many purely imaginary roots while the remaining roots have negative real parts. We will obtain conditions that assure the stability of differential-difference equations in this critical case.

Let us consider the initial value problem

$$(1) \qquad \sum_{i=0}^{n} \sum_{j=0}^{m} a_{ij} y^{(i)}(t - \tau_j) = 0$$

$$(2) \qquad y^{(i)}(t) = \varphi^{(i)}(t), \quad t \in [-\tau_m, 0], \qquad i = \overline{0, n}.$$

Here a_{ij}, $i = \overline{0, n}$, $j = \overline{0, m}$ are real constants, $a_{n0} \neq 0$, the constant retardations $0 < \tau_0 < \tau_1 < \ldots < \tau_m$ are commensurable (there exists a number q such that $\tau_i = qn_i$ with n_i integer; $i = \overline{1, m}$), $\varphi(t) \in C^n[-\tau_m, 0]$ is a given initial function asatisfying the following condition:

$$\sum_{i=0}^{n} \sum_{j=0}^{m} a_{ij} \varphi^{(i)}(-\tau_j) = 0.$$

With the above assumptions problem (1)-(2) admits a unique solution $y = y(t) \in C^n[0, \infty]$ for $t \geqslant 0$ which can be expanded into a series in terms of the basic solutions:

$$y(t) = \sum_{k=0}^{\infty} P_k(t) e^{z_k t},$$

where the z_k's are the roots of the characteristic equation

$$(3) \qquad D(z) = \sum_{i=0}^{n} \sum_{j=0}^{m} a_{ij} z^i e^{-\tau_j z} = 0,$$

and $P_k(t)$ is a polynomial of degree less than the multiplicity of the root z_k [2]. For the arrangement of the roots of the quasi-polynomial $D(z)$ corresponding to equation (1) of neutral type $\left(a_{n0} \neq 0, \ \sum_{j=1}^{m} |a_{nj}| \neq 0\right)$ two critical cases are possible:

1. Asymptotically critical case – all roots of equation (3) satisfy the condition

(4) $\operatorname{Re} z_k < 0$

and there exists an infinite sequence of roots approaching to the imaginary axis.

2. Very critical case: there exists an infinite sequence of purely imaginary roots and the remaining roots of (3) satisfy condition (4).

The asymptotic behavior of solutions of some differential-difference equations in the asymptotically critical case was considered in [6], [7], [8], [12], [14], [15], [18]. In these papers it was shown for various classes of equations of the form (1) that in the asymptotically critical case these equations have solutions that decrease, as $t \to \infty$, no faster than some power of t with negative exponent, besides the velocity of decrease of the solution depends on the smoothness of the solution and on the velocity with which the sequence of zeros of the characteristic polynomial approaches to the imaginary axis. It was also shown that in the case of multiple sequences of zeros approaching the imaginary axis and of high enough multiplicity, independently of (4), the solutions are non-stable in the sense of Liapunov as $t \to \infty$. Yet, none of these papers contain coefficient criteria for the stability or non-stability of solutions as $t \to \infty$.

In order to obtain such criteria for the stability of solutions of equation (1) as $t \to \infty$, we will follow the following way:

I. We find coefficient criteria for condition (4).

II. We find coefficient criteria for the existence of a sequence of roots of equation (3), approaching to the imaginary axis. We will also study the asymptotic behavior of such sequences

III. We apply the theorems of the papers [8], [18].

There are many papers devoted to the first question – the problem of Routh – Hurwitz for quasi-polynomials. These papers can be divided into four classes according to their method of study:*

*References are given only to papers containing results used in the present work. A more detailed bibliography can be found in [4] and [5].

1. The generalization of the Routh – Hurwitz determinant criterion for quasi-polynomials [4], [5].

2. The method of L.S. Pontragin [3], [2].

3. The method of N.G. Čebotaŕev [4].

4. The method of D-divisions and its modifications [5], [1].

Coefficient criteria for the existence of a sequence of zeros approaching to the imaginary axis and the asymptotic behaviour of these chains can be found in [9], [13]. Using the results of the paper [8], one can obtain various coefficient criteria for the stability of solutions of problem (1)-(2) as $t \to \infty$. Let us recall some of them.

First of all we state some theorems that will be necessary in the sequel. Replacing the exponential expression in (3) by its series expansion and writing $-iz$ instead of z $(i = \sqrt{-1})$, we obtain:

$$D(iz) = \sum_{l=0}^{\infty} (-1)^l c_{2l} z^{2l} + i \sum_{l=0}^{\infty} (-1)^l c_{2l+1} z^{2l+1} =$$

$$= f(z) + ig(z),$$

where

$$c_\nu = \sum_{j=0}^{m} a_{\nu j} + \sum_{p=1}^{\nu} (-1)^p \sum_{j=1}^{m} a_{\nu-p,j} \frac{\tau_j^p}{p!}, \qquad \nu \leqslant n$$

$$c_{n+k} = \sum_{p=0}^{n} (-1)^p \sum_{j=1}^{m} a_{pj} \frac{\tau_j}{(r+k-p)!}, \qquad k = 1, 2, \ldots .$$

Let us consider the Routh – Hurwitz determinants

$$(5) \qquad \Delta_0 = 1, \Delta_\mu = \begin{vmatrix} c_1 c_3 & \ldots & c_{2\mu-1} \\ c_0 c_2 & \ldots & c_{2\mu-2} \\ 0\, c_1 & \ldots & c_{2\mu-3} \\ 0\, c_0 & \ldots & c_{2\mu-4} \\ \ldots & \ldots & . \\ 0\;0 & \ldots & c_\mu \end{vmatrix}, \qquad \mu = 1, 2, \ldots$$

Theorem of Čebotarev [4]. *If the even* $(f(z))$ *and odd* $(g(z))$ *constituants of the quasi-polynomial* $D(iz)$ $(D(0) > 0)$ *have no common zeros, then the positiveness of the determinants* Δ_μ, $\mu = 1, 2, \ldots$ *is a necessary and sufficient condition in order that all the zeros of* $D(z)$ *satisfy* (4).

Let us also consider the following infinite matrix

$$
B_2 \left\| \begin{array}{cccccc}
c_1 & -c_3 & c_5 & -c_7 & \cdots \\
c_0 & -c_2 & c_4 & -c_6 & \cdots \\
0 & c_1 & -c_3 & c_4 & \cdots \\
0 & c_0 & -c_2 & c_4 & \cdots \\
0 & 0 & c_1 & -c_3 & \cdots \\
0 & 0 & c_0 & -c_2 & \cdots \\
\cdot & \cdot & \cdot & \cdot & \cdots
\end{array} \right\|
$$

Denote by $B_2, B_4, \ldots, B_{2n}, \ldots$ the diagonal minors of sizes $2 \times 2, 4 \times 4, \ldots, 2n \times 2n, \ldots$ starting from the left upper corner of the above matrix and set $\det B_{2n} = \Delta^*_{2n}$.

In this way we obtain an infinite sequence of Hermite – Hurwitz determinants

(6) $\qquad 1, -\Delta^*_2, \Delta^*_4, \ldots, (-1)^j \Delta^*_{2j},$

constructed for the functions $f(z)$ and $g(z)$. We can analogously construct the sequence of Hermite – Hurwitz determinants for the functions $f(z)$ and $-f'(z)$, the only difference is that in the infinite matrix the coefficients of $g(z)$ have to be replaced by those of $-f'(z)$.

Theorem of Naimark [5]. *In order that the quasi-polynomial* $D(z)$ *should not have zeros on the right hand side of the imaginary axis, it is necessary and sufficient that in the sequences of Hermite – Rademacher determinants constructed for the functions* $f(z), g(z)$ *and for the functions* $f(z), -f'(z)$ *there be no changes of sign.*

The application of the above criteria requires the determination of the sign of infinitely many determinants, which is not easy to do, even if the quasi-polynomials are numerically given. However, in [5] there can be found a special method which is effective in the case of quasi-polynomials.

In what follows we will use the following metrics of the space C^m:

$$\| y(t) \| = \max_{-\tau_m \leqslant s \leqslant 0} \{ |y(t+s)|, |y'(t+s)|, \ldots, |y^{(m)}(t+s)| \}.$$

We will say that the solution $y_{\varphi_0}(t)$ of problem (1)-(2) with initial function $\varphi_0(t)$ is stable (asymptotically stable, exponentially stable, etc.) in the space C^m, if it is stable (asymptotically stable, exponentially stable, etc.) in the above metrics of the space C^m, i.e., if for any $\epsilon > 0$, $t_0 > 0$ there exists $\delta(\epsilon, t_0) > 0$ such that for every solution $y_0(t)$ of problem (1)-(2), corresponding to the initial function $\varphi(t)$ and satisfying the inequality $\| \varphi(t) - \varphi_0(t) \| < \delta$ for $t \geqslant t_0$, we have $\| y_\varphi(t) - y_{\varphi_0}(t) \| < \epsilon$. The following theorems are true.

Suppose that in equation (1) we have $a_{n0} \neq 0$, $a_{ni} = 0$, $i \neq 0$ (equation of retarded type).

Theorem 1. *Assume that*

(1) $\varphi^{(n-1)}(t)$ *is absolutely continuous on the initial domain;*

(2) *the even and odd constituants* $f(z)$ *and* $g(z)$ *of the quasi-polynomial* $D(iz)$ *have no common roots;*

(3) *all determinants* $\Delta\mu$, $\mu = 1, 2, \ldots$ (5) *are positive.*

Then the solutions of problem (1)-(2) *are exponentially stable in the space* C^{n-1} *as well as in the space* C^n *as* $t \to \infty$.

Indeed, assumptions (2) and (3) imply (4) and the theorem follows from [1], p. 116.

Suppose that in equation (1) we have $a_{n0} \neq 0$, $\sum_{i=1}^{m} a_{ni} \neq 0$, i.e., equation (1) is of neutral type (not asymptotically critical case).

Theorem 2. *Assume that*

(1) $\varphi^{(n)}(t)$ *is continuous on the initial domain.*

(2) *The even and odd constituants* $f(z)$ *and* $g(z)$ *of the quasi-polynomial* $D(iz)$ *have no common zeros.*

(3) *All the determinants* Δ_μ $\mu = 1, 2, \ldots$ (5), *are positive.*

(4) *The polynomial* $P(w) = a_{n0} + a_{n1} w^{n_1} + \ldots + a_{nm} w^{n_m}$ *has no root of modulus* 1.

Then the solutions of problem (1)-(2) *are exponentially stable in the space* C^{n-1} *as* $t \to \infty$.

Because of condition (4), the roots of the quasi-polynomial are separated from the imaginary axis [13], thus the proof of Theorem 2 goes along the same lines as that of Theorem 1.

Concerning the asymptotically critical case, we have the following results.

Theorem 3. *Assume that*

(1) $\varphi^{(n)}(t)$ *is continuous and Lipschitz on the initial domain.*

(2) *The even and odd constituants* $f(z)$ *and* $g(z)$ *of the quasi-polynomial* $D(iz)$ *have no common zero.*

(3) *All the determinants* Δ_μ, $\mu = 1, 2, \ldots$ *are positive.*

(4) *One of the following conditions is satisfied:*

(a) $|a_{n0}| = |a_{ni_0}|$, $a_{nj} = 0$, $j \neq 0, i_0$

(b) *the polynomial* $P(w) = a_{n0} + a_{n1} w^{n_1} + \ldots + a_{nm} w^{n_m}$ *has at least one simple root of modulus* 1.

Then the solutions of problem (1)-(2) *are uniformly asymptotically stable and polynomially decreasing in the space* C^{n-1}.

Proof. Condition (4) implies [9], [13] that the quasi-polynomial $D(z)$ has at least one simple sequence of zeros polynomially approaching to the imaginary axis. Hence our theorems follows from [8].

Theorem 4. *Let* $p > 0$, $q \geqslant 0$ *and* r *real numbers. Then the solutions of the equation*

$$ry''(t-\tau) + y''(t) + py'(t) + qy(t) = 0$$

are uniformly asymptotically stable as $t \to \infty$, *provided that* $p^2 - 2q \geqslant 0$ *and* $-1 \leqslant r \leqslant 1$. *Moreover, the stability is exponential if* $|r| \neq 1$, *and the solutions decrease polynomially if* $|r| = 1$.

In fact, all roots of the characteristic function satisfy condition (4) ([2], p. 491) and if $|r| = 1$, then there exists a simple sequence of zeros polynomially approaching to the imaginary axis [9]. It remains only to apply the results of the paper [8].

We are now going to obtain some criteria for the stability of solutions of equation (1) in the very critical case. The asymptotic behaviour of solutions as $t \to \infty$ in the very critical case has already been studied in [10], [12], [16] and [18].

Let us consider equation (1) with $m = 1$:

(7) $$\sum_{i=0}^{n} [a_i y^{(i)}(t) + b_i y^{(i)}(t-\tau)] = 0, \quad \tau > 0.$$

Then the characteristic equation has the form

(8) $$D_1(z) = P(z) + Q(z)e^{-\tau z} = 0$$

where $P(z) = \sum_{i=0}^{n} a_i z^i$, $Q(z) = \sum_{i=0}^{n} b_i z^i$. A necessary and sufficient condition for the existence of infinite sequences of purely imaginary zeros of equation (8) is that [17]

(9) $$|P(i\omega)|^2 = |Q(i\omega)|^2$$

(ω real, $i = \sqrt{-1}$).

Theorem 5. *If all Hermite – Hurwitz determinants* (6), *constructed for the functions* $f(z), g(z)$ *and* $f(z), -f'(z)$ *of the quasi-polynomial* $D_1(iz)$ *are positive and* $|P(z)| \equiv |Q(z)|$, *then the solutions of equation* (7) *are stable of Liapunov as* $t \to \infty$.

It follows from Naimark's theorem that the characteristic function $D_1(z)$ has no zero with positive real part, furthermore, condition (9) assures the existence of a simple infinite chain of purely imaginary roots of the form $\frac{2k\pi}{\tau}i, \frac{2k+1}{\tau}\pi i,\ k = 0, \pm 1, \ldots$ [11]. Theorem 5 then follows by using the same method as in [10].

Let us denote by R the domain of those sequences of coefficients $a_0, a_1, \ldots, a_n, b_0, b_1, \ldots, b_n$ of equation (7) which satisfy condition (9).

Theorem 6. *If all Hermite – Hurwitz determinants* (6) *constructed for the functions* $f(z), g(z)$ *and* $f(z), -f'(z)$ *of the quasi-polynomial* $D_1(iz)$ *are positive,* $|P(z)| \not\equiv |Q(z)|$ *and condition* (9) *is satisfied, then the set of those sequences of coefficients for which equation* (7) *is not stable in the sense of Liapunov and* (9) *holds is dense in* R.

REFERENCES

[1] L.E. El'sgol's – S.B. Norkin, *Introduction to the theory and application of differential equations with deviating arguments,* New York, Acad. Press, 1973.

[2] R. Bellman – K. Cooke, *Differential-difference equations,* Academic Press, 1963.

[3] L.S. Pontrjagin, On the roots of certain elementary transcendental functions, *Izv. A.N. SSSR, Ser. Mat.,* 6, 3 (1942), 115-134 (in Russian).

[4] N.G. Čebotarěv – N.N. Meǐman, *The Routh – Hurwitz problem for polynomials and integer functions, Trudy Mat. Inst. Steklova,* 26 1949 (in Russian).

[5] Yu.I. Naimark, *D*-decomposition of the space of quasi-polynomials (for the stability of linearized distributed systems), *P.M.M.,* 13, 4 (1946), 349-380 (in Russian).

[6] W. Hahn, Über Differential-differenzengleichungen mit anomalen Lösungen, *Math. Annalen,* 133, 3 (1957), 251-255.

[7] W. Snow, *Existence, uniqueness and stability for nonlinear differential-difference equations in the neutral case,* New York University, 1965, 1-47.

[8] P.S. Gromova, On the stability of solutions of nonlinear equations of neutral type in the asymptotically critical case, *Mat. Zam.*, 6 (1967), 715-726 (in Russian).

[9] P.S. Gromova, Asymptotics for roots having of large absolute value of quasi-polynomials in a vicinity of the imaginary axis, *Trudy Sem. po Teor. Diff. Urav. s Otk. Arg., U.D.N., Moscow,* 6 (1968), 109-125 (in Russian).

[10] P.S. Gromova, On the instability of solutions of imaginary differential-difference equations of the first order, ibid., 8 (1972), 28-37 (in Russian).

[11] P.S. Gromova, On the asymptotics of purely imaginary roots of quasi-polynomials, *ibid.*, 7 (1969), 194-195 (in Russian).

[12] P.S. Gromova, On the stability of solutions of linear differential-difference equations of *n*-th order in the supercritical case, *ibid.*, 9 (1973) (in Russian).

[13] P.S. Gromova, A condition for the existence of infinitely many roots approximating the imaginary axis in the case of quasi-polynomials, *ibid.*, 10 (1974-75) (in Russian).

[14] W.E. Brumley, On the asymptotic behavior of solutions of differential-difference equations of neutral type, *J. Diff. Equ.*, 7, 1 (1970), 175-188.

[15] D.C. Subramanyam, Ont he asymptotic behavior of the solutions of the linear homogeneous differential-difference equations with real constant coefficients in critical case, *Mat. Nachr.*, 40, 4-6 (1969), 201-223.

[16] P.S. Gromova – A.M. Zverkin, Trigonometric series whose sum is continuous and unbounded on the real line and is a solution of a differential-difference equation with retarded argument, *Diff. Urav.*, 4, 10 (1968), 1774-1784 (in Russian).

[17] L.E. El'sgol's, On certain properties of periodic solutions of quasilinear differential equations with retarded argument, *Vestnik M.G.U.*, 5 (1959), 229-237 (in Russian).

[18] P.S. Gromova, A study of stability of solutions of a system of linear differential-difference equations in singular cases, *Zagadnienia Niel. Kol.*, 14 (1973), 149-158.

P.S. Gromova

U.D.N. Kaf. Appl. Math. B-302 ul. Ordzonikidze 3, Moscow, USSR.

COLLOQUIA MATHEMATICA SOCIETATIS JÁNOS BOLYAI
15. DIFFERENTIAL EQUATIONS, KESZTHELY (HUNGARY), 1975.

ON THE BOUNDEDNESS OF THE SOLUTIONS OF $2n$-DIMENSIONAL SYSTEMS

L.V. GUŠČO

We consider the system

$$(1)\qquad \begin{aligned} \dot{x}_k &= \frac{\partial H(x, y)}{\partial y_k} \\ \dot{y}_k &= -\frac{\partial H(x, y)}{\partial x_k} + f_k(x, y) \end{aligned} \qquad (k = 1, \dots, n)$$

with Hamilton function

$$H(x, y) \equiv V(x) + \frac{1}{2} y^T A y,$$

where y is an n-dimensional column vector and y^T is the transpose of the vector y.

We assume that the following conditions are fulfilled.

I. The potential energy $V(x)$ is a positive definite and continuously differentiable function in the Euclidean space R^n. Moreover, $V(x) \to +\infty$ as $|x| \to +\infty$.

II. The kinetic energy $T(y) \equiv \frac{1}{2} y^T A y$ has a positive definite constant symmetric inertia matrix.

III. The perturbation $f(x, y) \equiv \{f_1(x, y), \ldots, f_n(x, y)\}$ is defined and continuous on R^{2n}.

The derivative of the Hamilton function is called the Rayleigh function; owing to system (1), it can be written in the form

$$\dot{H}(x, y) \equiv R(x, y) \equiv f(x, y) A y.$$

Let T^+ be a positive semi-trajectory of system (1) with apex $M\{x(t), y(t)\}$, defined on the semi-interval $[0, \tau)$, where τ is either a positive number (if the semi-trajectory is non-extendable) or the symbol $+\infty$ (if T^+ is extendable).

The semi-trajectory T^+ is said to be divergent ([1], pp. 24, 372), if for any increasing sequence of numbers $t_N \in [0, \tau)$ $(N = 1, 2, \ldots)$ tending to τ and points $M(t_N) \equiv \{x(t_N), y(t_N)\}$ of the semi-trajectory T^+ we have

$$\lim_{N \to \infty} \{x^2(t_N) + y^2(t_N)\} = +\infty.$$

We say that the semi-trajectory is unbounded, if

$$\overline{\lim_{t \to \tau - 0}} \{x^2(t) + y^2(t)\} = +\infty.$$

Lemma 1. *If conditions* I-III *are fulfilled, and in the cylinder*

$$G \equiv \{V(x) < g \mid x, y \in R^n\}$$

we have

$$\frac{f(x, y)}{y^2} \to 0 \quad as \quad |y| \to +\infty \tag{2}$$

uniformly for x, *then the unbounded semi-trajectory* T^+ *of system* (1) *leaves the closure* $\bar{G}$ *at a finite instant of time* t.

Proof. Let the unbounded semi-trajectory T^+ of system (1) be defined on the semi-interval $[0, \tau)$, where τ is either a positive number or the symbol $+\infty$.

Fix a natural number

(3) $$N \geqslant \max \{g, H(x_0, y_0)\},$$

where the number g is taken from the definition of the cylinder G, and x_0, y_0 are the coordinates of the initial point of the semi-trajectory T^+.

Let now N_1, N_2 be any pair of natural numbers such that

(4) $$N < N_1 < N_2,$$

where N satisfies inequality (3). The distance of the apex $M(t) \equiv \{x(t), y(t)\}$ of the semi-trajectory T^+ from the origin 0 is a continuous function on the semi-interval $[0, \tau)$. Since a continuous function of a continuous function is continuous, making use of conditions I, II and of the assumption concerning the unboundedness of the semi-trajectory T^+ we obtain that the Hamilton function $H(x, y)$ is defined, continuous and unbounded on the semi-trajectory T^+,

(5) $$\sup H(x, y) = +\infty \quad \text{for} \quad M(x, y) \in T^+.$$

In other words, the function $H\{x(t), y(t)\}$ is defined, continuous and unbounded on the semi-interval $[0, \tau)$. In particular, there exists a number $\theta \in [0, \tau)$ for which

(6) $$H\{x(\theta), y(\theta)\} > N_2.$$

From inequalities (3) and (4) we obtain

(7) $$H\{x(0), y(0)\} < N_1.$$

By the second Bolzano – Cauchy theorem, there exists a number $\xi \in (0, \theta)$ for which $H\{x(\xi), y(\xi)\} = \frac{1}{2}(N_1 + N_2)$. Owing to the continuity of $H\{x(t), y(t)\}$ as a function of t on the segment $[0, \theta]$, we can find positive numbers τ_1, τ_2 such that

(8) $$N_1 < H\{x(t), y(t)\} < N_2 \quad \text{for} \quad \xi - \tau_1 < t < \xi + \tau_2.$$

Inequalities (6) and (7) imply the existence of the least upper bounds $\bar{\tau}_1$ and $\bar{\tau}_2$ of those numbers τ_1 and τ_2, respectively, which satisfy inequalities (8); setting $t_1 = \xi - \bar{\tau}_1$ and $t_2 = \xi + \bar{\tau}_2$ we obtain the relations

(9) $$N_1 < H\{x(t), y(t)\} < N_2 \quad \text{for} \quad t_1 < t < t_2,$$

(10) $$0 < t_1 < \xi < t_2 < \theta,$$

(11) $$H\{x(t), y(t)\} = \begin{cases} N_1 & \text{for} \quad t = t_1, \\ N_2 & \text{for} \quad t = t_2. \end{cases}$$

Suppose now that the semi-trajectory T^+ we are considering does not leave the closure $\bar{G}$, i.e.

(12) $$V\{x(t)\} \leqslant g \quad \text{for} \quad 0 \leqslant t < \tau.$$

We wish to prove that our assumption (12) leads to a contradiction, which will show the falsity of (12) and the validity of the lemma.

Let the absolute radial velocity of the curve of the semi-trajectory T^+ be denoted by

(13) $$v \equiv + \sqrt{2T(y)} > 0,$$

where $T(y) = \frac{1}{2} y^T A y$ is the kinetic energy appearing in condition II, and let the Euclidean norm of the matrix A be denoted by α:

(14) $$\| A \| = \alpha.$$

Along the arc $M_1 M_2$ of the semi-trajectory T^* (where

$$M_1 = M\{x(t_1), y(t_1)\}, \quad M_2 = M\{x(t_2, y(t_2)\},$$

t_1 and t_2 being taken from inequality (10) and assumed to have the properties (9) and (11)), according to notation (13) and property (11), the identity

(15) $$v^2(t) \equiv y^T(t) A y(t), \qquad t \in [0, \tau)$$

holds. Hence it follows ([2], p. 34, item (I.5.25)) that

(16) $$\sqrt{\lambda}\, | y(t) | \leqslant v(t) \leqslant \sqrt{\Lambda}\, | y(t) |, \qquad t \in [0, \tau),$$

where λ and Λ stand for the least and the greatest eigenvalue of the matrix A.

Inequalities (16) yield

$$(17) \qquad \frac{|f\{x(t), y(t)\}|}{y^2(t)} \leqslant \frac{|f\{x(t), y(t)\}|}{\lambda y^2(t)}, \qquad t \in [t_1, t_2].$$

Property (12) implies that for an arbitrarily small number $\epsilon > 0$ there exists a large number N_ϵ such that in the region G, and therefore on the closure $\bar{G}$ too, we have

$$(18) \qquad \frac{|f(x, y)|}{y^2} < \frac{\epsilon\lambda\sqrt{\lambda}}{2\alpha(\sqrt{\lambda} + \sqrt{\Lambda})} \quad \text{for} \quad |y| > N_\epsilon$$

uniformly in x, where λ and Λ are taken from (16), and α from (14).

So far, in the selection of the natural numbers N_1 and N_2 we have only been constrained by inequalities (3) and (4). At the present moment we impose an additional restriction on the choice of these numbers: we require that besides inequalities (3) and (4) also the inequality

$$(19) \qquad N_1 > g + \frac{1}{2}\Lambda N_\epsilon^2 \equiv K$$

be satisfied, where g is taken from the assumption of Lemma 1, Λ from inequality (16), and N_ϵ from (18).

Obviously, the additional constraint (19) we have just introduced does not alter the course of the foregoing argument.

From relations (9)-(11) it follows that

$$N_1 \leqslant H\{x(t), y(t)\} \equiv V\{x(t)\} + \frac{1}{2} y^T A y \quad \text{for} \quad t_1 \leqslant t \leqslant t_2.$$

On account of inequalities (12), (16) and (19) we have

$$g + \frac{1}{2}\Lambda N_\epsilon^2 < N_1 < V\{x(t)\} + \frac{1}{2} y^T A y \leqslant$$

$$\leqslant g + \frac{1}{2}\Lambda |y(t)|^2, \qquad t_1 \leqslant t \leqslant t_2,$$

which implies that

$$(20) \qquad |y(t)| > N_\epsilon \quad \text{for} \quad t_1 \leqslant t \leqslant t_2.$$

Consequently, inequalities (17), (18) and (20) yield

$$(21) \qquad \frac{|f\{x(t), y(t)\}|}{v^2(t)} < \frac{\epsilon\sqrt{\lambda}}{2\alpha(\sqrt{\lambda} + \sqrt{\Lambda})} \quad \text{for} \quad t_1 \leqslant t \leqslant t_2.$$

Owing to condition I, the gradient $\nabla V(x)$ of the function $V(x)$ is a vector valued function defined and continuous on $\boldsymbol{R}^n$ and, in particular, on the compact set

$$\bar{G}_{y=\bar{0}} \equiv \{V(x) \leqslant g \,|\, x \in \boldsymbol{R}^n, \ y = \bar{0}\},$$

on which, by the theorem of Weierstrass, the Euclidean length $|\nabla V(x)|$ somewhere attains its least upper bound

$$(22) \qquad l \equiv \max_{V(x) \leqslant g} |\nabla V(x)| > 0$$

(if we had $l = 0$, then the assumption $V(x) > 0$ for $x \neq \vec{0}$ appearing in condition I would not be staisfied).

Since on the arc M_1M_2 inequality (9) holds,

$$N_1 < H\{x(t), y(t)\} \equiv V\{x(t)\} + \frac{1}{2}v^2(t);$$

in view of inequality (12) we have

$$(23) \qquad v(t) \geqslant \sqrt{2(N_1 - g)} > 0 \quad \text{for} \quad t_1 \leqslant t \leqslant t_2.$$

We now impose a third restriction on the selection of the natural numbers N_1 and N_2: henceforth we assume that simultaneously with inequalities (3), (4), (19) also the inequality

$$(24) \qquad N_1 > \frac{l\alpha(\sqrt{\lambda} + \sqrt{\Lambda})}{\epsilon\sqrt{\lambda}} + g \equiv L$$

holds, where l is taken from (22), α from (14), λ and Λ from (16). In other words, (3) and the inequalities

$$(25) \qquad N_2 > N_1 > \max(N, K, L)$$

must be fulfilled, where N occurs in (3), K in (19), L in (24). Inequalities (22), (23) and (24) yield

$$(26)\qquad \frac{|\nabla V(x)|}{\nu^2(t)} \leqslant \frac{|\nabla V(x)|}{2(N_1-g)} < \frac{l\epsilon\sqrt{\lambda}}{2l\alpha(\sqrt{\lambda}+\sqrt{\Lambda})} = \frac{\epsilon\sqrt{\lambda}}{2\alpha(\sqrt{\lambda}+\sqrt{\Lambda})}.$$

The arc M_1M_2, considered for $t_1 \leqslant t \leqslant t_2$ of the semi-trajectory T^+ is a smooth curve in R^{2n}. Consequently, as this is a trajectory of the system, its projection $x(t_1)x(t_2)$ to the hyperplane $y = \bar{0}$ is a smooth curve in R^n.

Let the parameter of the projection curve $x(t_1)x(t_2)$ be the length s defined by the line element $ds^2 \equiv dx^T A^{-1} dx$. s is the length of the arc $x(t_1)x(t) \subset x(t_1)x(t_2)$ for $t_1 \leqslant t \leqslant t_2$, i.e. for $0 \leqslant s \leqslant s_2$.

Let us prove that

$$(27)\qquad |x''(s)| \equiv \left|\frac{d^2x}{ds^2}\right| < \epsilon \quad \text{for} \quad 0 \leqslant s \leqslant s_2.$$

Indeed, in our case, for the Lagrange function

$$L(x,\dot{x}) \equiv \frac{1}{2}\dot{x}^T A^{-1}\dot{x} - V(x)$$

the Jacobian matrix

$$\nabla^2_{\dot{x}} L(x,\dot{x}) \equiv A^{-1}$$

is a non-singular and, consequently, for the points of the arc M_1M_2 we have the identity

$$(28)\qquad \dot{\nu}x' + \nu^2 x'' \equiv A[f(x,y) - \nabla V(x)],$$

where the dashes and dots denote differentiation with respect to s and t, respectively. This identity is not difficult to obtain from system (1), taking into account that $\frac{ds}{df} \equiv \nu$. Let us estimate $|x'| \equiv \left|\frac{dx}{ds}\right|$. The identity $ds^2 \equiv dx^T A^{-1} dx$ yields $\frac{dx^T}{ds} A^{-1} \frac{dx}{ds} \equiv 1$, i.e.

$$(29)\qquad \sqrt{\lambda} \leqslant \left|\frac{dx}{ds}\right| \leqslant \sqrt{\Lambda},$$

where λ and Λ are taken from inequalities (16).

Carrying the first term of identity (28) to the right-hand side, we divide term by term both sides of the identity by ν^2. Since

$$\frac{d\nu}{dt} \equiv \frac{d\sqrt{y^T A y}}{dt} \equiv \frac{\dot{y}^T A y}{\nu},$$

therefore

$$x'' \equiv A\,\frac{f(x, y)}{\nu^2} - A\,\frac{\nabla V(x)}{\nu^2} - \frac{x'}{\nu^2}\,\frac{\dot{y}^T A y}{\nu}.$$

For the length of the vector x'' we have the estimation

$$|x''| < 1 + \frac{|x'|\,|y|}{\nu}\;\|A\|\,\frac{\epsilon\sqrt{\lambda}}{\alpha(\sqrt{\lambda} + \sqrt{\Lambda})},$$

which follows from the inequalities (21) and (26). Applying the estimations (16) and (29), we obtain (27).

We have the obvious identity

$$\text{(30)} \qquad [x^2(s)]'' \equiv 2\{[x'(s)]^2 + x(s)x''(s)\},$$

valid along the arc $x(t_1)x(t_2)$ of the projection for $0 \leqslant s \leqslant s_2$, where s and s_2 are the lengths of the arcs $x(t_1)x(t)$ and $x(t_1)x(t_2)$, respectively. Let us estimate identity (30) from below. Making use of inequalities (29), (27) and (12), we find

$$\text{(31)} \qquad [x^2(s)]'' \geqslant 2(\lambda - m\epsilon),$$

where

$$\text{(32)} \qquad \epsilon < \frac{\lambda}{m}$$

is an arbitrarily small positive number and

$$m = \max_{V(x) \leqslant g} |x|.$$

Integrating inequality (31) with respect to s twice and taking into account that

$$(x^2)'(0) \equiv 2x(0)x'(0) \leqslant 2m\sqrt{\Lambda},$$

we obtain

$$x^2(s) \geqslant (\lambda - m\epsilon)s^2 - 2m\sqrt{\Lambda}s.$$

The right-hand side of the latter inequality is a quadratic polynomial of s, which will not be less than m^2 if the length $s_2 \equiv s\{x(t_1)x(t_2)\}$ satisfies the inequality

$$(33) \qquad s_2 > \frac{3m\sqrt{\Lambda}}{\lambda - m\epsilon} \equiv s_1;$$

in this case we have

$$(34) \qquad |x(s)| > m \quad \text{for} \quad s_1 < s \leqslant s_2,$$

contrary to the assumption (12).

Let us prove that the length of the arc M_1M_2 of the semi-trajectory T^+ is greater than s_1. For this purpose, consider the differential $dH = vdv + dV$ of the Hamilton function $H(x, y)$. According to system (1), $dH \equiv f(x, y)\,dx$. This identity implies that the increment of the total energy of system (1) along the arc M_1M_2 of the semi-trajectory T^+ we are considering can be expressed by the line integral

$$\int_{M_1M_2} f(x, y)\,dx = N_2 - N_1,$$

where the numbers N_1 and N_2 are taken from inequalities (25). Passing to line integrals of the first kind, we obtain

$$\int_0^{s_2} f\{x(s), y(s)\}x'(s)\,ds = N_2 - N_1$$

or

$$(35) \qquad \int_0^{s_2} y^2(s)\,\frac{f\{x(s), y(s)\}}{y^2(s)}\,x'(d)\,ds = N_2 - N_1.$$

If the positive number ϵ satisfies, besides (32), also the inequality

$$\epsilon < \frac{\alpha\lambda\sqrt{\lambda}}{m\sqrt{\Lambda}(3\sqrt{\Lambda}+\alpha)},$$

where α is taken from (14), λ and Λ from (16), further

$$m = \max_{V(x)\leqslant g} |x|,$$

then inequalities (18), (20) yield

$$\frac{|f\{x(s), y(s)\}|}{y^2(s)} < \frac{\lambda}{4s_1\sqrt{\Lambda}},$$

where s_1 is taken from identity (33). Making use of the latter estimate, we find the following upper bound for the integral appearing in equation (35):

$$\frac{\lambda}{4s_1} y_m^2 s_2 > N_2 - N_1,$$

where

$$y_m^2 = \max_{0\leqslant s\leqslant s_2} y^2(s).$$

If inequality (33) is no fulfilled, then

$$\text{(36)} \qquad \frac{\lambda}{4s_1} y_m^2 s_1 > N_2 - N_1.$$

Let the maximum of $y(s)$ on the arc M_1M_2 be attained at the point $M(x_m, y_m)$. At this point the Hamilton function takes the value N_m, i.e.

$$V(x_m) + \frac{1}{2} y_m^T A y_m = N_m.$$

Since we have $M_m(x_m, y_m) \in M_1M_2$, where the relations (9)-(11) hold, therefore $N_m \leqslant N_2$ or $V(x_m) + \frac{1}{2} y_m^T A y_m \leqslant N_2$. Hence, on account of inequality (36), we obtain the inequalities $N_2 > \frac{1}{2} y_m^T A y_m \geqslant \frac{1}{2} y_m^2 \lambda \geqslant \geqslant 2N_2 - 2N_1$, which cannot be valid if

$$\text{(37)} \qquad N_2 > 2N_1.$$

Since natural numbers N_2 satisfying (37) are admissible in our reasoning, the assumption that inequality (33) cannot be fulfilled leads to the contradiction just indicated. But inequality (33), as already mentioned, contradicts our assumption (12); thus the latter is false, and the statement of Lemma 1 is true, which was to be proved.

Remark. Under the conditions of Lemma 1 every divergent semi-trajectory T^+ of system (1) leaves the closure $\bar{G}$ of the cylinder G at a fintie instant of time, as T^+ is unbounded by definition.

We say that the unstable (in the sense of Lagrange) semi-trajectory T^+ of system (1) with apex $M(t) \equiv \{x(t), y(t)\}$ is oscillating, if

$$\underline{\lim}_{t \to \tau - 0} |x(t)| < \overline{\lim}_{t \to \tau - 0} |x(t)| = +\infty.$$

Lemma 2. *If conditions* I-III *are fulfilled, if* (2) *holds for any* $g > 0$, *and if for a vector valued function* $y \in \boldsymbol{R}^n$ *defined and continuous on the curve* $0x$ *we have*

$$\overline{\lim}_{|x| \to +\infty} \int_{0x} f(x, y)\, dx < +\infty, \tag{38}$$

then any divergent semi-trajectory of system (1) *is oscillating.*

Proof. If the semi-trajectory T^+ of system (1) is divergent then, according to the Remark, the conclusion of Lemma 1 holds, i.e. with each number $g > 0$ a number $t_g \in (0, \tau)$ can be associated so that the apex M_g of the semi-trajectory T^+ does not belong to the closure of the cylinder $\bar{G}$ appearing in condition (2). Since this inclusion holds for arbitrarily large numbers g, condition I yields

$$\overline{\lim}_{t \to \tau - 0} |x(t)| = +\infty.$$

Therefore the divergent semi-trajectories T^+ of system (1) can be divided into two classes:

$$\underline{\lim}_{t \to \tau - 0} |x(t)| < \overline{\lim}_{t \to \tau - 0} |x(t)| = +\infty \tag{E$_1$}$$

(such semi-trajectories are oscillating by definition),

$$(E_2) \qquad \varliminf_{t\to\tau-0} |x(t)| = \varlimsup_{t\to\tau-0} |x(t)| = +\infty.$$

Let us prove that the class (E_2) is empty.

In fact, for all semi-trajectories T^+ of class (E_2) we have

$$\lim_{t\to\tau-0} |x(t)| = +\infty$$

and, consequently, in this case condition (38) implies the existence of a number $K > 0$ such that

$$H\{x(t), y(t)\} - H\{x(0), y(0)\} \equiv \int_{0x} f(x, y)\, dx \equiv$$

$$\equiv \int_0^x f\{x(t), y(t)\}\dot{x}(t)\, dt < K \quad \text{for} \quad t \in [0, \tau).$$

Thus the divergent semi-trajectory T^+ is embedded in the compact set characterized by the relation $H(x, y) < K + H\{x(0), y(0)\}$. $x, y \in \boldsymbol{R}^n$. This contradicts the definition of divergent semi-trajectory. Lemma 2 is proved.

Theorem. *If the conditions* I-III, (2) *and* (38) *are satisfied for every* $g > 0$, *then each semi-trajectory* T^+ *of system* (1) *is either oscillating or stable in the sense of Lagrange.*

Proof. The set of all semi-trajectories T^+ of system (1) that are unstable in the sense of Lagrange is the union of the following three classes:

(A_1) oscillating semi-trajectories, for which we have

$$\varliminf_{t\to\tau-0} |x(t)| < \varlimsup_{t\to\tau-0} |x(t)| = +\infty;$$

(A_2) divergent semi-trajectories, which are, under the present assumptions, also oscillating by Lemma 2;

(A_3) semi-trajectories for which there is a constant $c_0 > 0$, its own for each such, that

$$(39) \qquad \varlimsup_{t\to\tau-0} |x(t)| = c_0 < +\infty$$

and, at the same time,

(40) $$\varliminf_{t \to \tau - 0} |y(t)| \leqslant \varlimsup_{t \to \tau - 0} |y(t)| = +\infty.$$

Let us prove that the latter class is empty.

There exists a number $t_0 \in (0, \tau)$ such that

(41) $$|x(t)| < c_0 + 1 = c \quad \text{for} \quad t \in (t_0, \tau),$$

where c_0 is taken from relation (39). In view of condition I,

$$V\{x(t)\} \leqslant g \quad \text{for} \quad t \in (t_0, \tau),$$

where

$$g = \max_{|x| \leqslant c} V(x).$$

Consequently, beginning with a moment t_0, the semi-trajectories T^+ of class (A_3) are completely embedded in the cylinder defined in the $2n$-dimensional Liénard space $\{x, y\}$ by the inequality (41). On the other hand, owing to (40), they are unbounded. This contradicts Lemma 1. The class (A_3) is empty. The theorem is proved.

It can be shown that there exist systems satisfying the conditions of our Theorem and having oscillating semi-trajectories as well as systems, also satisfying these condition, with semi-trajectories state in the sense of Lagrange.

For instance, the equation

$$\ddot{x} + f(x)\dot{x} + g(x) = 0$$

may have, depending on the choice of the functions $f(x)$ and $g(x)$, only bounded semi-trajectories or only oscillating semi-trajectories. The Van der Pol equation $\ddot{x} + \mu(1 - x^2)\dot{x} + x = 0$ has, for $\mu < 0$, semi-trajectories of both kinds.

REFERENCES

[1] V.V. Nemyckiĭ – V.V. Stepanov, *The qualitative theory of differential equations.* Gostehizdat, 1947 (in Russian).

[2] B.P. Demidovič, *Lectures on the mathematical theory of stability.* Nauka, 1967 (in Russian).

L.V. Guščo

Moscow 117437, ul. Volgina d. 25, k. 2, kv. 81, USSR.

COLLOQUIA MATHEMATICA SOCIETATIS JÁNOS BOLYAI
15. DIFFERENTIAL EQUATIONS, KESZTHELY (HUNGARY), 1975.

ON ASYMPTOTICALLY ORDINARY FUNCTIONAL DIFFERENTIAL EQUATIONS

I. GYŐRI

1. Several authors have investigated the asymptotic behavior of functional differential equations having small retardments. The purpose of these investigations was to find a function or in case of a system, n functions such that the solutions can be asymptotically characterized by these functions. For example, in the articles of V.B. Uvarov [1], E. Kozakievicz [2], [3], Yu.A. Ryabov [4] these functions are solutions of the functional differential equations. On the other hand, K.L. Cooke [5], [6], J. Kato [7] and the author [8] obtained results in which these functions are solutions of a suitable ordinary differential equation.

In this article we introduce the concept of asymptotically ordinary functional differential equation and prove some theorems concerning this new concept. Our results contain some well-known theorems as special cases.

2. The linear space of n-dimensional column vectors x will be denoted by R^n, and $|\cdot|$ denotes the norm.

Denote by $R^n \times R^n$ the linear space of n by n matrices with norm $\|A\| = \sup_{|x|=1,\, x\in R^n} |Ax|$.

In this article we study the linear functional differential equation

$$\text{(LFDE)} \quad \dot{x}(t) = \int_{-r}^{0} [d_s Q(t,s)]x(t+s), \qquad t \geq t_0.$$

We suppose that, for some positive number r, the following conditions are fulfilled:

(A) The function $Q(t;s)$: $[t_0, \infty) \times [-r, 0] \to R^n \times R^n$ is of bounded variation with respect to s on $[-r, 0]$ for every fixed $t \in [t_0, \infty)$, the function $Q(t, 0)$ is continuous, and we have $Q(t, -r) = 0$ on $[t_0, \infty)$.

(B) The function

$$l(t, \varphi) = \int_{-r}^{0} [d_s Q(t,s)]\varphi(s)$$

maps $[t_0, \infty) \times C([-r, 0], R^n)$ continuously into R^n.

A function $x(t)$ is said to be a solution of (LFDE) if $x(t)$ is continuous on $[t_0 - r, \infty)$, differentiable on $[t_0, \infty)$, and satisfies the (LFDE) there.

The solution of (LFDE) belonging to the initial point $t = t_0$ and to the initial function $\varphi \in C = C([t_0 - r, t_0], R^n)$ is the solution $x(t)$ with $x(t) = \varphi(t)$ for $t_0 - r \leq t \leq t_0$. Let us denote this solution by $x(t_0, \varphi)(t)$.

Definition 2.1. The equation (LFDE) will be called asymptotically ordinary on $[t_0, \infty)$, if there exists a function $V(t) \in C([t_0 - r, \infty), R^n \times R^n)$ (to be called asymptotical fundamental system) such that

(a) $\det V(t) \neq 0 \qquad (t_0 - r \leq t < \infty)$,

(b) every solution $x(t)$ of (LFDE) has the form

$$\text{(2.1)} \qquad x(t) = V(t)[c[x] + o(1)], \qquad t \to +\infty,$$

where $c[x]$ is a vector depending on $x(t)$.

(c) for every $c \in R^n$ there exists a solution $x(t)$ of (LFDE) such that $c[x] = c$.

Theorem 2.1. *(LFDE) is asymptotically ordinary if and only if there exists a* $U(t)$ *solution of the matrix differential equation*

$$(2.2) \qquad \dot{U}(t) = \int_{-r}^{0} [d_s Q(t, s)]\, U(t + s)$$

which is continuous on $[t_0 - r, \infty)$ *and has the properties*

(i) $\det U(t) \neq 0$ *for* $t \geqslant T$, *where* $T \geqslant t_0$, *is an appropriate real number;*

(ii) *every solution* $x(t)$ *of (LFDE) can be represented in the form*

$$(2.3) \qquad x(t) = U(t)[c[x] + o(1)], \qquad t_0 \leqslant t \to +\infty,$$

where $c[x]$ *is a vector depending on* $x(t)$.

Proof. Suppose that $U(t)$ is a solution of (2.2) with the properties (i) and (ii). Then it is easy to see that the equation (LFDE) is asymptotically ordinary and $V(t)$ is an asymptotical fundamental system of (LFDE), where $V(t) \equiv U(T)$ for $t_0 - r \leqslant t \leqslant T$ and $V(t) = U(t)$ for $T \leqslant t < \infty$.

To prove necessity, suppose that the equation (LFDE) is asymptotically ordinary on $[t_0, \infty)$ and $V(t)$ is an asymptotical fundamental system. Then, according to Definition 2.1 (c), there exist solutions $x_i(t)$ of (LFDE) with

$$(2.4) \qquad \lim_{t \to +\infty} V^{-1}(t) x_i(t) = e_i, \qquad i = 1, 2, \ldots, n;$$

where the i-th component of e_i is equal to 1, the others are zero. The matrix function $U(t) = (x_1(t), \ldots, x_n(t))$ is continuous on $[t_0 - r, \infty)$ and satisfies the equation (2.2).

Furthermore, from (2.4)

$$(2.5) \qquad \lim_{t \to +\infty} V^{-1}(t) U(t) = (e_1, \ldots, e_n) = I.$$

Therefore we can find a real number $T \geqslant t_0$ such that $\det U(t) \neq 0$ for $t \geqslant T$. According to (2.1) and (2.5), for every solution $x(t)$ of (LFDE) we have

$$\lim_{t\to+\infty} U^{-1}(t)x(t) = \lim_{t\to+\infty} U^{-1}(t)V(t)V^{-1}(t)x(t) = c[x]$$

i.e. $x(t)$ is of the form (2.3). The proof is complete.

Using Theorem 2.1, we are going to present examples for not asymptotically ordinary equations.

Example 2.1. Consider the scalar differential equation

(2.6) $$\dot{u}(t) = -u(t-r), \qquad t \geqslant t_0,$$

where $r > 0$. It is known (see [9]) that for $r > e^{-1}$ every solution of (2.6) is oscillatory on $[t_0, \infty)$. Since (2.6) is a special case of (LFDE), applying Theorem 2.1 with $r > \frac{1}{e}$ it follows that (2.6) is not asymptotically ordinary.

Example 2.2. Consider the example constructed by J.A. Yorke [6]:

(2.7) $$\dot{u}(t) = -u(t-\tau(t)), \qquad t \geqslant t_0,$$

where $t_0 = -2$ and

$$\tau(t) = \begin{cases} t + \sqrt{-2t} & \text{for } -2 \leqslant t \leqslant 0 \\ 0 & \text{for } t \geqslant 0. \end{cases}$$

Every solution $u(t)$ of (2.7) that is continuous on $[-2, \infty)$ can be written in the form

(2.9) $$u(t) = \begin{cases} \alpha \dfrac{t^2}{4} & \text{for } -2 \leqslant t \leqslant 0, \\ 0 & \text{for } t \geqslant 0, \end{cases}$$

where α is a real number. Therefore (2.7) is not asymptotically ordinary.

3. The next theorem is concerned with the existence of a non-singu-

lar solution of the matrix differential equation belonging to the equation (LFDE). This problem is in close connection with the concept introduced above.

Suppose that the equation (LFDE) is scalar and the function $Q(t, s)$ is non-increasing for every fixed t. Then, as it was proved by A. D. Myškis (see [9], Theorem 39), for $-r \cdot \sup_{t \geq t_0} Q(t, 0) < \frac{1}{e}$ there exists a positive solution of (LFDE) on $[t_0 - r, \infty)$. The following theorem is a generalization of A. D. Myškis' result for systems.

Theorem 3.1. *Suppose that*

(i) *the function* $Q(t, s)$: $[t_0, \infty) \times [-r, 0] \to R^n \times R^n$ *satisfies* (A) *and* (B):

(ii) $r \cdot \sup_{t \geq t_0} m(t) < \frac{1}{e}$, *where* $m(t)$ *denotes the total variation of the function* $Q(t, s)$ *on* $-r \leq s \leq 0$ *for every fixed* $t \in [t_0, \infty)$.

Then the solution $U(t)$ *of the equation*

$$(3.1) \qquad \dot{U}(t) = \int_{-r}^{0} [d_s Q(t, s)]\, U(t + s), \qquad t \geq t_0,$$

$$U(t_0 + s) = I, \qquad -r \leq s \leq 0,$$

is a non-singular matrix for every $t \in [t_0 - r, \infty)$. *Furthermore,*

$$(3.2) \qquad \max_{-r \leq s \leq 0} |U(t + s) U^{-1}(t)| < e, \qquad t \geq t_0,$$

and

$$(3.3) \qquad |U(t + s) U^{-1}(t) - I| \leq e^2 \int_{t+s}^{t} m(\tau)\, d\tau$$

for $(t, s) \in [t_0, \infty) \times [-r, 0]$.

Proof. Since $U(t_0) = I$ and $U(t)$ is continuous, $U(t)$ is non-singular in a sufficiently small neighbourhood of t_0 and satisfies (3.2).

Assume now that, for some $\bar{t} \in [t_0, \infty)$, $U(\bar{t})$ is singular. Then, in view of $U(t_0) = I$, the set

$$\text{(3.5)} \qquad \Gamma = \{t;\ t \in [t_0, \infty),\ \max_{-r \leqslant s \leqslant 0} |U(t+s)U^{-1}(t)| \geqslant e\}$$

is not empty and we have $t_1 = \inf \Gamma > t_0$. Therefore the function $U(t)$ is non-singular on $[t_0 - r, t_1]$, and

$$\text{(3.6)} \qquad \begin{aligned} &\max_{-r \leqslant s \leqslant 0} |U(t_1+s)U^{-1}(t_1)| = e, \\ &\max_{-r \leqslant s \leqslant 0} |U(t+s)U^{-1}(t)| < e, \end{aligned}$$

$t_0 \leqslant t \leqslant t_1$. Relation (3.2) yields

$$\text{(3.7)} \qquad \begin{aligned} &\dot{U}(t) = A(t)U(t), \qquad t_0 \leqslant t \leqslant t_1, \\ &U(t_0+s) = I, \qquad -r \leqslant s \leqslant 0, \end{aligned}$$

where

$$\text{(3.8)} \qquad A(t) = \int_{-r}^{0} [d_s Q(t,s)][U(t+s)U^{-1}(t)].$$

From the theory of ordinary differential equations it is known (see [10], Lemma 2.4) that for the solution of equation (3.7) we have

$$\text{(3.9)} \qquad \begin{aligned} &\max_{-r \leqslant s \leqslant 0} |U(t_1+s)U^{-1}(t_1)| = \max_{\bar{t} \leqslant \tau \leqslant t_1} |U(\tau)U^{-1}(t_1)| \leqslant \\ &\leqslant \max_{\bar{t} \leqslant \tau \leqslant t_1} \exp\Big(\int_{\tau}^{t_1} |A(s)|\,ds\Big) \leqslant \exp\Big(\int_{\bar{t}}^{t_1} |A(s)|\,ds\Big), \end{aligned}$$

where $\bar{t} = \max\{t_0, t_1 - r\}$.

From the definition of $A(t)$ by the aid of (3.6) we obtain

$$\text{(3.10)} \qquad |A(t)| \leqslant m(t) \cdot \max_{-r \leqslant s \leqslant 0} |U(t+s)U^{-1}(t)| < e \cdot m(t)$$

for every $t_0 \leqslant t \leqslant t_1$ with $m(t) \neq 0$, If $m(t) \equiv 0$, then $U(t)$ is constant on $[t_0, t_1]$; this case is not interesting. If $m(t) \not\equiv 0$, from (3.9), (3.10) and (ii) we get

$$\max_{-r\leqslant s\leqslant 0} |U(t_1+s)U^{-1}(t_1)| < \exp\left(e\int_{\bar{t}}^{t_1} m(s)\,ds\right) \leqslant e.$$

The latter inequality contradicts (3.6). Thus Γ is empty, $U(t)$ is non-singular on $[t_0, \infty)$ and satisfies (3.2).

To verify (3.3), consider the equation (3.7), valid for $t \geqslant t_0$. Applying a well-known result (see [10]) and taking into account relations (3.8) and (3.10), we find:

$$|U(t+s)U^{-1}(t) - I| \leqslant \exp\left(\int_{t+s}^{t} |A(\tau)|\,d\tau\right) - 1 \leqslant$$

$$\leqslant \exp\left(e\int_{t+s}^{t} m(\tau)\,d\tau\right) - 1 \leqslant e^2 \int_{t+s}^{t} m(\tau)\,d\tau.$$

The proof is complete.

In the next theorems sufficient conditions on an equation to be asymptotically ordinary are given.

Theorem 3.2. *Let the assumptions of Theorem* 3.1 *be fulfilled. Moreover, let* $Q(t_1, s_1)Q(t_2, s_2) = Q(t_2, s_2)Q(t_1, s_1)$ *for* $(t_1, s_1), (t_2, s_2) \in [t_0, \infty) \times [-r, 0]$. *Then the equation is asymptotically ordinary on* $[t_0, \infty)$ *and the solution* $U(t)$ *of* (3.1) *is an asymptotical fundamental system of (LFDE).*

In the proof we will use the following

Lemma 3.1. *Let the assumptions of Theorem* 3.2 *be fulfilled. Then* $Q(t_1, s)$ *and the solution* $U(t)$ *of* (3.1) *are commutative matrices for* $t \in [t_0, \infty)$ *and* $(t_1, s) \in [t_0, \infty) \times [-r, 0]$.

The proof of this lemma is obvious.

Proof of Theorem 3.2. Let $x(t)$ be any solution of (LFDE). By Theorem 3.1, $U(t)$ is non-singular on $[t_0 - r, \infty)$. Hence

$$y(t) = U^{-1}(t)x(t), \qquad t \geqslant t_0 - r,$$

is a solution of the equation

$$(3.11)\qquad \dot{y}(t) = -\int_{-r}^{0}[d_sQ(t,s)]U^{-1}(t)U(t+s)[y(t)-y(t+s)],\qquad t\geq t_0.$$

We have made use of the fact that $U(t)$ and $Q(t_1, s)$ as well as $U^{-1}(t)$ and $Q(t_1, s)$ are commutative matrices.

Put

$$v(t) = \begin{cases} \max\limits_{t_0-r\leq t_1,t_2\leq t_0} |y(t_2)-y(t_1)| + \max\limits_{-r\leq s\leq 0} |y(t_0+s)|, \\ \qquad\qquad t_0 - r \leq t \leq t_0, \\ v(t_0) + \int_{t_0}^{t} |\dot{y}(s)|\, ds, \qquad t\geq t_0. \end{cases}$$

Then $\max\limits_{-r\leq s\leq 0} [y(t)-y(t+s)| \leq v(t)-v(t-r)$, so that by (3.11)

$$\dot{v}(t) \leq m(t)\cdot \max |U^{-1}(t)U(t+s)|\cdot [v(t)-v(t-r)].$$

This inequality and (3.2) yield

$$(3.12)\qquad \dot{v}(t) \leq q[v(t)-v(t-r)] = q\int_{t-r}^{t}\dot{v}(s)\,ds,\qquad t\geq t_0+r,$$

where $q = e\cdot\sup\limits_{t\geq t_0} m(t)$. Leaning on (3.12) it is easy to see that

$$(3.13)\qquad \dot{v}(t) < c\cdot (rq)^{\frac{t}{r}},\qquad t\geq t_0,$$

where $c > 0$ is a suitable real number.

According to (ii), $rq < 1$. Therefore, by (3.13), $\dot{v}(t)$ is bounded. Since $v(t)$ is increasing,

$$\lim_{t\to+\infty} v(t) = v(t_0) + \lim_{t\to+\infty}\int_{t_0}^{t}|\dot{y}(s)|\,ds$$

and

$$(3.14)\qquad \lim_{t\to+\infty} y(t) = \lim_{t\to+\infty} U^{-1}(t)\times(t)$$

exist and are finite.

As $x(t)$ was an arbitrary solution of (LFDE), it follows that (3.14) or the equivalent relation

$$x(t) = U(t)[c[x] + o(1)], \qquad t \to +\infty,$$

is valid, where $c[x]$ depends on $x(t)$ but not on t. Using Theorem 2.1, the proof is complete.

Theorem 3.3. *Let the assumptions of Theorem 3.2 be fulfilled. Let*

$$(3.15) \qquad \int_{t_0+r}^{\infty} \Big\{ \int_{-r}^{0} \Big[\int_{t+s}^{t} m(\tau)\, d\tau \Big] ds\, m(t, s) \Big\} dt < \infty,$$

where $m(t, s)$ *denotes the total variation of the function* $Q(t, \tau)$ *on* $-r \leqslant \tau \leqslant s$ *for every fixed* $(t, s) \in [t_0, \infty) \times [-r, 0]$, *and* $m(t) = m(t, 0)$.

Then the equation (LFDE) is asymptotically ordinary on $[t_0, \infty)$, *and the solution of the problem*

$$(3.16) \qquad \begin{aligned} &\dot{V}(t) = Q(t, 0)\, V(t), && t_0 \leqslant t, \\ &V(t_0 + s) = 1, && -r \leqslant s \leqslant 0, \end{aligned}$$

is an asymptotical fundamental system of (LFDE).

Proof. Consider the solution $U(t)$ of (3.1) and the function

$$(3.17) \qquad T(t) = U^{-1}(t)\, V(t), \qquad t \geqslant t_0 - r.$$

Since $Q(t, s)\, U^{-1}(t_1) = U^{-1}(t_1)\, Q(t, s)$, from (3.1) and (3.16) we get $T(t_0) = I$ and

$$(3.18) \qquad \dot{T}(t) = \int_{-r}^{0} [d_s Q(t, s)][U^{-1}(t)\, U(t + s) - I]\, T(t).$$

In view of (3.3) we have

$$\Big| \int_{-r}^{0} [d_s Q(t, s)][U^{-1}(t)\, U(t + s) - I] \Big| \leqslant$$

$$\leqslant e^2 \int_{-r}^{0} \Big[\int_{t+s}^{t} m(\tau)\, d\tau \Big] d_s m(t, s),$$

i.e. according to (3.15)

$$\int_{t_0+r}^{\infty} \Big| \int_{-r}^{0} [d_s Q(t, s)][U^{-1}(t)\, U(t+s) - I] \Big|\, dt < \infty.$$

Then (see [10]), for the solution $T(t)$ of (3.18) the limit

$$P = \lim_{t \to +\infty} T^{-1}(t) = \lim_{t \to +\infty} V^{-1}(t)\, U(t)$$

exists and is nonsingular. By Theorem 3.2, for every solution $x(t)$ of (LFDE) we obtain

$$x(t) = U(t)[c[x] + o(1)] = V(t)[Pc[x] + o(1)], \qquad t \to +\infty.$$

Hence, applying the nonsingularity of P, we deduce that $V(t)$ is an asymptotical fundamental system of (LFDE). The proof is complete.

Apply now the theorems to the equation

$$\dot{x}(t) = Ax(t - \tau(t)), \qquad t \geqslant t_0, \tag{3.19}$$

where A is an n by n constant matrix, $\tau(t)$ is a continuous function on $[t_0, \infty)$, and r is a positive constant.

Put $Q(t, s) = Ae(s + \tau(t))$, where $e(s) = 0$ $(s < 0)$, $e(s) = 1$ $(s \geqslant 0)$. Then the equation (LFDE) turns into the equation (3.19), so that according to Theorems 3.2 and 3.3 we obtain:

Corollary 3.1. Assuming that $|A| \cdot r < \frac{1}{e}$, equation (3.19) is asymptotically ordinary on $[t_0, \infty)$ and the solution of the equation

$$\dot{U}(t) = AU(t - \tau(t)), \qquad t \geqslant t_0,$$

$$U(t_0 + s) = I, \qquad -r \leqslant s \leqslant 0,$$

is an asymptotical fundamental system of (3.19). Furthermore, if (3.20)

$$\int_{t_0}^{\infty} \tau(t)\, dt < \infty, \tag{3.20}$$

then also the function

$$V(t) = e^{At}$$

is an asymptotical fundamental system of (LFDE).

Remark 3.1. If $A = [a]$ and $r|a| < \frac{1}{e}$, where a is a real number, then under condition (3.20) we obtain a theorem of K.L. Cooke [5].

Remark 3.2. If $A = [-1]$ and $\tau(t) = r$ $(t_0 \leqslant t)$, then from Corollary 3.1 we obtain that for $r < \frac{1}{e}$ the equation

$$\dot{x}(t) = -x(t-r), \qquad t \geqslant t_0,$$

is asymptotically ordinary on $[t_0, \infty)$.

Example 2.1 shows that this equation is not asymptotically ordinary for $r > \frac{1}{e}$.

Remark 3.3. As Example 2.2 shows, the condition $\int_{t_0}^{\infty} \tau(t)\, dt < \infty$ is not sufficient for equation (3.19) to be asymptotically ordinary on $[t_0, \infty)$.

Remark 3.4. If (LFDE) is a scalar equation and $Q(t, s)$: $[t_0, \infty) \times [-r, 0] \to R$ is non-decreasing for fixed t, then in Theorems 3.1-3.3 it is not necessary to make the assumption involving the bound $\frac{1}{e}$. Results of this type can be found e.g. in [3], [8] and [9].

REFERENCES

[1] V.B. Uvarov, Asymptotic properties of the energy distribution of neutrons, absorbed in an infinite medium, *Žurnal Vyč. Mat.*, 7 (1967), 836-851 (in Russian).

[2] E. Kozakiewicz, Über das asymptotische Verhalten der nichtschwindgenden Lösungen einer linearen Differentialgleichung mit nacheilendem Argument. *Wiss. Z. Humboldt Univ. Berlin, Math.-Nat. R.*, 13:4 (1964), 577-589.

[3] E. Kozakiewicz, Über lineare Differential-Functional-relationen erster Ordnung vom instabilen. *Typ. Math. Nachr.*, 40: 1-3 (1969), 61-78.

[4] Yu.A. Ryabov, An application of the small parameter method of Lyapunov – Poincaré in the theory of retarded systems, *Inž. Žurn.*, 1: 2 (1961), 3-15 (in Russian).

[5] K.L. Cooke, *Functional-differential equations: Some models and perturbation problems, in differential equations and dynamical systems* (J.K. Hale and J.P. La Salle, eds.) Academic Press, New York (1967), 167-183.

[6] K.L. Cooke, Functional differential equations with asymptotically vanishing lag, *Rendiconti del Circelo Matematico di Palermo*, 16 (1967), 39-55.

[7] J. Kato, On the existence of a solution approaching zero for functional differential equations, *Proceedings United States – Japan Seminar on differential and functional equations, W.A. Benjamin, INC., New York, Amsterdam* (1967), 153-169.

[8] I. Győri, Asymptotic behaviour of solutions of unstable-type first-order differential equations with delay, *Stud. Sci. Math. Hungarica,* 8 (1973), 125-132.

[9] A.D. Myškis, *Linear differential equations with retarded argument,* Nauka, Moscow (1972).

[10] Yu.L. Deleckiĭ – M.G. Kreĭn, *Stability of solutions of differential equations in Banach spaces,* Nauka, Moscow (1970).

I. Győri

6720 Szeged, Somogyi Béla út 4, Hungary.

COLLOQUIA MATHEMATICA SOCIETATIS JÁNOS BOLYAI
15. DIFFERENTIAL EQUATIONS, KESZTHELY (HUNGARY), 1975.

A CONDITION FOR NON-ASYMPTOTICAL STABILITY

L. HATVANI – A. KRÁMLI

0. A generalized conservative mechanical system (i.e. a mechanical system having time-independent Hamiltonian function) can not have asymptotically stable equilibrium position. This follows from the fact that the Hamiltonian function is constant along the phase trajectories (cf. [1]).

In this paper there is given a sufficient condition for non-attractivity of the zero solution of a general non-autonomous differential equation. From this result it follows that a general non-stationary Hamiltonian system can not have asymptotically stable equilibrium position. Further we study the perturbations, under which a general Hamiltonian system preserves this property.

1. For any column vector $x = \operatorname{col}(x_1, x_2, \dots, x_n)$ in the Euclidean n-space R^n we denote by $\|x\|$ the norm of x.

Consider the differential equation

$$(1.1) \qquad \dot{x} = f(t, x),$$

where $t \in I = [0, \infty)$; the function $f(t, x)$, defined on the set

$$\Gamma = \{(t, x):\ t \in I,\ \|x\| < H\} \qquad (0 < H \leqslant \infty),$$

is continuous and has continuous first partial derivatives with respect to the components of x.

Denote by $x(t; t_0, x_0)$ the unique solution of (1.1) passing through the point (t_0, x_0). Assume that this solution exists for all $t \geqslant t_0$, provided that $\|x_0\|$ is sufficiently small. In addition, $f(t, 0) \equiv 0$ for $t \in I$, so that $x = 0$ is a solution of (1.1), called the zero solution.

The zero solution of (1.1) is said to be

(i) *stable* (in the sense of Lyapunov): if given any $\epsilon > 0$ and any $t_0 \in I$ there exists a $\delta(\epsilon, t_0) > 0$ such that $\|x_0\| < \delta$ implies $\|x(t; t_0, x_0)\| < \epsilon$ for $t \geqslant t_0$.

(ii) *attractive:* if given any $t_0 \in I$ there exists a $\sigma(t_0) > 0$ such that $\|x_0\| < \sigma$ implies $\|x(t; t_0, x_0)\| \to 0$ as $t \to \infty$.

(iii) *asymptotically stable:* if it is stable and attractive.

For fixed t_0 and t $(t_0 \in I,\ t \geqslant t_0)$ the function $x(t; t_0, x_0)$ defines a diffeomorphism of a certain ball $S_\lambda = \{x:\ \|x\| < \lambda\}$ into R^n. Let the Jacobian determinant of $x(t; t_0, x_0)$ be denoted by

$$J(x_0; t_0, t) = \frac{D(x_1(t; t_0, x_0), \ldots, x_n(t; t_0, x_0))}{D(x_{01}, \ldots, x_{0n})}.$$

2. In the proof of the main theorem the following slight generalization of the well known Liouville's formula [2] will be needed:

Lemma 2.1. *The Jacobian determinant* $J(x_0; t_0, t)$ *can be given by the expression*

$$J(x_0; t_0, t) = \exp\Big[\int_{t_0}^{t} \Big(\sum_{i=1}^{n} \Big(\frac{\partial f_i(s, x)}{\partial x_i}\Big)_{x = x(s; t_0, x_0)}\Big)\, ds\Big].$$

Proof. It is known [2] that the vector $\dfrac{\partial x(t; t_0, x_0)}{\partial x_{0j}}$ $(1 \leqslant j \leqslant n)$, as function of t, is the solution of the initial value problem

$$\dot{z}_i = \sum_{k=1}^{n} \left(\frac{\partial f_i}{\partial x_k}\right)_{x=x(t;\, t_0,\, x_0)} z_k;\; z_i(t_0) = \delta_{ij} \qquad (i = 1, 2, \ldots, n);$$

hence

$$\left(\frac{\partial}{\partial t} J(x_0;\, t_0,\, t)\right)_{t=t_0} = \sum_{i=1}^{n} \left(\frac{\partial f_i(t_0, x)}{\partial x_i}\right)_{x=x_0}. \tag{2.1}$$

Taking into consideration the rule

$$J(x_0;\, t_0,\, t) = J(x(s;\, t_0,\, x_0);\, s,\, t) J(x_0;\, t_0,\, s) \qquad (t \geqslant t_0,\;\; s \geqslant t_0),$$

from (2.1) we obtain the differential equation

$$\left(\frac{\partial}{\partial t} J(x_0;\, t_0,\, t)\right)_{t=s} =$$

$$= \sum_{i=1}^{n} \left(\frac{\partial f_i(s, x)}{\partial x_i}\right)_{x=x(s;\, t_0,\, x_0)} \cdot J(x_0;\, t_0,\, s)$$

for $s \geqslant t_0$. The integration of this equation gives the statement of the lemma.

Introduce the notations

$$x(t;\, t_0,\, F) = \{x(t;\, t_0,\, x_0):\; x_0 \in F \subset R^n\},$$

$$D_t = x(t;\, 0,\, \{x_0:\; \| x_0 \| \leqslant k\}) \qquad (t \in I),$$

where k is a sufficiently small positive number such that the solution $x(t;\, t_0,\, x_0)$ of (1.1) exists for all $t \in I$ provided that $\| x_0 \| \leqslant k$.

Theorem 2.1. *Assume that*

$$\limsup_{t \to \infty} \int_0^t \min \left\{ \sum_{i=1}^{n} \frac{\partial f_i(s, x)}{\partial x_i} :\; x \in D_s \right\} ds > -\infty. \tag{2.2}$$

Then the zero solution of (1.1) *is not attractive, moreover the Lebesgue measure* $\mu(E)$ *of the set*

$$E = \{x_0:\; \| x_0 \| \leqslant k,\; \lim_{t \to \infty} \| x(t;\, 0,\, x_0) \| = 0\}$$

is equal to 0.

Proof. By assumption (2.2) there exists a sequence $\{t_m\}$ and a constant K such that $t_m \to \infty$ as $m \to \infty$, and for every measurable subset F of D_0 we have

$$\mu(x(t_m; 0, F)) = \int_{x(t_m; 0, F)} \cdots \int dx_1 \ldots dx_n =$$

$$= \int_F \cdots \int J(x_0; 0, t_m)\, dx_{01} \ldots dx_{0n} \geqslant e^K \mu(F) \tag{2.3}$$

$$(m = 1, 2, \ldots).$$

Let us now suppose that the statement of the theorem is not true, i.e. $\mu(E) > 0$. Then by Egorov's theorem (see [3]) there exists a measurable

$$E^* \subset E \quad \text{such that} \quad \mu(E^*) > \frac{\mu(E)}{2} \quad \text{and} \quad \| x(t_m; 0, x_0) \| \to 0$$

as $m \to \infty$ uniformly with respect to $x_0 \in E^*$. Hence for every $\eta > 0$ there exists a number $n(\eta)$ such that the set $x(t_{n(\eta)}; 0, E^*)$ is contained in a ball G_η of measure less than η. Introducing the notation $G_\eta^{-1} = x(0; t_{n(\eta)}, G_\eta)$, from (2.3) we get the estimation

$$\eta \geqslant \mu(x(t_{n(\eta)}; 0, G_\eta^{-1})) \geqslant e^K \mu(G_\eta^{-1}) > e^K \frac{\mu(E)}{2},$$

which contradicts to the fact that η is arbitrarily small.

The theorem is proved.

Consider a general non-autonomous Hamiltonian system

$$\dot{q}_i = \frac{\partial H(t, q, p)}{\partial p_i}, \quad \dot{p}_i = -\frac{\partial H(t, q, p)}{\partial q_i} \quad (i = 1, 2, \ldots, n), \tag{2.4}$$

where the function $H(t, q, p)\colon I \times R^n \times R^n \to R$ is continuous and has cotninuous second partial derivatives with respect to q_i, p_j. Assume further that $q = p = 0$ is an equilibrium position of (2.4).

Corollary 2.1. *The equilibrium position of the Hamiltonian system (2.4) can not be attractive, and therefore it can not be asymptotically stable.*

3. Small oscillation of a dynamical system about the equilibrium position is described in normal coordinates by the Lagrangian equations

$$\ddot{q}_i + \lambda_i q_i = 0 \qquad (i = 1, 2, \ldots, n),$$

where the frequencies λ_i are constant. Kelvin proved (see [4], pages 103-104) that if the equilibrium position $q = 0$ is stable (i.e. $\lambda_i > 0$ for all $i = 1, 2, \ldots, n$), then by addition of definite dissipative forces it becomes asymptotically stable, i.e. the solution $q = 0$ of the equations

$$\ddot{q}_i + \lambda_i q_i = -\frac{\partial}{\partial \dot{q}_i} R(\dot{q}) \qquad (i = 1, 2, \ldots, n)$$

is asymptotically stable provided that the dissipation function $R(\dot{q}) = \sum_{i,j=1}^{n} b_{ij}\dot{q}_i\dot{q}_j$ ($b_{ij} = \text{const}$) is a positive definite quadratic form. On the other hand, if the additional dissipative forces are not definite (i.e. the quadratic form $R(\dot{q})$ is positive semidefinite), then the equilibrium position does not become asymptotically stable.

Let us now consider a non-stationary mechanical system described by the Lagrangian equations

$$(3.1) \qquad \frac{d}{dt}\frac{\partial T}{\partial \dot{q}_i} - \frac{\partial T}{\partial q_i} = -\frac{\partial V}{\partial q_i} + \sum_{j=1}^{n} g_{ij}\dot{q}_j - \frac{\partial R}{\partial \dot{q}_i} \qquad (i = 1, 2, \ldots, n),$$

where we used the following notations: $V = V(t, q)$ is the potential energy ($V(t, 0) \equiv 0$), the symmetrical quadratic form $T = T(t, q, \dot{q}) = \frac{1}{2}\sum_{i,j=1}^{n} a_{ij}(t, q)\dot{q}_i\dot{q}_j = \frac{1}{2}(\mathscr{A}(t, q)\dot{q}, \dot{q})$ is the kinetic energy of the system; the dissipation function $R = R(t, q, \dot{q}) = \frac{1}{2}\sum_{i,j=1}^{n} b_{ij}(t, q)\dot{q}_i\dot{q}_j = \frac{1}{2}(\mathscr{B}(t, q)\dot{q}, \dot{q})$ is symmetrical and positive semidefinite. ($\mathscr{A}(t, q)$ and $\mathscr{B}(t, q)$ are symmetrical $n \times n$ matrices, $[g_{ij}(t, q)]_{i,j=1,\ldots,n}$ is an antisymmetrical one.)

The first statement of Kelvin's theorem was extended to systems of type (3.1) by V. Matrosov [5]. He proved that under certain restrictions the equilibrium position of (3.1) is uniformly asymptotically stable provided that $R(t, q, \dot{q})$ is positive definite. In the following theorem we generalize the second statement of Kelvin's theorem.

Denote by $\mathscr{M}^{-1}$ the inverse, and by $\mathrm{tr}\,[\mathscr{M}]$ the trace of a $n \times n$-matrix $\mathscr{M}$.

Theorem 3.1. *Suppose that the equilibrium position* $q = 0$ *of system* (3.1) *is stable with respect to* q *(see* [6]*).*

If there exists a constant $r > 0$ *such that*

$$\liminf_{t\to\infty} \int_0^t \max\{\mathrm{tr}\,[\mathscr{B}(s, q)\mathscr{A}^{-1}(s, q)] : \| q \| \leqslant r\}\, ds < \infty, \tag{3.2}$$

then the equilibrium position is not asymptotically stable, i.e. for every $\epsilon > 0$ *and every* $t_0 \in I$ *there are* $q_0, \dot{q}_0$ *such that* $\| q_0 \|^2 + \| \dot{q}_0 \|^2 < \epsilon^2$ *and*

$$\limsup_{t\to\infty} (\| q(t; t_0, q_0, \dot{q}_0)\|^2 + \|\dot{q}(t; t_0, q_0, \dot{q}_0)\|^2) > 0.$$

Proof. Introducing the conjugate momenta $p_i = \dfrac{\partial T}{\partial \dot{q}_i}$ and the Hamiltonian function $H(t, q, p) = T + V = \frac{1}{2}(\mathscr{A}^{-1}(t, q)p, p) + V(t, q)$, system (3.1) can be written in the canonical form

$$\begin{aligned} \dot{q}_i &= \frac{\partial H}{\partial p_i} \\ \dot{p}_i &= -\frac{\partial H}{\partial q_i} + \sum_{j=1}^n g_{ij} \frac{\partial H}{\partial p_j} - \sum_{j=1}^n b_{ij} \frac{\partial H}{\partial p_j} \qquad (i = 1, 2, \ldots, n). \end{aligned} \tag{3.3}$$

It is sufficient to show that (3.3) satisfies assumption (2.2) of Theorem 2.1.

Since $[g_{ij}(t, q)]$ is antisymmetrical we have

$$\begin{aligned} &\sum_{i=1}^n \frac{\partial^2 H}{\partial p_i \partial q_i} - \sum_{i=1}^n \frac{\partial^2 H}{\partial q_i \partial p_i} + \sum_{i,j=1}^n g_{ij} \frac{\partial^2 H}{\partial p_i \partial p_j} - \sum_{i,j=1}^n b_{ij} \frac{\partial^2 H}{\partial p_i \partial p_j} = \\ &= -\sum_{i,j=1}^n b_{ij} \frac{\partial^2 H}{\partial p_i \partial p_j} = -\,\mathrm{tr}\,[\mathscr{B}(t, q)\,\mathscr{A}^{-1}(t, q)]. \end{aligned} \tag{3.4}$$

The equilibrium position $q = 0$ of (3.1) is stable with respect to q, and therefore for $r > 0$ there exists a $k > 0$ such that $\| q_0 \|^2 + \| \dot{q}_0 \|^2 <$

$< k^2$ implies $\| q(t; 0, q_0, \dot{q}_0) \| < r$. Quantity (3.4) is independent of p; hence we can replace the set D_s in (2.2) by $\{q\colon q \in R^n, \ \|q\| \leqslant r\}$ for every $s \in I$.

Thus (3.2) guarantees that the assumption (2.2) holds for (3.3). The application of Theorem 2.1 completes the proof.

REFERENCES

[1] F. Gantmacher, *Lectures in analytical mechanics,* Mir Publishers, Moscow, 1970.

[2] E.A. Coddington – N. Levinson, *Theory of ordinary differential equations,* McGraw-Hill Book Co., New York – Toronto – London, 1955.

[3] B.Sz.-Nagy, *Introduction to real functions and orthogonal expansions,* Oxford Univ. Press, New York and Akadémiai Kiadó, Budapest, 1965.

[4] N. Chetayev, *Stability of motion,* Russian; Moscow, 1955.

[5] V. Matrosov, On the stability of motion, *Prikl. Mat. Meh.*, 26 (1962), 885-895, Russian; transl. *Appl. Math. Mech.*, 26, 1337-1353.

[6] A. Oziraner – V. Rumyancev, Method of Lyapunov's functions for the problem of stability of motion with respect to a part of variables, *Prikl. Mat. Meh.*, 36 (1972), 364-384 (Russian).

[7] W. Hahn, *Stability of motion,* Springer Verlag, Berlin – Heidelberg – New York, 1967.

[8] J.M. Bownds, Stability implications on the asymptotic behavior of second order differential equations, *Proc. Amer. Math. Soc.*, 39 (1973), 169-172.

L. Hatvani

University of Szeged, Institute Bolyai, H-6720 Szeged, Aradi vértanúk tere 1, Hungary.

A. Krámli

Hungarian Academy of Sciences, Computing Center, H-1111 Budapest, Kende u. 13-17, Hungary.

COLLOQUIA MATHEMATICA SOCIETATIS JÁNOS BOLYAI
15. DIFFERENTIAL EQUATIONS, KESZTHELY (HUNGARY), 1975.

ON A TWO-SIDED ITERATIVE METHOD

J. HEGEDŰS

Since 1919, when the key-work of S. A. Čapligin was published (see [1]), several authors constructed function-sequences approaching from two sides to the solutions of initial-value problems for differential equations (see [6], [7] and their bibliographies).

Sometimes we need such approximations $\{z_i(x)\}$ and $\{w_i(x)\}$, which have the property (A):

$$\text{(A)} \qquad \begin{cases} w_1^{(s)}(x) \leqslant w_2^{(s)}(x) \leqslant \ldots \leqslant y^{(s)}(x) \leqslant \ldots \leqslant z_2^{(s)}(x) \leqslant z_1^{(s)}(x) \\ w_i^{(s)}(x), \ z_i^{(s)}(x) \longrightarrow y^{(s)}(x) \end{cases}$$

where $y(x)$ is the solution of the problem for the differential equation, under consideration $a \leqslant x \leqslant b$; $s = 0, \ldots, n-1$ (n is the order to the equation).

Although such approximations with property (A) are known for some special initial-value and boundary-value problems (see [2]) and the References of [3]), but for more general problems only a few construction was given.

In this article we shall study the equation

$$(1.1) \qquad y = \mathscr{A} y$$

on a partially ordered, metric, complete set $\mathscr{M}$, where $\mathscr{A}$ is a given operator mapping $\mathscr{M}$ into itself in a contractive manner, i.e. for every $h, g \in \mathscr{M}$ we have $\mathscr{A} h, \mathscr{A} g \in \mathscr{M}$ and

$$\rho(\mathscr{A} h, \mathscr{A} g) \leqslant \theta \rho(h, g) \qquad (0 < \theta < 1,\ \theta = \text{const.}).$$

On the metric (ρ) and the ordering $(\leqslant)$ we assume that they are compatible in the following sense:

$$h \leqslant g \leqslant l \Rightarrow \rho(h, g), \rho(g, l) \leqslant \rho(h, l) \qquad (h, g, l \in \mathscr{M})$$

Finally, suppose that for every monotone sequence $\{h_p\}$ from $\mathscr{M}$ tending in metric to a given $h \in \mathscr{M}$, h_p and h are in ordering relation (from $\{h_p\} \searrow$ or $\{h_p\} \nearrow$ it follows obviously, that $h_p \geqslant h$, $h_p \leqslant h$ resp.).

1.a. THE CASE OF MONOTONE INCREASING OPERATORS

Suppose that for arbitrary two elements $h \leqslant g$ of $\mathscr{M}$ we have $\mathscr{A} h \leqslant \mathscr{A} g$. In the sequel denote by y the solution of the equation (1.1).

Theorem 1. *If there exist two elements* $z_1, w_1 \in \mathscr{M}$ *such that*

(i) $\mathscr{A} z_1 \leqslant z_1$, $w_1 \leqslant \mathscr{A} w_1$

then the sequences $\{z_p\}, \{w_p\}$ *given by*

$$z_{p+1} = \mathscr{A} z_p, \quad w_{p+1} = \mathscr{A} w_p \qquad (p = 1, 2, \ldots)$$

have the property (B)

$$\text{(B)} \qquad \begin{cases} w_1 \leqslant w_2 \leqslant \ldots \leqslant y \leqslant \ldots \leqslant z_2 \leqslant z_1 \\ z_p, w_p \xrightarrow{\rho} y. \end{cases}$$

Let us mention a few applications of the theorem.

I. *Initial-value problem for a retarded differential equations:*

$$y^{(n)}(x) = f[y] \equiv f(x, y(x), \ldots, y^{(n-1)}(x); y(g_0(x)), \ldots$$

$$\ldots, y^{(n-1)}(g_{n-1}(x))) \qquad (0 \leqslant x \leqslant 1;\ n \geqslant 2),$$

$$y(0) = \ldots = y^{(n-1)}(0) = 0, \qquad y|_E = Q(x),$$

where $f(x, u_0, \ldots, u_{n-1}; v_0, \ldots, v_{n-1})$ is a continuous function and

$$\frac{\partial f}{\partial u_i}, \frac{\partial f}{\partial v_i} \geqslant 0 \qquad (i = 0, \ldots, n-1).$$

II. *Boundary-value problem for a retarded differential equation:*

$$y^{(n)}(x) = f[y] \quad (0 \leqslant x \leqslant 1;\ n \geqslant 2), \quad y|_E = Q(x),$$

$$L_i y = \sum_{k=0}^{n-1} (a_{ik} y^{(k)}(0) + b_{ik} y^{(k)}(1)) = 0 \qquad (i = 0, \ldots, n-1)$$

(the constants a_{ik}, b_{ik} are such that the problem

$$y^{(n)}(x) = 0 \quad (0 \leqslant x \leqslant 1,\ n \geqslant 2), \quad L_i y = 0 \quad (i = 0, \ldots, n-1)$$

has only the trivial solution $y \equiv 0$) if the Green function corresponding to the problem and the function f satisfy

$$\frac{\partial^i G(x,t)}{\partial x^i} \leqslant 0; \quad \frac{\partial f}{\partial u_i}, \frac{\partial f}{\partial v_i} \leqslant 0 \qquad (i = 0, \ldots, n-1).$$

We assume that the continuous functions $g_i(x) \leqslant x$ are such that the boundary-value problem has solution in the usual sense (see e.g. [3]).

II′. *A special case of* II.

It is taken from [3], where

$$L_i y = y^{(i)}(0) \quad (i \leqslant n-2), \quad L_{n-1} y = y^{(n-1)}(1),$$

$$\frac{\partial^i G(x,t)}{\partial x^i} \leqslant 0; \quad \frac{\partial f}{\partial u_i}, \frac{\partial f}{\partial v_i} \leqslant 0 \qquad (i = 0, \ldots, n-1).$$

III. *Initial-value or boundary-value problem for a functional-differential equation*

$$y^{(n)}(x) = f[y] \equiv f(x, y(x), \dots, y^{(n-1)}(x); g_{0,x}(y), \dots,$$

$$\dots, g_{n-1,x}(y)) \qquad (0 \leqslant x \leqslant 1;\ n \geqslant 2), \quad y|_E = Q(x),$$

$$y_i(0) = 0 \quad \text{or} \quad L_i y = 0 \qquad (i = 0, \dots, n-1)$$

where $\dfrac{\partial f}{\partial u_i}, \dfrac{\partial f}{\partial v_i} \geqslant 0$ $(i = 0, \dots, n-1)$, the $g_{i,x}$ are continuous, monotone non-decreasing functionals on $C^{n-s}[\lambda, 1]$ (for initial-value problem $s = 1$, for boundary-value problem $s \geqslant 2$, $[\lambda, 0] = E$, $\lambda < 0$ constant), namely in the case of boundary-value problem we suppose that

$$\frac{\partial^i G(x, t)}{\partial x^i} \geqslant 0 \qquad (i = 0, \dots, n-1).$$

$E = [\lambda, 0]$ denotes always the initial set. The domain of the function $f(x, u_0, \dots, u_{n-1}; v_0, \dots, v_{n-1})$ is either the prism

$$\mathscr{D}_K\colon\ 0 \leqslant x \leqslant 1, \quad |u_i| \leqslant K, \quad |v_i| \leqslant K \qquad (i = 0, \dots, n-1)$$

(in this case Q and $g_{i,x}$ can not be arbitrary (see e.g. [3])) and let $|f| \leqslant$ $\leqslant L(K)$ be (where the constant $L(K)$ is so small, that $\mathscr{A}h \in \mathscr{M}$ for every $h \in \mathscr{M}$) * or the prism.

$$\mathscr{D}_\infty\colon\ 0 \leqslant x \leqslant 1, \quad |u_i| < \infty, \quad |v_i| < \infty \qquad (i = 0, \dots, n-1).$$

We assume the continuity of the function f as well as the existence and continuity of the first partial derivatives with respect to u_i and v_i and

$$\left|\frac{\partial f}{\partial u_i}\right|, \left|\frac{\partial f}{\partial v_i}\right| \leqslant N \qquad (i = 0, \dots, n-1),$$

where the constant N is so small, that the operator $\mathscr{A}$ corresponding to the problem would be a contraction.

Problems I, . . . , III are equivalent to the problem

*For problem II′ $L(K) = K$.

$$y(x) = \mathscr{A} y = \begin{cases} Q(x), & x \in E, \\ \int_0^1 G(x, t) f[y(t)]\, dt, & x \in [0, 1], \end{cases}$$

in the space of sufficiently smooth functions, where $G(x, t)$ is the Green function of the boundary-value problem, or in case of initial-value problem

$$G(x, t) = \begin{cases} \dfrac{(x-t)^{n-1}}{(n-1)!} & t \leqslant x, \\ 0 & t \geqslant x. \end{cases}$$

We note that several more general problems may be reduced to I, . . . , III; by the transforming the variable x and the unknown y and that Theorem 1 as well as the following theorems may be applied also for ordinary differential equations.

For initial-value problem we define $\mathscr{M}$ as follows:

$$\mathscr{M} = \{z(x)\colon\ z \in C^{n-1}[\lambda, 1],\ z^{(i)}(0) = Q^{(i)}(0) = 0,\ z|_E = Q\}$$

$$(i = 0, \ldots, n-1)$$

if the domain of f is the prism $\mathscr{D}_\infty$, in the case of $\mathscr{D}_K$ put

$$\mathscr{M} = \{z(x)\colon\ z \in C^{n-1}[\lambda, 1],\ z^{(i)}(0) = Q^{(i)}(0) = 0,$$

$$z|_E = Q,\ |z^{(i)}| \leqslant K\} \qquad (i = 0, \ldots, n-1).$$

For boundary-value problems we can not require that for $z \in \mathscr{M}$ $z^{(i)}(0) = Q^{(i)}(0) = 0$ holds for every i and we can replace the condition $z \in C^{n-1}[\lambda, 1]$ with conditions

$$z \in C^{n-s}[\lambda, 1] \quad (s > 1), \quad z|_E \in C^{n-1}[\lambda, -0],$$

$$z|_{[0,1]} \in C^{n-1}[+0, 1].$$

For $z, w \in \mathscr{M}$ we can define the ordering by the formula

$$z \leqslant w \Leftrightarrow z^{(i)}(x) \leqslant w^{(i)}(x) \qquad (0 \leqslant x \leqslant 1;\ i = 0, \ldots, n-1)$$

and the metric by

$$\rho(z, w) = \| z - w\| = \sum_{i=0}^{n-1} \max_{[0,1]} | z^{(i)}(x) - w^{(i)}(x)|.$$

A sufficient condition for the contractivity of the operator $\mathscr{A}$ corresponding to the problems I, II, II′ is for instance the following

$$N\Big\{\sum_{i=0}^{n-1} \Big[\max_x \int_0^1 \Big| \frac{\partial^i G(x, t)}{\partial x^i}\Big| dt\Big] +$$

$$+ \max_j \sum_{i=0}^{n-1} \max_x \Big[\int_{M_j} \Big| \frac{\partial^i G(x, t)}{\partial x^i}\Big| dt\Big]\Big\} < 1$$

where

$$M_j = \{x\colon\ g_j(x) > 0\} \qquad (j = 0, \ldots, n-1).$$

Under similar conditions also the operator $\mathscr{A}$ of problem III is a contraction.

Remark 1.1. Instead of (i) in the Theorem 1, it is sufficient to require the existence of two elements $z_1, w_1 \in \mathscr{M}$ having the property that with suitable natural numbers k, l

$$\mathscr{A}^l z_1 \leqslant z_1, \quad \mathscr{A}^k w_1 \geqslant w_1.$$

In this case the sequences $\{z_{pl+1}\}$, $\{w_{pk+1}\}$ have the property (B).

Remark 1.2. If there exist maximal and minimal element in $\mathscr{M}$, then these obviously satisfy (i).*

Remark 1.3. If the operator $\mathscr{A}$ and the ordering in $\mathscr{M}$, can be extended to $M \supset \mathscr{M}$ such that $\mathscr{A}$ remains monotone increasing in M and $\mathscr{A} M \subseteq \mathscr{M}$, then in certain cases it is reasonable to look for z_1 and w_1 among the elements of M.

For example in problem II′ (for $\mathscr{D}_K$) if $f[y] \equiv f(y)$, $n = 2$ ($\frac{\partial f}{\partial y}$ may change sign) and the ordering in $\mathscr{M}$ is defined by

$$z \leqslant w \Leftrightarrow z(x) \leqslant w(x) \qquad (0 \leqslant x \leqslant 1),$$

*z is maximal element if $z \geqslant m$ for every $m \in \mathscr{M}$.

then for M we can choose the set

$$\{z(x):\ z \in C[0, 1],\ |z(x)| \leqslant K\}$$

and the functions $z_1 = K$, $w_1 = -K$ obviously satisfy condition (i).

Remark 1.4. In problem II′ in the case of $\mathscr{D}_K$ the functions

$$z_1(x) = \begin{cases} Q(x), & x \in E, \\ K\left(\dfrac{x^{n-1}}{(n-1)!} - \dfrac{x^n}{n!}\right), & 0 \leqslant x \leqslant 1, \end{cases}$$

$$w_1(x) = \begin{cases} Q(x), & x \in E, \\ -K\left(\dfrac{x^{n-1}}{(n-1)!} - \dfrac{x^n}{n!}\right), & 0 \leqslant x \leqslant 1, \end{cases}$$

satisfy the condition (i) (the monotonicity of $\mathscr{A}$ is not required), for the prism $\mathscr{D}_\infty$ the functions

$$z_1^{(n)}(x) = f[z_1] + \delta_1, \quad w_1^{(n)}(x) = f[w_1] + \delta_2 \qquad (z_1, w_1 \in \mathscr{M})$$

are the solutions of the problems $(\delta_1 \leqslant 0,\ \delta_2 \geqslant 0)$.

1.b. THE CASE OF MONOTONE DECREASING OPERATORS

Suppose that the operator $\mathscr{A}$ of problem (1.1) is monotone decreasing, i.e. for arbitrary $h \leqslant g$; $h, g \in \mathscr{M}$ we have $\mathscr{A}h \geqslant \mathscr{A}g$.

Furthermore assume that there exist elements $z_1, w_1 \in \mathscr{M}$ such that

(i) $\mathscr{A}z_1 \leqslant z_1$, $\mathscr{A}w_1 \geqslant w_1$,

(ii) $\mathscr{A}^2 z_l \leqslant z_1$, $\mathscr{A}^2 w_1 \geqslant w_1$,

and consider the sequences $\{z_p\}, \{w_p\}$ defined by $z_{p+1} = \mathscr{A}z_p$, $w_{p+1} = \mathscr{A}w_p$. Then for the solution y of (1.1) and for these sequences the following theorem holds.

Theorem 2. *The sequences* $\{z_{2l+1}\}$, $\{w_{2l+1}\}$ *and* $\{w_{2l}\}$, $\{z_{2l}\}$ *have the property* (B).

Remark 1.5. Observe that if $\mathscr{A}$ is monotone decreasing, then $\mathscr{A}^2$ is monotone increasing. Therefore from (ii) and Remark 1.1 we obtain that the sequences $\{z_{2l+1}\}$, $\{w_{2l+1}\}$ have the property (B).

Remark 1.6. If there exist $z_1, w_1 \in \mathscr{M}$ such that $\mathscr{A}^{2l} z_1 \leqslant z_1$, $\mathscr{A}^{2k} w_1 \geqslant w_1$ ($l, k \geqslant 1$ natural numbers) then as $\mathscr{A}^{2l}$, $\mathscr{A}^{2k}$ are isotonic (monotone increasing) operators, the sequences $\{z_{2lp+1}\}$, $\{w_{2kp+1}\}$ ($p = 0, 1, \ldots$) have the property (B).

Remark 1.7.

(a) The functions $z_1 = K$, $w_1 = -K$ which appeared in the problem of Remark 1.3 obviously satisfy the conditions $\mathscr{A}^2 z_1 \leqslant z_1$, $\mathscr{A}^2 w_1 \geqslant w_1$.

(b) In the problem II′ (in the case of $\mathscr{D}_\infty$) if $\dfrac{\partial f}{\partial u_i}, \dfrac{\partial f}{\partial v_i} \geqslant 0$ ($i = 0, \ldots, n-1$) hold then it is easy to construct functions z_1, w_1 satisfying (i) and at the same time these functions satisfy (ii), if the following inequalities are valid

$$N\Big[2 - \frac{1}{n!} + \int_{M_{n-1}} dt + \sum_{i=0}^{n-2} \int_{M_i} \Big| \frac{\partial^i G(x,t)}{\partial x^i} \Big| dt\Big] \leqslant \frac{M_{\alpha_1}}{m_{\alpha_1}}, \frac{m_{\beta_1}}{M_{\beta_1}}$$

where

$$0 < m_{\beta_1} = \min_{[0,1]} \beta_1(x), \quad M_{\beta_1} = \max_{[0,1]} \beta_1(x)$$

$$0 > M_{\alpha_1} = \max_{[0,1]} \alpha_1(x), \quad m_{\alpha_1} = \min_{[0,1]} \alpha_1(x),$$

$$\alpha_1(x) = z_1^{(n)}(x) - f[z_1] < 0, \quad \beta_1(x) = w_1^{(n)}(x) - f[w_1] > 0.$$

(c) Similar condition may be imposed for problem II, if

$$\frac{\partial^i G(x,t)}{\partial x^i} \leqslant 0; \quad \frac{\partial f}{\partial u_i}, \frac{\partial f}{\partial v_i} \geqslant 0 \qquad (i = 0, \ldots, n-1)$$

or for problem I, if $\dfrac{\partial f}{\partial u_i}, \dfrac{\partial f}{\partial v_i} \leqslant 0$ ($i = 0, \ldots, n-1$).

2. A METHOD FOR REDUCING PROBLEMS TO THE MONOTONE CASE

In the sequel suppose that $\mathcal{M}$ is a subset of a linear vector space V, and suppose, that for every,

$$h, g \in \mathcal{M} \quad \text{implies} \quad \frac{1}{2}(h+g) \in \mathcal{M}.$$

Put

$$\mathcal{M}^\circ = \left\{\alpha(h-g) \mid h, g \in \mathcal{M};\ \alpha = \frac{1}{2}, 1\right\} \qquad (h - h = 0 \in \mathcal{M}^\circ).$$

Let us denote by $\widetilde{\mathcal{M}}$ the following set:

$$\widetilde{\mathcal{M}} = \{h \pm q \mid h \in \mathcal{M},\ q \in \mathcal{M}^\circ\} \qquad (V \supseteq \widetilde{\mathcal{M}} \supseteq \mathcal{M}).$$

Let be given in $2\mathcal{M}^\circ$ a metric (ρ) (suppose that $2\mathcal{M}^\circ$ * is complete) and a partial ordering relation $(\leqslant)$ which are compatible and for which holds that the terms of monotone, convergent sequences can be compared with the limit.

We define the ordering and the distance in $\mathcal{M}$ by the formulas

$$a \leqslant b \Leftrightarrow a - b \leqslant 0, \quad \rho(a, b) = \rho(a-b, 0) = \| a - b \|$$

$$(a, b \in \mathcal{M}).$$

The ordering and the distance may be defined similarly in $\widetilde{\mathcal{M}}$ and between the elements of $\widetilde{\mathcal{M}}$ and $\mathcal{M}$.

Suppose that the set $\mathcal{M}$ has the property

(2.1) $$[h_1 \leqslant g,\ g \leqslant h_2\ (h_1, h_2 \in \mathcal{M};\ g \in \widetilde{\mathcal{M}})] \Rightarrow g \in \mathcal{M}.$$

Consider now the equation (1.1) $y = \mathcal{A} y$, where $\mathcal{A}$ is a non-necessarily monotone operator, mapping $\mathcal{M}$ into itself contractively. The solution of (1.1) is denoted by y.

Let be given an operator $\mathcal{B}\colon \mathcal{M} \times \mathcal{M} \to 2\mathcal{M}^\circ$ with

*$2\mathcal{M}^\circ = \{2e \mid e \in \mathcal{M}^\circ\}$

$$\|\mathscr{B}(h, g)\| \leqslant \theta \| h - g\| \qquad (0 < \theta < 1,\ \theta = \text{const.}),$$

and let $\mathscr{B}(h, g)$ be non-increasing in h and non-decreasing in g.

Suppose that if the operators $\mathscr{F}$ and Φ are defined by

$$\begin{aligned} \mathscr{F}(h, g) &= \frac{1}{2}(\mathscr{A}h + \mathscr{A}g) + \frac{1}{2}\mathscr{B}(h, g), \\ \Phi(h, g) &= \frac{1}{2}(\mathscr{A}h + \mathscr{A}g) - \frac{1}{2}\mathscr{B}(h, g), \end{aligned} \qquad (\mathscr{F}, \Phi\colon \mathscr{M} \times \mathscr{M} \to \widetilde{\mathscr{M}}),$$

then $\mathscr{F}$ resp. Φ is increasing with respect to g resp. h and decreasing with respect to h resp. g.

Under the above conditions we have

Theorem 3. *If there exist elements* $z_1, w_1 \in \mathscr{M}$ *such that for* $z_2 = \mathscr{F}(z_1, w_1)$ *and* $w_2 = \Phi(z_1, w_1)$ *the inequalities*

(i) $z_2 \leqslant z_1,\ \ w_1 \leqslant w_2,$

(ii) $\mathscr{F}(z_2, w_2) \leqslant z_1,\ \ \Phi(z_2, w_2) \geqslant w_1,$

hold, then the subsequences $\{z_{2l-1}\}, \{w_{2l-1}\}; \{w_{2l}\}, \{z_{2l}\}$ $(l = 1, 2, \ldots)$ *of the sequences defined by*

(2.2) $$z_{p+1} = \mathscr{F}(z_p, w_p), \quad w_{p+1} = \Phi(z_p, w_p) \qquad (p = 1, 2, \ldots)$$

have the property (B).

The proof of this theorem will be published elsewhere.

Remark 2.1. It is easy to see, that Theorem 3 may be applied for problems I, . . . , III if $\dfrac{\partial^i G(x, t)}{\partial x^i} \leqslant 0\ (\leqslant 0)$ $(i = 0, \ldots, n-1)$.

Remark 2.2. If all terms of (2.2) remain in $\mathscr{M}$ and $z_{2k+1} \leqslant z_1$, $w_{2l+1} \geqslant w_1$ are valid for some suitable k and l, then the sequences $\{z_{2(kl)p+1}\}$, $\{w_{2(kl)p+1}\}$ admit property (B).

Remark 2.3. If $\mathscr{M}^\circ$ is a linear, partially ordered, complete, metric space, and $\mathscr{M} = \widetilde{\mathscr{M}}$, then the assumptions may be weakened considerably.

Remark 2.4.

(1) If there exist maximal and minimal element in $\mathcal{M}$ (z_1 resp. w_1) and $\mathcal{F}, \Phi$ map $\mathcal{M} \times \mathcal{M}$ into $\mathcal{M}$, then (i) and (ii) are satisfied.

(2) In problem II′ the functions z_1, w_1 (in the case of $\mathcal{D}_K$) of the Remark 1.4 obviously satisfy (i), while for $\mathcal{D}_\infty$ concrete procedure is given in [3] for the construction of the functions z_1, w_1 satisfying (i). This latter construction is applicable also for problem II. In these problems

$$\mathcal{B}(z, w) = \begin{cases} 0 & x \in E, \\ \int_0^1 G(x, t)\,\Delta(t)\,dt, & 0 \leqslant x \leqslant 1, \end{cases}$$

where

$$\Delta(t) = N \sum_{i=0}^{n-1} [z^{(i)}(t) - w^{(i)}(t) + z^{(i)}(g_i(t)) - w^{(i)}(g_i(t))]$$

$$(0 \leqslant t \leqslant 1),$$

and $\mathcal{M}^\circ$ coincides with $\mathcal{M}$ belonging to $Q \equiv 0$.

Remark 2.5.

(1) If $\mathcal{F}$ and Φ are monotone operators and $\mathcal{F}, \Phi \colon M \times M \to \mathcal{M}$ (where $M \supset \mathcal{M}$), then we can look for elements z_1, w_1 satisfying (ii) in M.

In the problem of Remark 1.3 the functions $z_1 = K$, $w_1 = -K$ satisfy (ii).

(2) In the problem II (for $\mathcal{D}_\infty$) the functions z_1, w_1 obtained by construction mentioned in the Remark 2.4, satisfy (ii) if

$$N \max_x \sum_{i=0}^{n-1} \Big[\int_0^1 \Big| \frac{\partial^i G(x, t)}{\partial x^i} \Big| dt + \int_{M_i} \Big| \frac{\partial^i G(x, t)}{\partial x^i} \Big| dt \Big] \leqslant$$

$$\leqslant \min \Big(\frac{M_{\alpha_1}}{m_{\alpha_1} - S_{\beta_1}}, \frac{s_{\beta_1}}{S_{\beta_1} - m_{\alpha_1}} \Big),$$

where

$$m_{\alpha_1} = \min_x \alpha_1(x), \quad M_{\alpha_1} = \max_x \alpha_1(x),$$
$$(m_{\alpha_1}, M_{\alpha_1} < 0;\ s_{\beta_1}, S_{\beta_1} > 0)$$
$$s_{\beta_1} = \min_x \beta_1(x), \quad S_{\beta_1} = \max_x \beta_1(x),$$

$$\alpha_1(x) = z_1^{(n)}(x) - \frac{1}{2} f[z_1] - \frac{1}{2} f[w_1] - \frac{1}{2} \Delta_1(x)$$
$$(0 \leqslant x \leqslant 1)$$
$$\beta_1(x) = w_1^{(n)}(x) - \frac{1}{2} f[z_1] - \frac{1}{2} f[w_1] + \frac{1}{2} \Delta_1(x)$$

$$\Delta_1(x) = N \sum_{i=0}^{n-1} [z_1^{(i)}(x) + z_1^{(i)}(g_i(x)) - w_1^{(i)}(x) - w_1^{(i)}(g_i(x))].$$

Remark 2.6. Under conditions similar to those of Theorem 3, the sequences $\{z_p\}, \{w_p\}$ obtained by the iterative procedure $z_{p+1} = \mathscr{A} z_p + \mathscr{B}(z_p, w_p)$, $w_{p+1} = \mathscr{A} w_p - \mathscr{B}(z_p, w_p)$ have the property (B).

Corollary 2.7. For the problem II in the special case

$$y^{(n)}(x) = f[y] \equiv f(x, y(x), \ldots, y^{(n-1)}(x)) \qquad (0 \leqslant x \leqslant 1,\ n \geqslant 2)$$
(2.3)
$$L_i y = 0 \qquad (i = 0, \ldots, n-1)$$

if there exists differentiable function $\widetilde{G}(x, t)$ with

$$N \sum_{i=0}^{n-1} \Big[\max_x \int_0^1 \Big| \frac{\partial^i \widetilde{G}(x, t)}{\partial x^i} \Big| dt\Big] = \theta_1 < 1$$

and

$$\frac{\partial^i \widetilde{G}(x, t)}{\partial x^i} \leqslant - \Big| \frac{\partial^i G(x, t)}{\partial x^i} \Big| \qquad (i = 0, \ldots, n-1) \tag{2.4}$$

then the property (A) is valid on the operators $\mathscr{F}$, Φ given by the aid of

$$\mathscr{B}(h, g) = N \int_0^1 \widetilde{G}(x, t) \sum_{i=0}^{n-1} [h^{(i)}(t) - g^{(i)}(t)]\, dt$$

and $\mathscr{A}$, and on the subsequences of the sequences $\{z_p\}, \{w_p\}$ (but in this case, differently from the previous, z_p and w_p do not belong to $\mathscr{M}$).

If (2.4) holds and $\widetilde{G}$ is differentiable (with respect to x) except a finite set of x's, then also the inequality-chains of the property (A) are valid except for this finite set of x's. On the other hand in several cases it is easy to find function $\widetilde{G}$ with such property and therefore also for boundary value problems, which do not have the properties $\frac{\partial^i G}{\partial x^i} \leqslant 0$ or $\frac{\partial^i G}{\partial x^i} \geqslant 0$ $(i = 0, \dots, n-1)$, one can construct approximations with property (A). E.g. for the problems of the articles [4] and [5] it is relatively simple to construct such function $\widetilde{G}$ (consequently we obtain sharpenings of the constructions of [4] and [5]).

Finally we mention that the above methods may be applied also for partial differential equations, for systems and also for equations given in an implicite form.

REFERENCES

[1] S.A. Čapligin, *Collected Papers,* I, Gostekhizdat, Moscow – Leningrad, 1948 (in Russian).

[2] T. Amankulov, Approximate solution of systems of integro-differential equation by majorant-minorant functions, *Izv. V.U.Z. Mat.*, 40 (70): 1968, 6-12 (in Russian).

[3] Yu.I. Kovács – J. Hegedűs, On a two-sided iteration process for solving boundary value problems with delay, *Acta. Sci. Math.*, 36 (1974), 69-89 (in Russian).

[4] Yu.I. Kovács – L.I. Savčenko, Solving a boundary value problem for a nonlinear system of ordinary differential equations of second order, *Ukrainskiĭ Mat. Žurnal,* 20, 1 (1968), 34-44 (in Russian).

[5] A. Averna, Studio, con un procedimento di quasilinearizzazione, del problema: $\begin{cases} u^{(IV)} = f(x, u, u', u'', u'''), \\ u(0) = u'(0) = 0, \quad u(a) = u'(a) = 0, \end{cases}$ Matematiche (Catania), 24 (1969), 66-91.

[6] V. Lakshmikantham – S. Leela, *Differential and integral inequalities,* Volume 1, 2, New York – London, Academic Press, 1969.

[7] W. Walter, *Differential- und Integral-ungleichungen,* Volume 2, Springer-Verlag, 1964.

J. Hegedűs

JATE Bolyai Intézet, 6720 Szeged, Aradi Vértanúk tere 1, Hungary.

COLLOQUIA MATHEMATICA SOCIETATIS JÁNOS BOLYAI
15. DIFFERENTIAL EQUATIONS, KESZTHELY (HUNGARY), 1975.

INVESTIGATING THE SOLUTIONS OF CERTAIN ELLIPTIC DIFFERENTIAL EQUATIONS BY COMPLEX ANALYTIC METHODS

G. JANK

1. INTRODUCTION

Constructive methods in the theory of partial differential equations have recently become more and more important. First of all, methods using linear operators which take holomorphic functions into the set of solutions of a given differential equation have been intensively studied. Such operators were already introduced by I.N. Vekua [6] and St. Bergman [4] (as integral operators), and by E. Peschl and K.W. Bauer [3] (as differential operators).

Complex analytic methods are especially useful when investigating differential operators. For example, there exists a surjective differential operator for the class of differential equations

$$(1) \qquad F(z, \bar{z}) w_{z\bar{z}} - n(n+1) w = 0 \qquad (n \in N),$$

where F is an arbitrary solution of the nonlinear differential equation

$$(2) \qquad F(\log F)_{z\bar{z}} + 2 = 0$$

(cf. [2]). The solutions of (2) may be regarded as known, since an appropriate transformation reduces (2) to the Liouville differential equation

$$\Theta_{z\bar{z}} - e^{2\Theta} = 0.$$

In what follows we consider the differential equation

$$(3) \qquad w_{z\bar{z}} - \frac{m+1}{z-\bar{z}}\, w_z + \frac{n+1}{z-\bar{z}}\, w_{\bar{z}} = 0 \qquad (n, m \in N_0)$$

which, in the case $n = m$ belongs to the class (1). There exist very simple representation theorems for solutions of (3) defined in a simply connected region of the upper half-plane or in a punctured neighbourhood in the upper half-plane. By the aid of these representation theorems it is possible to construct a class of solutions of (3) defined (apart from isolated singular points) in the whole upper half-plane and remaining bounded near the real axis. These functions admit analytic continuations across the real axis. Using one of these functions a fundamental solution in the large can be obtained. For $n = m$ this is a real valued function widely applied in the theory of supersonic flows when dealing with generalized Tricomi equations (cf. [1]). Since, for $n = m$, equation (3) can be transformed so as to become invariant under all automorphisms of the upper half-plane, one can find (as it was shown in [5]) a connection between automorphic solutions of (3), the theory of Eichler integrals and, on the other hand, the theory of automorphic forms.

2. REPRESENTATION THEOREMS

For our simple equation (3) the following representation theorem holds.

Theorem 1.

(i) *For each solution* w *of equation* (3) *defined in a simply connected region* G *of the upper half-plane there exist functions* $f(z)$ *and* $g(z)$, *holomorphic in* G, *such that*

$$(4) \qquad w = \frac{\partial^{n+m}}{\partial z^n \partial \bar{z}^m} \left(\frac{f(z) + \overline{g(z)}}{z - \bar{z}} \right).$$

(ii) *Conversely, for any* $f(z)$ *and* $g(z)$, *holomorphic in* G, *the function* w *defined by* (4) *is a solution of* (3).

(iii) *Given a solution* w *of* (3) *defined in* G, *the functions*

$$\frac{d^{n+m+1}}{dz^{n+m+1}} f(z), \quad \frac{d^{n+m+1}}{dz^{n+m+1}} g(z)$$

are uniquely determined and holomorphic in G:

$$(5) \qquad \begin{aligned} f^{(n+m+1)}(z) &= \frac{d^{m+1}[(z-\bar{z})^{n+1}w]}{m!\,(z-\bar{z})^{n+m+2}}, \\ g^{(n+m+1)}(z) &= -\frac{d^{n+1}[(z-\bar{z})^{m+1}\bar{w}]}{n!\,(z-\bar{z})^{n+m+2}}, \end{aligned}$$

where $d = (z-\bar{z})^2 \frac{\partial}{\partial z}$, $d^{k+1} = dd^k$.

(iv) *The solution* $w \equiv 0$ *is generated by the functions*

$$f_0(z) = \sum_{k=0}^{n+m} a_k z^k, \quad g_0(z) = -\sum_{k=0}^{n+m} \bar{a}_k z^k, \qquad a_k \in C.$$

These statements can be proved by the methods of [2].

If the region G is not simply connected, say it is a punctured disc entirely contained in the upper half-plane H, then we have the following representation theorem.

Theorem 2. *Let* w *be a solution of equation* (3) *defined in* $\dot{U}(z_0) = \{z \mid 0 < |z - z_0| < r\} \subset H$. *Then the representation* (4) *still holds with the multi-valued generators*

$$\begin{aligned} f(z) &= f_1(z) + p_{n+m}(z) \log(z - z_0), \\ g(z) &= g_1(z) + \overline{p_{n+m}(\bar{z})} \log(z - z_0), \end{aligned}$$

where f_1 *and* g_1 *are holomorphic in* $U(z_0)$ *and* $p_{n+m}(z)$ *belongs to* π_{n+m}, *the set of all polynomials of degree* $\leqslant n+m$.

For, given a solution w defined in $\dot{U}(z_0)$, the $n+m+1$-th derivatives of the generators, holomorphic in $\dot{U}(z_0)$, are uniquely determined.

Integrating $n + m + 1$ times and taking into account that w itself is holomorphic in $\dot{U}(z_0)$ we obtain the desired conclusion.

3. A GENERALIZATION OF RIEMANN'S THEOREM ON REMOVABLE SINGULARITIES AND A REPRESENTATION THEOREM FOR SOLUTIONS BOUNDED ON THE REAL AXIS

Theorem 3. *Let* w *be a solution of* (3) *in a punctured neighbourhood* $\dot{U}(z_0) \subset H$ *of* z_0, *with generators*

$$f(z) = \sum_{-\infty}^{+\infty} \alpha_k (z - z_0)^k + p_{n+m}(z) \log (z - z_0),$$

$$g(z) = \sum_{-\infty}^{+\infty} \beta_k (z - z_0)^k + \overline{p_{n+m}(\bar{z})} \log (z - z_0),$$

suppose there exists a positive constant c *such that* $|w(z)| < c$ *for* $z \in \dot{U}(z_0)$. *Then* f *and* g *have holomorphic extensions* $U(z_0)$, *i.e.* $\alpha_k = \beta_k = 0$, $k = -1, -2, \ldots$ *and* $p_{n+m}(z) \equiv 0$.

Proof. By assumption, z_0 is a removable singularity of w. From Theorem 1 it follows that, say, the function $f^{(n+m+1)}(z)$ is bounded in $\dot{U}(z_0)$ and therefore it has a holomorphic extension to $U(z_0)$.

Next we are interested in representing solutions defined in H and bounded on the real axis. For the sake of brevity, set

$$E_{nm} f = \frac{\partial^{n+m}}{\partial z^n \partial \bar{z}^m} \frac{f(z)}{z - \bar{z}}.$$

Denote by B^{nm} the set of those solutions of (3) which are defined in H with the possible exception of a finite number of points and bounded near the real axis. Denote by H^{nm} the set of holomorphic functions of the form

$$f(z) = g(z) + \sum_{l=1}^{k} s_l(z) \log (z - z_l),$$

where $g(z)$ is holomorphic on the whole plane with the possible exception of a finite number of isolated non-real singular points, s_l belongs to π_{n+m}

and the z_l's are fixed points of the upper half-plane, $l = 1, 2, \ldots, k$. We have

Theorem 4.

(i) *For* $w \in B^{nm}$ *there exists a function* $f \in H^{nm}$ *such that*

$$w = E_{nm}(f(z) - f(\bar{z})). \tag{6}$$

(ii) *Conversely, for any* $f \in H^{nm}$, *the function* w *defined by* (6) *belongs to* B^{nm}.

For the proof we shall need the following

Lemma. *If the function* $w \in B^{nm}$ *has no singularities in* H, *then there exists an entire function* F *such that* $w = E_{nm}(F(z) - F(\bar{z}))$.

For denote by a_{rs} the coefficients of the power series expansion of w in the disk $|z - z_0| < \mathrm{Im}\,(z_0) < 1$, where $z_0 \in H$.

The series

$$\Theta(z, \bar{z}) = \sum_{r=0}^{\infty} \sum_{s=0}^{\infty} \frac{a_{rs}(z - z_0)^{r+n}(\bar{z} - \bar{z}_0)^{s+m}}{(r+1)\ldots(r+n)(s+1)\ldots(s+n)}$$

converges absolutely and uniformly in the disk $|z - z_0| < \mathrm{Im}\,(z_0)$, and we have

$$w = \frac{\partial^{n+m}}{\partial z^n \partial \bar{z}^m}\,\Theta.$$

Moreover, Θ remains bounded on the circle of convergence. It follows from Theorem 1 and the assumption of the lemma that there exist functions $f(z)$ and $g(z)$, holomorphic in H, such that Θ coincides with the power series expansion of $\dfrac{f(z) + \overline{g(z)}}{z - \bar{z}}$ around z_0. Hence $(f(z) + \overline{g(z)})|_{z \in R} = 0$. The function

$$U(z) = \begin{cases} f(z) + \overline{g(z)}, & z \in H \\ 0, & z \in R \\ -f(\bar{z}) - \overline{g(\bar{z})}, & z \in \bar{H} \end{cases}$$

is harmonic on the whole plane, so there exists a pair F, G of entire functions such that $U = F + \bar{G}$. In view of the relation $U | R = 0$ the assertion follows.

Proof of Theorem 4.

Let $w \in B^{nm}$, and denote by z_l $(l = 1, 2, \ldots, k)$ the singularities of w in H. By Theorem 2, in $U(z_l)$ there exists a representation $w = E_{nm}[f_l + {}_lp_{n+m}(z) \log(z - z_l) + \bar{g}_l + \overline{{}_lp_{n+m}(\bar{z}) \log(z - z_l)}$. Denote by f_l^* and g_l^* the principal parts of the Laurent series of f_l and g_l, respectively, and consider the function

$$w_1 = w - E_{nm}[\alpha(z) - \alpha(\bar{z})]$$

where

$$\alpha(z) = \sum_{l=1}^{k} [f_l^*(z) - \overline{g_l^*(\bar{z})} + {}_lp_{n+m} \log(z - z_l) - $$

$$- \overline{{}_lp_{n+m} \log(\bar{z} - z_l)}] \in H^{nm}.$$

Suppose for a moment that assertion (ii) of our theorem is already proved. Then w_1 belongs to B^{nm} and has no singularities in H. By the Lemma, there exists an entire function F such that $w_1 = E_{nm}(F(z) - F(\bar{z}))$. Together with α also $\alpha + F$ belongs to H^{nm} This yields the required representation for w. For proving (ii) we recall that f is holomorphic on the real axis. Substituting the power series expansion of f into (6) we get

$$\lim_{\substack{z \to x \\ \operatorname{Im} z > 0}} E_{nm}(f(z) - f(\bar{z})) = f^{(n+m+1)}(x).$$

Thus w is bounded on R.

4. EXTENSION OF FUNCTIONS OF THE CLASS B^{nm} AND FUNDAMENTAL SOLUTIONS IN THE LARGE

As already mentioned, the functions of the class B^{nm} have real analytic extensions to the lower half-plane.

Theorem 5. *The real analytic extension of a function* $w \in B^{nm}$ *to the lower half-plane exists. If for* $x \in R$, $\lim\limits_{\substack{z \to x \\ \operatorname{Im} z > 0}} w(z, \bar{z})$ *is real, this extension is defined by the relation* $w(z, \bar{z}) = \overline{w(z, \bar{z})}$, $\operatorname{Im} z < 0$.

Proof. Since $w(z, \bar{z}) \in B^{nm}$ there exists a function $f \in H^{nm}$ such that

$$w(z, \bar{z}) = \frac{\partial^{n+m}}{\partial z^n \partial \bar{z}^m} \frac{f(z) - f(\bar{z})}{z - \bar{z}}.$$

$\dfrac{f(z) - f(\bar{z})}{z - \bar{z}}$ has a real analytic extension over the real axis, what proves the first assertion. The further supposition gives, in view of the last formula of Section 3 and the reflection principle, $f(z) = \overline{f(\bar{z})} - q(z)$, $q \in \pi_{n+m}$. This gives the result.

An interesting result in this direction is provided by

Theorem 6. *If* $w \in B^{nm}$ *is a solution corresponding to a holomorphic generator* $f \in H^{nm}$, *then* $w \equiv 0$.

By Theorem 4, $w = E_{nm}(f(z) - f(\bar{z}))$, and by the assumption $E_{nm} f(\bar{z}) \equiv 0$. By Theorem 1 this implies $f(\bar{z}) = \overline{q_{n+m}(z)}$, where $q_{n+m} \in \pi_{n+m}$. Again by Theorem 1, $E_{nm}(f(z) - f(\bar{z})) = E_{nm}(\overline{q_{n+m}(\bar{z})} - \overline{q_{n+m}(z)}) \equiv 0$.

For $n = m$ (3) has also real valued solutions. Since $\frac{\partial}{\partial z} = \frac{1}{2}\left(\frac{\partial}{\partial x} - i\frac{\partial}{\partial x}\right)$, (3) can be transformed into the equation

$$\text{(7)} \qquad \Theta_{xx} + \Theta_{yy} + \frac{2(m+1)}{y}\Theta_y = 0 \qquad m = 0, 1, \ldots$$

for the new unknown function $\Theta(x, y) = w(z, \bar{z})$. Equation (7) can be obtained also from the generalized Tricomi equation

$$\eta\psi_{\xi\xi} + \psi_{\eta\eta} + \frac{6m-5}{2\eta}\psi_\eta = 0$$

by the transformation $x = \xi$, $y = \frac{2}{3}\eta^{\frac{3}{2}}$ ($\eta > 0$ is the elliptic half-plane)

$\Theta(x, y) = \psi(\xi, \eta)$, see [1]. In the hyperbolic half-plane $(\eta < 0)$ a similar transformation leads to the (hyperbolic) differential equation

$$\chi_{xx} - \chi_{tt} - \frac{2(m+1)}{t} \chi_t = 0. \tag{8}$$

In [1] a single global fundamental solution containing no integrals is obtained which is defined on the whole upper half-plane apart from a logarithmic singularity. This solution is then extended to the hyperbolic half-plane. For constructing (in the case $n = m$) a real valued fundamental solution we first need a general representation of all real valued solutions of class B^{nm}. By Theorem 4, for a real valued $w \in B^{nm}$ there exists a generator $f \in H^{nn}$ with $f(z) = \overline{f(\bar{z})}$ such that

$$w = E_{nn} f(z) + \overline{E_{nn} f(z)}. \tag{9}$$

Conversely, any $f(z)$ with the above property defines a real valued function of class B^{nn}.

Choosing the generator

$$f(z) = ic \log \frac{z - ib}{z + ib}, \qquad c \in \boldsymbol{R},\ b > 0, \tag{10}$$

which is of class H^{nn} $(n \in N_0)$, we get a fundamental solution $\gamma(x, y)$ with a logarithmic singularity at ib, satisfying

$$\gamma(x, 0) = \alpha_1(x),$$

$$\gamma_x(x, 0) = \alpha_2(x),$$

$$\gamma_y(x, 0) = 0,$$

where the explicitly computable functions α_1, α_2 are bounded on $\boldsymbol{R} \cup \{\infty\}$ and depend also on b and n [1].

In [1] a generator similar to (10) was used on the hyperbolic half-plane. This provided an extension, with continuous first derivatives, of the fundamental solution the hyperbolic half-plane.

5. A CONNECTION WITH THE EICHLER INTEGRALS

Let Γ be a Fuchsian group with limit circle $R \cup \{\infty\}$. A singular Eichler integral (of degree $n+1$, $n \in N$) is a function f holomorphic in $H \cup \bar{H}$ with the exception of isolated singular points that has no more than a finite number of singularities in $F \cup \bar{F}$ (F denotes the fundamental domain of Γ) and is such that for all $A \in \Gamma$ the function

$$p_A(z) = f(z) - (A'(z))^{-n} f(A(z))$$

coincides with the restriction to $H \cup \bar{H}$ of a polynomial belonging to π_{2n}. Thus any polynomial of the class π_{2n} is itself an Eichler integral. Two Eichler integrals whose difference belongs to π_{2n} are considered equivalent. We denote by $E_n(\Gamma)$ the vector space consisting of the equivalence classes so obtained. The equivalence classes themselves are called Eichler classes.

The substitution $w = (z - \bar{z})^{-n-1} v$ transforms the differential equation (3) into

$$v_{z\bar{z}} + \frac{n-m}{z-\bar{z}} v_z + \frac{m(n+1)}{(z-\bar{z})^2} v = 0.$$

The case interesting for our present purpose is again $n = m$ yielding the equation

$$(11) \qquad (z-\bar{z})^2 v_{z\bar{z}} + n(n+1)v = 0,$$

since this equation is invariant with respect to all automorphisms of the upper half-plane. By Theorem 1, a solution v defined in H can be represented by the formula

$$v = (z-\bar{z})^{n+1} \frac{\partial^{2n}}{\partial z^n \partial \bar{z}^n} \frac{f(z) + \overline{g(z)}}{z - \bar{z}},$$

where f and g are holomorphic in H. We can rewrite this representation in the form

$$(12) \qquad v = (H_n f)(z) + \overline{(H_n g)(z)},$$

$$(H_n f)(z) = (z - \bar{z})^{n+1} \frac{\partial^{2n}}{\partial z^n \partial \bar{z}^n} \frac{f}{z - \bar{z}} =$$

$$= \sum_{k=0}^{n} \frac{(2n-k)!}{k!\,(n-k)!} (\bar{z} - z)^{k-n} \frac{d^k}{dz^k} f.$$

The generators f and g can be determined in the same manner as in Theorem 1:

$$f^{(2n+1)}(z) = \frac{d^{n+1} v}{(z - \bar{z})^{2n+2}}, \quad g^{(2n+1)}(z) = \frac{d^{n+1} \bar{v}}{(z - \bar{z})^{2n+2}}.$$

Moreover, $H_n f + \overline{H_n g} \equiv 0$ if and only if $g \in \pi_{2n}$ and $f(z) = (-1)^{n+1} \overline{g(\bar{z})}$. Solutions of (11) holomorphic in a punctured neighbourhood $\dot{U}(z_0) \subset H$ can be obtained from (12) by the aid of the generators

$$\text{(13)} \quad \begin{aligned} f(z) &= f_0(z) + p(z) \log (z - z_0), \\ g(z) &= g_0(z) + (-1)^{n+1} \overline{p(\bar{z})} \log (z - z_0), \end{aligned}$$

where f_0 and g_0 are holomorphic in $\dot{U}(z_0)$ and $p \in \pi_{2n}$.

We denote by $A_n(\Gamma)$ the vector space of all solutions v of (11) which satisfy the following conditions:

(a) v is defined in H apart from isolated non-logarithmic singularities. Hence for the corresponding generators (13) we have $p \equiv 0$.

(b) b has no more than finite number of singularities in F.

(c) $v(A(z)) = v(z)$ for all $z \in H$, and $A \in \Gamma$.

In [5] the following theorem was proved:

Theorem 7. *A solution* v *of* (11) *belongs to* $A_n(\Gamma)$ *if and only if* $v = G_n f = H_n f(z) + (-1)^{n+1} \overline{H_n f(\bar{z})}$, *where* f *is a suitable element of an Eichler class of* $E_n(\Gamma)$. *For* $v \in A_n(\Gamma)$ *the corresponding Eichler class is uniquely determined by*

$$(z - \bar{z})^{2n+2} f^{(2n+1)}(z) = \begin{cases} d^{n+1} v(z) & z \in H \\ d^{n+1} [(-1)^{n+1} v(\bar{z})], & z \in \bar{H}. \end{cases}$$

Hence there is a one-to-one correspondence γ_n between the elements of $A_n(\Gamma)$ and $E_n(\Gamma)$. Actually, there exists an isomorphism

$$\gamma_n: A_n(\Gamma) \to E_n(\Gamma).$$

Thus it is possible to translate assertions regarding solutions of (11) into assertions for Eichler integrals (cf. [5]).

REFERENCES

[1] K.W. Bauer, Differentialoperatoren bei einer Klasse verallgemeinerter Tricomi-Gleichungen, *ZAMM* (to appear).

[2] K.W. Bauer – G. Jank, Differentialoperatoren bei einer inhomogenen elliptischen Differentialgleichung, *Rend. Ist. Mat. Univ. Trieste,* 3 (1971), 1-29.

[3] K.W. Bauer – E. Peschl, Ein allgemeiner Entwicklungssatz für die Lösungen der Differentialgleichung $(1 + \epsilon z\bar{z})^2 w_{z\bar{z}} + \epsilon n(n+1)w = 0$ in der Nähe isolierter Singularitäten, *S.-ber. d. Bayer. Akad. d. Wiss., math.-nat. Klasse, S* (1965), 113-146.

[4] St. Bergman, *Integral Operators in the Theory of Linear Partial Differential Equations,* Erg. d. Math. u. Grenzgeb., Berlin, Göttingen, Heidelberg (1961).

[5] I. Haeseler – St. Ruscheweyh, Singuläre Eichlerintegrale und Eisensteinreihen, *Math. Ann.*, 203 (1973), 251-259.

[6] I.N. Vekua, *New Methods for Solving Elliptic Equations,* Amsterdam (1967).

G. Jank

I. Lehrkanzel und Institut für Mathematik an der Technischen Hochschule, A-8010 Graz, Kopernikusgasse 24, Austria.

COLLOQUIA MATHEMATICA SOCIETATIS JÁNOS BOLYAI
15. DIFFERENTIAL EQUATIONS, KESZTHELY (HUNGARY), 1975.

ON RYABOV'S SPECIAL SOLUTIONS OF FUNCTIONAL DIFFERENTIAL EQUATIONS

J. JARNÍK – J. KURZWEIL

1. INTRODUCTION

Let R^n denote the n-dimensional Euclidean space, $R = R^1$. Let us denote for $\tau > 0$

$$C_\tau = \{x \mid x\colon\ [-\tau, 0] \to R^n, x \text{ continuous}\},$$

$$\|x\| = \sup_{\sigma \in [-\tau, 0]} |x(\sigma)|.$$

Given $x\colon [t-\tau, t] \to R^n$, denote by x_t the function mapping $[-\tau, 0]$ into R^n, defined by $x_t(\sigma) = x(t+\sigma)$ for $\sigma \in [-\tau, 0]$.

Our results concern the equation

$$\frac{dx}{dt} = F(t, x_t) \tag{1}$$

where $F\colon R \times C_\tau \to R^n$ is continuous in both variables,

$$|F(t,u) - F(t,v)| \leqslant L\|u - v\| \tag{2}$$

for all $t \in R$, $u, v \in C_\tau$ (L independent of t) and

(3) $$|F(t, 0)| \leqslant A = \text{const}.$$

We assume further

(4) $$L\tau e < 1.$$

It is easy to see that the assumption (4) guarantees that

(5) $$\Theta = L\tau e^{\Theta}$$

has two positive roots Θ_1, Θ_2 (depending on L, τ) and $L\tau < \Theta_1 < L\tau e < < 1 < \Theta_2$.

Yu. A. Ryabov [1; cf. also 2] showed that under assumptions (1)-(4), for every $t_* \in R$, $x_* \in R^n$ there exists a solution $\bar{x}$ of (1) passing through (t_*, x_*), defined on the whole real line and satisfying

(6) $$\sup_{t < t_*} e^{\frac{t}{\tau}} |\bar{x}(t)| < \infty$$

and this solution is unique. Following Ryabov, we call it special solution and denote it by $\bar{x}(t_*, x_*)$.

Further, Ryabov introduced the assumption

(4*) $$L\tau e^{1 + L\tau + \Theta_1} < 1$$

and proved that under (1)-(3), (4*) every solution of (1) approaches asymptotically one of the special solutions if $t \to +\infty$, more precisely, to any solution x of (1) defined on $[T - \tau, +\infty)$ and any $t_0 \geqslant T$ there exists a special solution $\bar{x}(t_0, x_0)$, $x_0 \in R^n$ such that

(7) $$\lim_{t \to +\infty} [x(t) - \bar{x}(t_0, x_0)(t)] = 0.$$

2. GENERALIZATION OF RYABOV'S RESULTS

It is possible to show that it is not necessary to strengthen the assumption (4) in order to obtain (7). Furthermore, it can be shown that for every solution x of (1) there is exactly one special solution $\bar{x}$ which is approached by x most rapidly. In other words, denoting by $\bar{y}$ any other special solution, it holds

$$\frac{x(t) - \bar{x}(t)}{x(t) - \bar{y}(t)} \to 0$$

if $t \to +\infty$. This is the main result of [3] which may be formulated as

Theorem 1. *Let* (1)-(4) *be fulfilled and let* x *be a solution of* (1) *defined on* $[T - \tau, +\infty)$, $t_0 \geqslant T$. *Then the limit*

$$x_0 = \lim_{s \to +\infty} \bar{x}(s, x(s))(t_0) \tag{8}$$

exists and

$$\sup_{t > t_0} e^{\frac{t}{\tau}} |x(t) - \bar{x}(t_0, x_0)(t)| < \infty. \tag{9}$$

The crucial point in the proof of Theorem 1 is a lemma which asserts that two special solutions $\bar{x}, \bar{y}$ satisfy for all $t < t_1$ the inequality

$$|\bar{x}(t) - \bar{y}(t)| \leqslant e^{\frac{\Theta_1(t_1 - t)}{\tau}} |\bar{x}(t_1) - \bar{y}(t_1)|. \tag{10}$$

Here Θ_1 is the solution of (5) satisfying $\Theta_1 < 1$. The inequality (10) expresses the fact that two special solutions cannot approach each other too quickly (with increasing t). Among other, (10) immediately implies that (9) cannot be satisfied by more than one special solution.

3. NEW CLASSES OF SPECIAL SOLUTIONS

To generalize further our results, let us introduce in C_τ the norm

$$\| x \|_\vartheta = \sup_{\sigma \in [-\tau, 0]} e^{\vartheta(\tau + \sigma)} |x(\sigma)|$$

for arbitrary $\vartheta \geqslant 0$ (for $\vartheta = 0$ we have the usual supremum norm considered above).

Definition. Let $0 \leqslant \omega_1 \leqslant \omega_2$. Equation (1) is called (ω_1, ω_2)-special if to every $t_* \in \boldsymbol{R}$, $x_* \in \boldsymbol{R}^n$ there exists a solution $\bar{x}: \boldsymbol{R} \to \boldsymbol{R}^n$ of (1) passing through (t_*, x_*) and satisfying

$$\sup_{t<t_*} e^{\omega_1 t} |\bar{x}(t)| < \infty; \tag{11}$$

for any solution $y: \boldsymbol{R} \to \boldsymbol{R}^n$ of (1) passing through (t_*, x_*) and satisfying

$$\sup_{t<t_*} e^{\omega_2 t} |y(t)| < \infty \tag{12}$$

it is $y = \bar{x}$.

It is clear that if (1) is (ω_1, ω_2)-special, $0 \leqslant \omega_1 \leqslant \rho_1 \leqslant \rho_2 \leqslant \omega_2$ then (1) is also (ρ_1, ρ_2)-special. The solution $\bar{x}$ from the definition is also called (ω_1, ω_2)-special.

Theorem 2. *Let the equation* (1) *satisfy* (3),

$$|F(t, u) - F(t, v)| \leqslant L_\vartheta \| u - v \|_\vartheta, \tag{2'}$$

$$L_\vartheta \tau e < 1. \tag{4'}$$

Denote by Θ_1, Θ_2 *the roots of* (5) *with* $L = L_\vartheta$ *and suppose*

$$0 \leqslant \vartheta\tau < \Theta_2.$$

Choose ω_1, ω_2 *so that*

$$\max(\Theta_1, L_\vartheta \tau e^{\vartheta\tau}) < \omega_1 \tau \leqslant \omega_2 \tau < \Theta_2.$$

Then the equation (1) *is* (ω_1, ω_2)*-special.*

Remark. This theorem includes Ryabov's result if we put $\vartheta = 0$, $\omega_1 = \omega_2 = \dfrac{1}{\tau}$.

Example. The linear equation

$$\dot{x}(t) = ax(t) + bx(t - \tau)$$

(a, b constants) is transformed into (1) by putting

$$F(t, u) = au(0) + bu(-\tau).$$

Evidently $L_\vartheta = |a| e^{-\vartheta\tau} + |b|$ and hence (for $a \neq 0$) L_ϑ decreases with increasing ϑ.

Under the hypotheses of Theorem 2 it is possible to prove an inequality analogous to (10), namely, if x, y are (ω_1, ω_2)-special solutions $(\omega_1, \omega_2$ being chosen according to Theorem 2) then for $t < t_1$ we have

$$(10') \qquad |\bar{x}(t) - \bar{y}(t)| \leq e^{\frac{\nu(t_1 - t)}{\tau}} |\bar{x}(t_1) - \bar{y}(t_1)|$$

where $\nu = \max(\Theta_1, L_\vartheta \tau e^{\vartheta\tau})$.

If $\vartheta\tau \leq 1$ then it is possible to prove a theorem completely analogous to Theorem 1, i.e., to prove that the asymptotic behaviour of all solutions of (1) (under the assumptions of Theorem 2) for $t \to +\infty$ is determined by $\left(\frac{1}{\tau}, \frac{1}{\tau}\right)$-special solutions. If $1 < \vartheta\tau < \Theta_2$ then the problem remains open.

Remark. Simple examples show that ω_1, ω_2 cannot be chosen outside the interval (Θ_1, Θ_2) and that neither ω_1 nor ν in (11′) can be less than $L_\vartheta \tau e^{\vartheta\tau}$.

REFERENCES

[1] Yu.A. Ryabov, Certain asymptotic properties of linear systems with small time lag. *Trudy Sem. Teor. Diff. Uravnenii s Otklon. Argumentom Univ. Druzhby Narodov Patrisa Lumumby,* 3 (1965), 153-164.

[2] R.D. Driver, On Ryabov's asymptotic characterization of the solutions of quasi-linear differential equations with small delays. *SIAM Review,* 10, 3 (1968), 329-341.

[3] J. Jarník – J. Kurzweil, Ryabov's special solutions of functional differential equations. *Boll. U. Mat. Ital.*, (4) 11, Suppl. fasc., 3 (1975), 198-208.

[4] J. Kurzweil – J. Jarník, New classes of special solutions of functional differential equations. To appear in *Czechoslovak Math. Journ.*

J. Jarník – J. Kurzweil
Matematický ústav, ČSAV, Žitna 25, 11567 Praha 1, Czechoslovakia.

COLLOQUIA MATHEMATICA SOCIETATIS JÁNOS BOLYAI
15. DIFFERENTIAL EQUATIONS, KESZTHELY (HUNGARY), 1975.

ON THE ASYMPTOTIC BEHAVIOUR OF POSITIVE SOLUTIONS OF TWO DIFFERENTIAL INEQUALITIES WITH RETARDED ARGUMENT

E. KOZAKIEWICZ

Our purpose is the investigation of positive solutions of the differential inequalities

$$(1 \geqslant) \qquad y'(x) \geqslant a(x)y(x) - \int_0^\infty y(x-s)\,dr(x,s) \qquad (A \leqslant x < +\infty)$$

and $(1 \leqslant)$. We get $(1 \leqslant)$ from $(1 \geqslant)$ substituting "$\leqslant$" by "$\geqslant$". This convention will be used in the sequel too. We suppose that $r(x, s)$ satisfies the usual conditions while there is no restriction concerning the sign of $a(x)$. We omit here a more detailed discussion of the assumptions, since for the sake of sufficient generality we shall integrate relations $(1 \geqslant)$ and $(1 \leqslant)$. For our purposes this method requires more care, than a simple passing over from a differential equation to an integral equation.

1. THE INTEGRATED RELATIONS

In the sequel all integrals are taken in the sense of Lebesgue. A and B denote fixed points of the real line, $-\infty < A < B \leqslant +\infty$. A real function is said to be locally summable, if it is summable on any interval $A \leqslant x \leqslant C$ $(A < C < B)$. Denote $l(t)$ a measurable real function, defined for $A \leqslant t < B$, such that $|l(t)|$ is locally summable. For any x $(A \leqslant x < B)$, $C(x)$ denotes the space of continuous real functions with domain M_x, where $M_x \neq \phi$, $0 \leqslant \inf M_x$. Denote $E(P)$ $(A \leqslant P < B)$ the infimum of the set of points $x \leqslant P$ for which there exist s and t such that $x = t - s$, $t \geqslant P$, $s \in M_t$. We assume $-\infty < E(A)$ and $\lim_{x \to B-0} E(x) = B$. Note that in most places these last conditions can be omitted or substituted by weaker ones.

Let L_x be a real valued functional defined on $C(x)$, with the following properties:

1. For $f \in C(x)$, $g \in C(x)$, $L_x(f+g) = L_x f + L_x g$ holds.

2. For $f \in C(x)$, a real, $L_x(af) = aL_x f$ holds.

3. $f \in C(x)$, $f \geqslant 0$ (i.e. $f(s) \geqslant 0$ for $s \in M_x$) implies $L_x f \geqslant 0$.

4. $L_t g(t-s)$ is locally summable for any function $g(x)$, continuous on $E(A) \leqslant x < B$.

A function $u(x)$, continuous for $E(A) \leqslant x < B$ is said to be a solution of

$$(2 \geqslant) \qquad u(x) \geqslant u(\xi) + \int_{\xi}^{x} l(t)u(t)\,dt - \int_{\xi}^{x} L_t u(t-s)\,dt \qquad (A \leqslant \xi \leqslant x < B)$$

if the relation $(2 \geqslant)$ holds for all ξ and x with $A \leqslant \xi \leqslant x < B$.

2. A THEOREM ABOUT THE SIGN OF SOLUTIONS OF $(2 \geqslant)$

2.1. *Statement and proof of the theorem*

Theorem 1. *Let* $u(x)$ *be a solution of* $(2 \geqslant)$, *such that* $u(x) \leqslant 0$ *for* $E(A) \leqslant x < A$ *and* $u(A) = 0$. *Then there exists a* $C > A$ *such that* $u(x) \geqslant 0$ holds *for* $A \leqslant x < C$.

Proof. Suppose that the assertion does not hold. Then there are two possibilities:

(a) There exist right neighbourhoods of A with $u(x) \leqslant 0$ and each of them contains points x with $u(x) < 0$.

(b) In all right neighbourhoods of A there are points x with $u(x) < 0$ as well as points with $u(x) > 0$.

Case (a). We can choose T $(A < T < B)$ so that $u(x) \leqslant 0$ holds for $A \leqslant x \leqslant T$, and at the same time

$$\int_A^T |l(t)|\, dt < 1.$$

Denote

$$\rho = \min_{A \leqslant t \leqslant T} u(t).$$

For $A \leqslant x \leqslant T$ we have

$$u(x) \geqslant u(A) + \int_A^x l(t)u(t)\, dt - \int_A^x L_t u(t-s)\, dt \geqslant \int_A^x l(t)u(t)\, dt \geqslant$$

$$\geqslant \int_A^x |l(t)|\, u(t)\, dt \geqslant \rho \int_A^x |l(t)|\, dt \geqslant \rho \int_A^T |l(t)|\, dt > \rho.$$

This is a contradiction, since $u(x)$ attains its minimum value ρ in the interval $A \leqslant x \leqslant T$.

Case (b). Choose S $(A < S < B)$ so that

$$\int_A^S |l(t)|\, dt < \frac{1}{2}, \quad \int_A^S L_t 1\, dt < \frac{1}{2} \quad \text{and} \quad u(S) = 0$$

should hold. Denote $\sigma(S) = \max\limits_{A \leqslant t \leqslant S} u(t)$ and let P $(A < P < S)$ be such that $u(P) = \sigma(S)$. Denote by Q the first zero of $u(t)$ in the interval $P \leqslant t \leqslant S$. For $P \leqslant x \leqslant Q$ we have

$$u(x) \geqslant u(P) + \int_P^x l(t)u(t)\, dt - \int_P^x L_t u(t-s)\, dt \geqslant$$

$$\geqslant \sigma(S) - \int_P^x |l(t)|\, |u(t)|\, dt -$$

$$-\int_P^x L_t(u(t-s)-\sigma(S))\,dt - \int_P^x L_t\sigma(S)\,dt \geqslant$$

$$\geqslant \sigma(S) - \sigma(S)\int_P^x |l(t)|\,dt - \sigma(S)\int_P^x L_t 1\,dt \geqslant$$

$$\geqslant \sigma(S)\Big\{1 - \int_A^S |l(t)|\,dt - \int_A^S L_t 1\,dt\Big\} > 0$$

in contrary to $u(Q) = 0$. Hence Theorem 1 is proved.

2.2. *On the variable lower limit*

2.2.1. *The importance of the variable lower limit in the case of* $(2\geqslant)$

The simplest way to integrate relation $(1\geqslant)$ is to integrate from a fixed lower limit A to a variable limit x. Thus we obtain a relation of the form $(2\geqslant)$ with $\xi = A$. However, we do not use this simpler relation, since Theorem 1 fails to be true for relations $(2\geqslant)$, specialized this way, as it is shown by the following example. Let $A = \xi = 0$, $l(t) \equiv -1$, $L_t u(t-s) \equiv 0$, $B = +\infty$. Then $(2\geqslant 0)$ implies

(3) $$u(x) \geqslant u(0) - \int_0^x u(t)\,dt \qquad (0 \leqslant x < +\infty).$$

There exist solutions of (3), assuming in any neighbourhood of $x = 0$ as well positive, as negative values. For proving this we choose a function $u(x)$ with the following properties:

Let $u(x)$ be continuous for $0 < x < +\infty$, $u(0) = 0$, $u\left(\frac{1}{n}\right) = 0$ $(n = 1, 2, \ldots)$, and $u(x) \equiv 0$ $(1 < x < +\infty)$. Suppose that if $\frac{1}{n+1} \leqslant x \leqslant \frac{1}{n}$, then $-\frac{1}{4(n+1)^2} \leqslant u(x) \leqslant \frac{3}{n}$ holds.

Moreover, suppose

$$\int_{\frac{1}{n+1}}^{\frac{1}{n}} u(t)\,dt \geqslant \frac{1}{n^3}$$

and let there exist x_n $\left(\frac{1}{n+1} < x_n < \frac{1}{n}\right)$, such that $u(x_n) < 0$ holds for $n = 1, 2, \ldots$. It is easy seen that such fucntions exist. They are continuous for $0 \leqslant x < +\infty$. We shall show, that each of them satisfies inequality (3). Suppose $\frac{1}{n+1} \leqslant x \leqslant \frac{1}{n}$ $(n = 1, 2, \ldots)$. Then

$$-\int_0^x u(t)\,dt = -\int_0^{\frac{1}{n+1}} u(t)\,dt - \int_{\frac{1}{n+1}}^x u(t)\,dt \leqslant$$

$$\leqslant -\sum_{\nu=1}^{\infty} \int_{\frac{1}{n+\nu+1}}^{\frac{1}{n+\nu}} u(t)\,dt + \int_{\frac{1}{n+1}}^x \frac{1}{4(n+1)^2}\,dt \leqslant$$

$$\leqslant -\sum_{\nu=1}^{\infty} \frac{1}{(n+\nu)^3} + \frac{1}{4n(n+1)^3} \leqslant$$

$$\leqslant -\int_{n+1}^{\infty} \frac{1}{x^3}\,dx + \frac{1}{4n(n+1)^3} =$$

$$= -\frac{1}{2}\frac{1}{(n+1)^2} + \frac{1}{4n(n+1)^3} \leqslant -\frac{1}{4}\frac{1}{(n+1)^2} \leqslant u(x).$$

For $1 < x < +\infty$ (3) is also satisfied.

2.2.2. *Variable lower limit in* (2 =)

In 2.2.1 we have pointed out that the variable lower limit is essential for Theorem 1. However, if we consider (2 =) instead of (2 $\geqslant$), the variable lower limit doesn't matter, since (2 =) is equivalent to

$$y(x) = y(A) + \int_A^x l(t)y(t)\,dt - \int_A^x L_t y(t-s)\,dt \qquad (A \leqslant x \leqslant B). \tag{4}$$

Obviously (4) is a special case of (2 =). On the other hand, writing down (4) at x and ξ respectively, a subtraction yields (2 =).

Though by these remarks in the case of equations it would be sufficient to deal with fixed lower limit only, we use also in this case variable lower limit, because it is convenient to form the difference between an equation (2 =) and an inequality of the form (2 ⩾) or (2 ⩽).

2.2.3. *Return to differential inequalities*

From (2 ⩾) it follows

$$u(x) - u(\xi) \geqslant \int_{\xi}^{x} l(t)u(t)\,dt - \int_{\xi}^{x} L_k u(t-s) \qquad (A \leqslant \xi \leqslant x < B).$$

Suppose, that $l(t)$ and $L_t u(t-s)$ are continuous in t, and $u(x)$ is differentiable for $A \leqslant x < B$. Dividing by $x - \xi$ $(A \leqslant \xi < x < B)$ and letting $x \to \xi + 0$ yields

$$u'(\xi) \geqslant l(\xi)u(\xi) - L_{\xi} u(\xi - s) \qquad (A \leqslant \xi < B).$$

From (2 ⩽) we obtain an inequality with the opposite direction.

2.3. *The special case* $L_t \equiv 0$

Theorem 1′. *Let* $u(x)$ *be a solution of* (2 ⩾) *with* $L_t \equiv 0$ $(A \leqslant t < B)$, $u(A) \geqslant 0$, *then* $u(x) \geqslant 0$ *for* $A \leqslant x < B$.

Proof. Suppose conversely, that there exists an x $(A < x < B)$ such that $u(x) < 0$. Denote by T the infimum of the set of such x-values. Then $u(T) = 0$ while $u(x)$ assumes negative values in any right neighbourhood of T. Applying Theorem 1 with T as initial point, we get a point C $(T < C < B)$ such that $u(x) \geqslant 0$ $(T \leqslant x \leqslant C)$. This contradiction proves Theorem 1′.

2.4. *On the possibility of changing the sign of* $u(x)$ *under the assumptions of Theorem* 1

Theorem 1′ shows that in the case $L_t \equiv 0$ $(A \leqslant t < B)$ Theorem 1 holds with $C = B$. In general this is not true as it is shown by the solutions of $y'(x) = y(x-p)$ if $p > \frac{1}{e}$. However, an additional condition, which is often satisfied, implies $u(x) \geqslant 0$ $(A \leqslant x < B)$:

Theorem 1″. *Beyond the assumptions of Theorem* 1 *suppose that*

$(2\leqslant)$ *admits a solution* $y(x)$ *for which* $y(x)\geqslant 0$ *holds for* $E(A)\leqslant x<A$ *and* $y(x)>0$ *for* $A\leqslant x<B$. *Then* $u(x)\geqslant 0$ *for* $A\leqslant x<B$.

Proof. Suppose in the contrary that there exists an x_0 $(A<x_0<B)$ such that $u(x_0)<0$. Let $T=\inf\{x;\ u(x)<0,\ A\leqslant x<B\}$. We have $A<T<B$ and $u(T)=0$. By Theorem 1, $u(x)\equiv 0$ for $A\leqslant x\leqslant T$ would imply $u(x)\geqslant 0$ for $T\leqslant x\leqslant T+C$ $(C>0)$, contradicting to the definition of T. Hence $Q=\max\limits_{A\leqslant x\leqslant T}\dfrac{u(x)}{y(x)}>0$. Denote by P the largest real number from the interval $A\leqslant x\leqslant T$, at which this maximum value is attained. Clearly $P<T$. The function $v(x)=u(x)-Qy(x)$ satisfies $(2\geqslant)$. Moreover, $v(x)\leqslant 0$ $(E(P)\leqslant x\leqslant P)$ and $v(x)<0$ $(P<x<T)$ what contradicts to Theorem 1 and proves Theorem 1″.

3. A COMPARISON THEOREM

Theorem 2. *Let* $y(x), \bar{y}(x)$ *be solutions of* $(2\leqslant)$ *and* $(2\geqslant)$ *respectively, and suppose that for* $E(A)\leqslant x<A$ $0\leqslant y(x)$ *and* $\bar{y}(x)\leqslant y(x)$ *holds, moreover* $\bar{y}(A)=y(A)$, *and* $y(x)>0$ *for* $A\leqslant x<B$. *Then* $q(x)=\dfrac{\bar{y}(x)}{y(x)}$ *is nondecreasing for* $A\leqslant x<B$.

Proof. The difference $v(x)=\bar{y}(x)-y(x)$ satisfies $(2\geqslant)$, $v(x)\leqslant 0$ $(E(A)\leqslant x<A)$ and $v(A)=0$. Hence by Theorem 1″, $v(x)\geqslant 0$ or, equivalently $q(x)\geqslant 1$ for $A\leqslant x<B$. Now we shall prove

$$q(x_0)y(x)\leqslant\bar{y}(x)\qquad(A\leqslant x_0<x<B). \tag{5}$$

Let $T=\inf\{x;\ q(x_0)y(x)=\bar{y}(x),\ A\leqslant x<B\}$. Clearly $A\leqslant T\leqslant x_0$, $q(x_0)y(T)=\bar{y}(T)$, while $\bar{y}(x)\leqslant q(x_0)y(x)$ for $A\leqslant x\leqslant T$, $\bar{y}(x)\leqslant\leqslant y(x)\leqslant q(x_0)y(x)$ for $E(A)\leqslant x<A$. We apply Theorem 1″ for $w(x)==\bar{y}(x)-q(x_0)y(x)$ with the initial point T. We get $q(x_0)y(x)\leqslant\bar{y}(x)$ $(T\leqslant x<B)$. Since $T\leqslant x_0$, (5) holds. Dividing by $y(x)$ yields the assertion of Theorem 2.

4. THE EXISTENCE OF POSITIVE SOLUTIONS OF (2 =) WITH DIFFERENT ASYMPTOTIC BEHAVIOUR

In the sequel we assume $B = +\infty$. A solution $u(x)$ is said to be positive if $u(x) > 0$ holds for $E(A) \leqslant x < +\infty$. Suppose that (2 =) admits a positive solution. Then the solutions of (2 =) constitute a linear function system of penetration type.*

The notion of linear function system of penetration type was introduced in [1]. If there exists an interval $P \leqslant x < +\infty$ such that in this interval any solution of (2 =) can be represented as a fixed solution multiplied by a suitable real number, then the set of solutions of (2 =) is said to be finally one dimensional. Since we have just assumed the existence of a positive solution of (2 =) therefore in the case of final one dimensionality any solution of (2 =), for x large enough is scalar multiple of this positive solution. Now [1, Satz 9.] can be formulated as follows:

Theorem 3. *There exist two positive solutions* $y_1(x), y_2(x)$ *of* (2 =), *such that* $y_2(x) = o(y_1(x))$ $(x \to +\infty)$ *if and only if the set of solutions of* (2 =) *is not finally one dimensional.*

Thus assuming the existence of asymptotically distinguishable solutions of (2 =) we exclude only the uninteresting finally one dimensional case. Here we shall not investigate the problem of existence of at least one positive solution of (2 =).

5. SPLITTING THE SET OF POSITIVE SOLUTIONS OF (2 ≥). USING SOLUTIONS OF (2 =)

Theorem 4. *Let* $y_1(x)$ *and* $y_2(x)$ *be positive solutions of* (2 =) *such that* $y_2(x) = o(y_1(x))$ *as* $x \to +\infty$, *and let* $u(x)$ *be a positive solution of* (2 ≥). *Then either there exists a constant* $k > 0$ *such that* $ky_1(x) \leqslant u(x)$ *for* $A \leqslant x < +\infty$, *or there exists a constant* $K > 0$, *such that* $u(x) \leqslant Ky_2(x)$ *for* $A \leqslant x < +\infty$.

Proof. Choose K so that for $E(A) \leqslant x \leqslant A$ $u(x) \leqslant Ky_2(x)$ holds. If this inequality is satisfied for all x $(A \leqslant x < +\infty)$, then the assertion is

*In German: "lineares Funktionensystem vom Durchdringungstyp".

true. If not, then there exists $P > A$, such that $Ky_2(P) < u(P)$. Let $v(x) = Ky_2(x) + cy_1(x)$. Choose $c > 0$ so that $v(P) < u(P)$ and put $T = \inf\{x;\ v(x) = u(x),\ E(A) \leqslant x < +\infty\}$. We have $A < T < P$, $v(T) = u(T)$ and $v(x) \geqslant u(x)$ for $E(T) \leqslant x < T$. Theorem 2 yields $v(x) \leqslant y(x)$ $(T \leqslant x < +\infty)$ whence $cy_1(x) \leqslant u(x)$ $(T \leqslant x < +\infty)$. Decreasing c if necessary, one can find a constant k such that $ky_1(x) \leqslant u(x)$ $(A \leqslant x < +\infty)$. The proof is complete.

If $y_1(x), y_2(x)$ and $u(x)$ become positive only for x large enough, then Theorem 4 can be applied from a larger initial point. If $u(x)$ is not assumed to be positive, the assertion still remains true for x large enough, essentially by the same argument. It can be proved by the same method that if an oscillatory solution $u(x)$ of $(2\geqslant)$ once gets below $y_2(x)$ in an influence domain $E(P) \leqslant x \leqslant P$, then it can not intersect with $y_2(x)$ for $x \geqslant P$.

Theorem 4 provides an extension of the dichotomy principle for equation containing on their right hand side a term $l(x)y(x)$ with no restriction on the sign of $l(x)$. For a discussion of dichotomy cf. [2, §33]. Several nonlinear equations are also included, for example $u'(x) = l(x)u(x) - L_x u(x-s) + f(x,u)$ $(A \leqslant x < +\infty)$. Here $f(x,u)$ denotes a real valued functional of x and u, such that $f(x,u) \geqslant 0$ for $A \leqslant x < +\infty$.

6. ENCLOSURE OF POSITIVE SOLUTIONS OF $(2\leqslant)$ BY SOLUTIONS OF $(2=)$

Theorem 5. *Let $y_1(x)$ and $y_2(x)$ be positive solutions of $(2=)$, such that $y_2(x) = o(y_1(x))$ $(x \to +\infty)$. Let $u(x)$ be a positive solution of $(2\leqslant)$. Then there exist positive constants k and K such that $ky_2(x) \leqslant u(x) \leqslant Ky_1(x)$ $(A \leqslant x < +\infty)$.*

Proof. Choose $c > 0$ so that $cy_2(x) > y_1(x)$ $(E(A) \leqslant x \leqslant A)$ and define $v(x) = y_1(x) - cy_2(x)$. Denote by P the first zero of v. This exists surely, $A < P < +\infty$. Choose Q so that $v(Q) > 0$ holds, and choose d so that $dv(Q) \geqslant u(Q)$ holds and put $w(x) = dv(x)$. Let $T = \inf\{x;\ u(x) = w(x),\ A \leqslant x < +\infty\}$. Then $P < T < +\infty$, $w(x) \leqslant u(x)$

$(E(T) \leqslant x \leqslant T)$ and $w(T) = u(T)$. Since $w(x)$ is a solution of (2 =), Theorem 2 yields for $T \leqslant x < +\infty$ $u(x) \leqslant w(x) = dy_1(x) - cy_2(x) \leqslant dy_1(x)$. Increasing d if necessary, we get $u(x) \leqslant Ky_1(x)$ $(A \leqslant x < +\infty)$. Now, define $k > 0$ so that $ky_2(x) \leqslant u(x)$ for $E(A) \leqslant x \leqslant A$. We claim that this inequality holds for all $x \geqslant A$. Suppose conversely, that there exists an R with $ky_2(R) > u(R)$. If $z(x) = ky_2(x) - ly_1(x)$, where l is a sufficiently small positive constant, then $z(R) > u(R)$ still holds. Put $T = \inf\{x;\ z(x) = u(x),\ A \leqslant x < +\infty\}$. We have $A < T < +\infty$, $z(x) \leqslant u(x)$ $(E(T) \leqslant x \leqslant T)$ and $z(T) = u(T)$. By Theorem 2 $u(x) \leqslant z(x)$ $(T \leqslant x < +\infty)$ what leads to a contradiction since $u(x)$ remains positive for all x. This proves Theorem 5.

7. INTEGRABLE CASES OF (2 =)

In some special cases the comparison functions $y_1(x)$ and $y_2(x)$ occurring in Theorems 4 and 5 can explicitly be determined. Consider in differentiated form the equation

$$y'(x) = ay(x) - by(x - c) \qquad (A \leqslant x < +\infty).$$

Here a, b, c are constants $b, c > 0$. The Euler type substitution yields the characteristic equation

$$a - \lambda = be^{-\lambda c}. \tag{6}$$

Put $D = \frac{1}{c}(1 + \ln bc)$. Then it is easy to see that

1. For $a < D$ no real zero exists.

2. If $a = D$ then one has a real double root $\lambda_{12} = \frac{1}{c} \ln bc$.

3. For $a > D$ there are two simple real zeros λ_1, λ_2, moreover

$$-\infty < \lambda_2 < \lambda_{12} < \lambda_1 < +\infty.$$

In case 2 we can choose $y_1(x) = (x - d)e^{\lambda_{12}x}$ $(d < A - c)$ and $y_2(x) = e^{\lambda_{12}x}$, in case 3 $y_1(x) = e^{\lambda_1 x}$ and $y_2(x) = e^{\lambda_2 x}$.

An example for equations with nonconstant coefficients is

$$y'(x) = \frac{a}{x}\, y(x) - \frac{b}{x}\, y(kx) \qquad (1 \leq x < +\infty).$$

Here a, b, k are constants, $b > 0$, $0 < k < 1$. The substitution $y = x^\lambda$ leads to the characteristic equation

$$\alpha - \lambda = bk^\lambda. \tag{7}$$

The substitution $k = e^{-c}$ transforms (7) into (6). Hence for the zeros of (6) the following holds. Denote

$$K = -\frac{1}{\ln k}\,(1 + \ln(-b \ln k))$$

1. If $a < K$, then no real zero exists.

2. If $a = K$, one has the double real root

$$\lambda_{12} = \frac{1}{-\ln k} \ln(-b \ln k).$$

3. If $a > K$, then there exist two simple real roots

$$-\infty < \lambda_2 < \lambda_{12} < \lambda_1 < +\infty$$

In case 2 one can choose $y_1(x) = (\ln x - d)x^{\lambda_{12}}$ $(d < \ln k)$ and $y_2(x) = x^{\lambda_{12}}$.

In case 3: $y_1(x) = x^{\lambda_1}$, $y_2(x) = x^{\lambda_2}$.

REFERENCES

[1] E. Kozakiewicz, Über lineare Funktionensysteme vom Durchdringungstyp, *Periodica Mathematica Hungarica,* Vol. 7 (1), (1976).

[2] A.D. Myškis, *Linear differential equations with retarded argument,* Moscow, 1972.

E. Kozakiewicz

Humboldt-Universität zu Berlin, Sektion Mathematik, 108 Berlin, Unter den Linden 6, GDR.

COLLOQUIA MATHEMATICA SOCIETATIS JÁNOS BOLYAI
15. DIFFERENTIAL EQUATIONS, KESZTHELY (HUNGARY), 1975.

SUFFICIENT CONDITIONS FOR THE OSCILLATION OF NONLINEAR DELAY DIFFERENTIAL EQUATIONS

P. MARUŠIAK

We consider the nonlinear delay differential equation

$$(1) \qquad [r(t)y^{(n-1)}(t)]' + F(t, y[h_1(t)], \dots, y^{(n-2)}[h_m(t)]) = 0, \quad n \geqslant 2$$

where

$$(2) \qquad \begin{aligned} &r \in C[R_+ \equiv [0, \infty), (0, \infty)], \; r(t) \quad \text{is nondecreasing for} \\ &t \geqslant t_0 \in R_+, \quad \int^{\infty} \frac{ds}{r(s)} = \infty \end{aligned}$$

$$(3) \qquad h_i \in C[R_+, R], \quad h_i(t) \leqslant t \quad \text{for} \quad t \in R_+, \qquad i = 1, 2, \dots, m$$

$$(4) \qquad F \in C[D \equiv R_+ \times R^{m(n-1)}, R], \quad y_{10}y_{i0} > 0 \qquad (i = 1, \dots, m)$$

implies $y_{10}F(t, y_{10}, \dots, y_{m,n-2}) > 0$ for all $t \in R_+$.

We shall assume that under the initial conditions

$$y^{(k)}(t) \equiv \Phi^{(k)}(t), \quad t \leqslant t_0, \quad (k = 0, 1, \dots, n-2),$$

$$y^{(n-1)}(t_0) = y_0^{(n-1)}$$

equation (1) admits a solution defined for all $t \geqslant t_0 \in \boldsymbol{R}_+$. A solution $y(t)$ of (1) is called oscillatory if the set of zeros of $y(t)$ is not bounded from the right. Otherwise a solution is called nonoscillatory. A nonoscillatory solution is said to be strongly monotone, if it tends monotonically to zero as $t \to \infty$ together with its first $n-1$ derivatives.

The purpose of this paper is to give sufficient conditions for all solutions of (1) to be oscillatory in the case n is even and to be either oscillatory or strongly monotone when n is odd. Our result generalizes Theorem 1 [2], Corollary of Theorem 2 [3] and Theorem 1 [5].

Let us denote $R_k[h(t), h(u)] = \int_{h(u)}^{h(t)} \dfrac{[x - h(u)]^k}{r(x)}\, dx.$

Theorem 1. *Let the functions in* (1) *satisfy* (2)-(4) *and, in addition, suppose that there exist function* h, p, f *and a positive constant* $\alpha \neq 1$, *such that*

$$\begin{aligned} &h \in C^1[\boldsymbol{R}_+, \boldsymbol{R}], \quad h(t) \leqslant h_i(t), \quad (i = 1, \dots, m), \\ &h'(t) \geqslant 0, \quad t \geqslant t_0 \in \boldsymbol{R}_+ \quad \lim_{t\to\infty} h(t) = \infty \end{aligned} \tag{5}$$

$$p \in C[\boldsymbol{R}_+, (0, \infty)] \tag{6}$$

$$f \in C[\boldsymbol{R}, \boldsymbol{R}], \quad xf(x) > 0 \quad \textit{for} \quad x \neq 0, \tag{7}$$

$f(x)$ *is nondecreasing on* $\boldsymbol{R}$,

$$\liminf_{|x|\to\infty} \frac{|f(x)|}{|x|^\alpha} > 0 \tag{7'}$$

$$\begin{aligned} &|F(t, y_{10}, \dots, y_{m,n-2})| \geqslant p(t)\,|f(y_{10})| \\ &\textit{for} \quad (t, y_{10}, \dots, y_{m,n-2}) \in D. \end{aligned} \tag{8}$$

Then

$$\int^{\infty} (R_{n-2}[h(t), h(T)])^{\beta} p(t)\, dt = \infty, \tag{9}$$

where $\beta = \min\{1, \alpha\}$, $T \in \boldsymbol{R}_+$

is a sufficient condition for all solutions of (1) to be oscillatory if n is even and to be either oscillatory or strongly monotone when n is odd.

Proof. Let us suppose that there exists a nonoscillatory solution $y(t)$ of (1). Let $y(t) > 0$ for $t \geqslant t_0 \in R_+$. The case $y(t) < 0$ can be treated similarly. In view of (5), there exists $t_1 \geqslant t_0$ such that $y[h(t)] > 0$, $y[h_i(t)] > 0$, $i = 1, 2, \ldots, m$. Then from (1), concerning (4), (8), we have

$$(10) \qquad [r(t)y^{(n-1)}(t)]' \leqslant -p(t)f(y[h_1(t)]) < 0, \qquad t \geqslant t_1$$

which implies that $r(t)y^{(n-1)}(t)$ is nonincreasing. We claim that $y^{(n-1)}(t) \geqslant 0$ for $t \geqslant t_1$. For, if $y^{(n-1)}(\bar{t}) < 0$ for some $\bar{t} \geqslant t_1$, then $r(t)y^{(n-1)}(t) \leqslant r(\bar{t})y^{(n-1)}(\bar{t}) < 0$, $t \geqslant \bar{t}$. Integrating the last inequality between $\bar{t}$ and t, we obtain

$$(11) \qquad y^{(n-2)}(t) \leqslant y^{(n-2)}(\bar{t}) + r(\bar{t})y^{(n-1)}(\bar{t})R_0[t, \bar{t}].$$

Now as $t \to \infty$, the right hand side of (11), in view of (2), tends to $-\infty$, which is contradiction, since $y(t) > 0$ on $[t_1, \infty)$.

Integrating (10) from t $(t > t_1)$ to ∞ and neglecting $r(\infty)y^{(n-1)}(\infty)(\geqslant 0)$, we get

$$r(t)y^{(n-1)}(t) \geqslant \int_t^\infty p(s)f(y[h_1(s)])\,ds.$$

From the last inequality, by the monotonicity of $r(t)y^{(n-1)}(t)$, we obtain

$$(12) \qquad y^{(n-1)}[h(t)] \geqslant \frac{1}{r[h(t)]} \int_t^\infty p(s)f(y[h_1(s)])\,ds, \qquad t \geqslant t_1.$$

$y^{(n-1)}(t) > 0$ implies either

(a), $y^{(n-2)}(t) \leqslant 0$ for $t \geqslant t_1$, or

(b), $y^{(n-2)}(t) > 0$ for $t \geqslant t_2 \geqslant t_1$.

(a) If $y^{(n-2)}(t) \leqslant 0$, $y(t) > 0$ on $[t_1, \infty)$, then, by a lemma of Ryder and Wend [4], exactly one of the following cases occurs:

(i) $y'(t), \ldots, y^{(n-2)}(t)$ tend monotonically to zero as $t \to \infty$,

(ii) there exists an odd integer k, $1 \leq k \leq n-3$ $(n > 3)$ such that $\lim_{t\to\infty} y^{(n-j)}(t) = 0$ for $2 \leq j \leq k+1$, $\lim_{t\to\infty} y^{(n-k-2)}(t) \geq 0$, $\lim_{t\to\infty} y^{(n-k-3)}(t) > 0$ and $\lim_{t\to\infty} y^{(n-i)}(t) = +\infty$, $i = k+4, \ldots, n$.

(iii) $y^{(n-i)}(t) > 0$, $(i = 0, 1, \ldots, n-1)$, $t \geq t_2$.

Choose $T \geq t_2$, so large that $h'(t) \geq 0$, $(i = 1, 2, \ldots, m)$, $r(t)$ is nondecreasing for $t \geq T$.

Suppose that case (i) holds.

Multiply (12) by $h'(t)$ and integrate the resulting inequality from t to ∞, $(t \geq T)$, we have

$$-y^{(n-2)}[h(t)] \geq \int_t^\infty R_0[h(s), h(t)]p(s)f[y[h_1(s)]]\,ds. \tag{13}$$

Repeating the above procedure, we get

$$(-1)^{n-2}y'[h(t)] \geq \int_t^\infty \frac{R_{n-3}[h(s), h(t)]}{(n-3)!}\,p(s)f(y[h_1(s)])\,ds. \tag{14}$$

I. If n is even then (14) implies $y'(t) \geq 0$ for $t \geq T$. Thus

$$y'[h(t)] \geq \int_t^\infty \frac{R_{n-3}[h(s), h(t)]}{(n-3)!}\,p(s)f(y[h(s)])\,ds, \tag{15}$$

since $h(t) \leq h_1(t)$ and $f(x)$ is nondecreasing in x.

Multiplying (15) by $h'(t)$ and integrating from T to t $(T < t)$, we have

$$\begin{aligned} y[h(t)] \leq{}& \int_T^t \frac{R_{n-2}[h(s), h(T)]}{(n-2)!}\,p(s)f(y[h(s)])\,ds\,+ \\ &+ \frac{R_{n-2}[h(t), h(T)]}{(n-2)!}\int_t^\infty p(s)f(y[h(s)])\,ds. \end{aligned} \tag{16}$$

Let $\alpha > 1$. Raising (16) (with the second neglected term on the right side) by $-\alpha$, multiplying by $\frac{R_{n-2}[h(t), h(T)]}{(n-2)!} p(t)f(y[h(t)])$ and integrating the resulting inequality from t_3 to t_4 $(T \leqslant t_3 < t_4)$, we get

$$\int_{t_3}^{t_4} \frac{R_{n-2}[h(s), h(T)]}{(n-2)!} p(s) \frac{f(y[h(s)])}{(y[h(s)])^\alpha} ds \leqslant \tag{17}$$

$$\leqslant \frac{1}{1-\alpha} \Big[\Big\{\int_T^t \frac{R_{n-2}[h(s), h(T)]}{(n-2)!} p(s)f(y[y(s)]\Big\}^{1-\alpha} ds\Big]_{t_3}^{t_4}.$$

For $t_4 \to \infty$ the right hand side of (17) is bounded and therefore the integral on the left hand of (17) converges.

Now, there are two possibilities: Either $\lim_{t\to\infty} y(t) = b > 0$, or $\lim_{t\to\infty} y(t) = \infty$. In both cases, by the assumptions (7), (7′), there exists $T_1 > t_3$ such that

$$\frac{f(y[h(t)])}{(y[h(t)])^\alpha} \geqslant d > 0, \qquad t \geqslant T_1.$$

Then (17) implies

$$\infty > d \int_{T_1}^{\infty} R_{n-2}[h(s), h(T)]p(s)\, ds,$$

which contradicts (9).

Let $\alpha < 1$. Raising (16) (with the first neglected term on the right hand) by $-\alpha$, we have

$$(y[h(t)])^{-\alpha}(R_{n-2}[h(t), h(T)])^\alpha \leqslant \Big\{\int_t^\infty \frac{p(s)f(y[h(s)])}{(n-2)!} ds\Big\}^{-\alpha}. \tag{18}$$

Multiplying (18) by $\frac{p(t)f(y[h(t)])}{(n-2)!}$ and integrating from t_3' to t_4' $(T \leqslant t_3' < t_4')$, we obtain

$$\int_{t_3'}^{t_4'} \frac{(R_{n-2}[h(s), h(T)])^\alpha}{(n-2)!} p(s) \frac{f(y[h(s)])}{(y[h(s)])^\alpha} ds \leqslant$$

$$\leqslant -\frac{1}{1-\alpha}\left[\left\{\int_t^{\infty}\frac{p(s)f(y[h(s)])}{(n-2)!}\,ds\right\}^{1-\alpha}\right]_{t_3'}^{t_4'}.$$

Now, exactly as in the case $\alpha > 1$, we can derive

$$\int_{t_3'}^{\infty}(R_{n-2}[h(s), h(T)])^{\alpha}p(s)\,ds < \infty,$$

which contradicts (9).

II. If n is odd, then (14) reduces to

$$(19)\qquad -y'[h(t)] \geqslant \int_t^{\infty}\frac{R_{n-3}[h(s), H(t)]}{(n-3)!}\,p(s)f(y[h_1(s)])\,ds$$

and this implies $y'(t) < 0$ for $t \geqslant T$. Hence $y(t)$ decreases to a limit $L \geqslant 0$. Suppose $L > 0$. We take T so large that $f(y[h_1(t)]) \geqslant \frac{1}{2}f(L)$ for $t \geqslant T$. Multiplying (19) by $h'(t)$ and integrating the resulting inequality from T to ∞, we have after the modification

$$y[h(T)] \geqslant L + \frac{f(L)}{2}\int_T^{\infty}\frac{R_{n-2}[h(s), h(T)]}{(n-2)!}\,p(s)\,ds,$$

which contradicts (9). Therefore $L = 0$ and so the nonoscillatory solution $y(t)$ must be strongly monotone.

Suppose case (ii) holds. Then there exists a $t_3 \geqslant T$ such that $y^{(j)}(t) > 0$ for $j = 0, 1, \ldots, n-k-3$. If we proceed as in case (i), we obtain

$$(20)\qquad y^{(n-k-2)}[h(t)] \geqslant \int_t^{\infty}\frac{R_k[h(s), h(t)]}{k!}\,p(s)f(y[h(s)])\,ds.$$

Multiplying (20) by $h'(t)$, integrating from t_3 to t $(t_3 < t)$, we have

$$y^{(n-k-3)}[h(t)] \geqslant \frac{R_{k+1}[h(t), h(t_3)]}{(k+1)!}\int_t^{\infty}p(s)f(y[h(s)])\,ds.$$

Repeating the above procedure and using $r(t)$ is nondecreasing for $t \geqslant T$, we obtain

(21) $$y'[h(t)] \geqslant \frac{R_{n-3}[h(t), h(t_3)]}{(n-3)!} \int_t^{\infty} p(s) f(y[h(s)])\, ds, \qquad t \geqslant t_3.$$

Multiplying (21) by $h'(t)$ and then integrating from t_3 to t, we get the following inequality

$$y[h(t)] \geqslant \int_{t_3}^{t} \frac{R_{n-2}[h(s), h(t_3)]}{(n-2)!} p(s) f(y[h(s)])\, ds +$$

$$+ \frac{R_{n-2}[h(t), h(t_3)]}{(n-2)!} \int_t^{\infty} p(s) f(y[h(s)])\, ds,$$

which is analogous to (16).

The proof of now proceeds exactly as in case (i).

Suppose case (iii) holds. Multiply (12) by $h'(t)$, integrate from T to t $(T < t)$, then use $y^{(n-2)}[h(T)] \geqslant 0$, $y[h_1(t)] \geqslant y[h(t)]$ and (6), we obtain

(22) $$y^{(n-2)}[h(t)] \geqslant \int_T^t R_0[h(s), h(T)] p(s) f(y[h(s)])\, ds + \\ + R_0[h(t), h(T)] \int_t^{\infty} p(s) f(y[h(s)])\, ds,$$

from which we get

$$y^{(n-2)}[h(t)] \geqslant R_0[h(t), h(T)] \int_t^{\infty} p(s) f(y[h(s)])\, ds$$

Repeating the above procedure $\overline{n-1}$ times, nearby using $r(t)$ is nondecreasing for $t \geqslant T$, we have the inequality analogous to (21). The proof now proceeds exactly as in case (ii). The proof of Theorem 1 is therefore complete.

Consider the equation

(23) $$[r(t) y^{(n-1)}(t)]' + p(t) |y[h(t)]|^{\gamma} \operatorname{sign} y[h(t)] = 0$$

where r, h and p satisfy conditions (2), (5) and (6), respectively, γ is a positive constant.

Corollary 1. *Assume that*

$$\int^{\infty} (R_{n-2}[h(t), h(T)])^{\nu} p(t)\, dt = \infty,$$

where

$$\nu = \begin{cases} 1 & \text{for} \quad \gamma > 1 \\ 1 - \epsilon;\ (0 < \epsilon < 1) & \text{for} \quad \gamma = 1 \\ \gamma & \text{for} \quad 0 < \gamma < 1 \end{cases}$$

Then all solutions of (23) *are oscillatory if* n *is even and they are either oscillatory or strongly monotone when* n *is odd.*

Remark. For $n = 2$, only case (iii) can occur. By virtue of (22), without the assumption that $r(t)$ is nondecreasing we can prove Theorem 1, because the proof proceeds exactly as in case (i).

Corollary 2. *Consider the equation*

(24) $$[r(t)y'(t)]' + p(t)\,|y[h(t)]|^{\gamma} \operatorname{sign} y[h(t)] = 0$$

where $r \in C[R_+, (0, \infty)]$, $R_0[t, T] \to \infty$ *as* $t \to \infty$, $(T \in R_+)$, h *and* p *satisfy* (5) *and* (6), γ *is a positive constant.*

If

$$\int^{\infty} (R_0[h(t), h(T)])^{\nu} p(t)\, dt = \infty,$$

where

$$\nu = \begin{cases} 1 & \text{for} \quad \gamma > 1 \\ 1 - \epsilon,\ (0 < \epsilon < 1) & \text{for} \quad \gamma = 1 \\ \gamma & \text{for} \quad 0 < \gamma < 1 \end{cases}$$

then all solutions of (24) *are oscillatory.*

Proof. Corollary 2 follows, considering the remark, from Corollary 1.

Corollary 2 extends results of Theorem 1 and Theorem 2 [1].

REFERENCES

[1] Ja.V. Bykov – L.Ja. Bykova – E.I. Šercov, Sufficient conditions for oscillations of nonlinear differential equations with deviating argument. *Differencial'nye Uravnenija*, 9 (1973), 1555-1560 (in Russian).

[2] T. Kusano – H. Onose, Oscillation of solutions of nonlinear differential delay equations of arbitrary Order. *Hiroshima Math. J.*, 2 (1972), 1-13.

[3] P. Marušiak, Oscillation of solutions of delay differential equations. *Czechoslovak Math. J.*, 2 (1974), 284-291.

[4] G.H. Ryder – D.V.V. Wend, Oscillation of solutions of certain ordinary differential equations of n-th order. *Proc. Amer. Math. Soc.*, 25 (1970), 463-469.

[5] V.N. Ševelo – N.V. Varech, On some properties of solutions of differential equations with delay. *Ukrain. Mat. Ž.*, 24 (1972), 807-813 (in Russian).

P. Marušiak

Katedra Matematiky, Marxa a Engelsa 25, 01088 Zilina, Czechoslovakia.

COLLOQUIA MATHEMATICA SOCIETATIS JÁNOS BOLYAI
15. DIFFERENTIAL EQUATIONS, KESZTHELY (HUNGARY), 1975.

ON THE CONDITIONS FOR THE NON-OSCILLATORY SOLUTIONS OF NON-LINEAR THIRD ORDER DIFFERENTIAL EQUATIONS

B. MEHRI

In this article we are concerned with asymptotic and nonoscillatory bounded solutions of the non-linear third order differential equation

(1) $$x''' + f(t, x) = 0.$$

We shall assume that the function $f(t, x)$ satisfies the Caratheodory Conditions in every finite rectangular region $0 \leqslant t < \infty$, $|x| < \infty$. Here

(2) $$x f(t, x) \geqslant 0$$

(3) $$|f(t, x_1)| < |f(t, x_2)|, \quad \text{if} \quad |x_1| < |x_2|, \quad x_1 \cdot x_2 > 0.$$

A solution $x(t)$ of (1) which exists in the future is said to be oscillatory if for every $T > 0$, there is a $t_0 > T$ such that $x(t_0) = 0$, non-oscillatory, otherwise.

Theorem 1. *In order for* (1) *to have a bounded nonoscillatory solution it is sufficient that the conditions*

$$(4) \qquad \int_{t_0}^{\infty} t^2 |f(t, c)| \, dt < \infty$$

be satisfied for some non-zero constant c.

Proof. We have to prove that if the condition

$$\int_{t_0}^{\infty} t^2 |f(t, c)| \, dt < \infty$$

is satisfied for some constant c, then (1) has at least one nonoscillatory bounded solution. Consider the integral equation

$$(5) \qquad x(t) = \frac{c}{2} + \frac{1}{2!} \int_t^{\infty} (s - t)^2 f(s, x(s)) \, ds$$

where $t > t_1 > 1$ is so large that

$$(6) \qquad \int_{t_1}^{\infty} s^2 |f(s, c)| \, ds < |c|.$$

Now consider the sequence $\{x_n(t)\}$, defined in the following manner:

$$(7) \qquad x_0(t) = \frac{c}{2}$$

$$x_n(t) = \frac{c}{2} + \frac{1}{2!} \int_t^{\infty} (s - t)^2 f(s, x_{n-1}(s)) \, ds.$$

Using (2), (3) and (7) we easily deduce from (7) that

$$(8) \qquad \frac{|c|}{2} < x_n(t) \operatorname{sign} c < |c|, \qquad n = 1, 2, \ldots .$$

Again, from (7), we have

$$(9) \qquad x_n'(t) = - \int_t^{\infty} (s - t) f(s, x_{n-1}(s)) \, ds,$$

hence

$$(10) \qquad |x_n'(t)| < \int_t^{\infty} s |f(s, c)| \, ds < \int_t^{\infty} s^2 |f(s, c)| \, ds < |c|.$$

for $t > t_1 > 1$. It follows from (9) and (10) that the sequence $\{x_n(t)\}$ defines a uniformly bounded and equicontinuous family on $[t_1, \infty)$, hence it follows from Arzela – Ascoli theorem that there exists a subsequence $\{x_{n_k}(t)\}$ uniformly convergent on every subinterval of $[t_1, \infty)$. Now a standard argument, see for example [2], yields a solution $x(t)$ which is a solution to (5), as easily checked, a solution of the differential equation (1). But on the other hand according to (8), $x(t)$ is bounded nonoscillatory.

Theorem 2. *Let the function f satisfy the conditions* (2), (3), *and in addition suppose*

(11) $$|f(t, x)| < a(t)|x|$$

where $a(t)$ is nonnegative function such that

(12) $$\int_{t_0}^{\infty} t^2 a(t)\, dt < \infty.$$

Then equation (1) *has solutions which are asymptotic to αt^2 $(\alpha \neq 0)$ as $t \to \infty$.*

Proof. Consider the following integral equation

(13) $$x(t) = c_0 + c_1 t + c_2 t^2 - \frac{1}{2!} \int_{t_0}^{t} (t - s)^2 f(s, x(s))\, ds.$$

From this and in view of (11) we get, for $t > t_0$.

(14) $$|x(t)| < t^2 \left(A + \int_{t_0}^{t} a(s)|x(s)|\, ds\right)$$

where A is a positive constant.

Define the function

$$F(t) = A + \int_{t_0}^{t} a(s)|x(s)|\, ds.$$

Then

(15) $$|x(t)| < t^2 F(t).$$

This implies

(16) $$F(t) < A + \int_{t_0}^{t} a(s)s^2F(s)\,ds.$$

Now using Gronwall's inequality and (12) it follows

$$F(t) < K < \infty.$$

The inequality (15) now becomes

(17) $$|x(t)| < Kt^2.$$

Integrating Eq. (1) from t_1 to t we get

(18) $$x''(t) = x''(t_1) - \int_{t_1}^{t} f(s, x(s))\,ds$$

using (11), (12), (17), the integral in (18) converges as $t \to \infty$ and therefore a finite $\lim_{t\to\infty} x''(t)$ exists. Choose t_1 so large that

$$1 - K\int_{t_1}^{\infty} s^2a(s)\,ds > 0$$

holds. If a solution $x(t)$ satisfies $x''(t_1) = 1$, then this solution $x(t)$ obeys the given asymptotic property.

REFERENCES

[1] D.V. Izymova, On the conditions of oscillation and nonoscillation of non-linear second order differential equations, *Differencial, nye uravn,* 2 (1966), 1572-1586.

[2] J.S.W. Wong, On second order nonlinear oscillation, *Math. Research Center University of Wisconsin, Technical Report No.* 836, also *Funkcial. Ekvac.*, 11 (1969), 207-234.

[3] P. Marusiak, Note on the Ladas' paper on oscillation and asymptotic behavior of solutions of differential equations with retarded argument, *Journal of Differential Equations,* 13 (1973), 150-156.

B. Mehri

Univ. of Arya-Mehr, Tehran, Iran.

COLLOQUIA MATHEMATICA SOCIETATIS JÁNOS BOLYAI
15. DIFFERENTIAL EQUATIONS, KESZTHELY (HUNGARY), 1975.

DECOMPOSITION OF THE SPACE OF SOLUTIONS OF RETARDED DIFFERENTIAL EQUATIONS AND THE OSCILLATION OF SOLUTIONS

S.B. NORKIN

§1. DECOMPOSITION OF THE SPACE OF SOLUTIONS OF A LINEAR SYSTEM

Consider the system of retarded linear differential equations

$$(1.1) \qquad x'(t) = \int_0^{\infty} [dr(t, s)]x(t - s), \qquad t \geq t_0$$

where $x \in R^n$, $r(t, s)$ is an $n \times n$ matrix, with entries satisfying the usual conditions of A. D. Myškis [1], integration with respect to s is understood in the sense of Stieltjes. We assume, that the set of s's which are places of change of the kernel $r(t, s)$, satisfying $t - s < t_0$ for some $t \geq t_0$, is non-void, i.e. the initial set E_{t_0} does not reduce to the initial point t_0. Then the fundamental initial value problem for (1.1) is the following: given a continuous n-vector $\Phi(t)$, defined on the initial set E_{t_0}, a solution of (1.1) satisfying the initial condition

$$(1.2)\qquad x(t-s) \equiv \Phi(t-s), \qquad t-s \in E_{t_0},\ t \geqslant t_0$$

should be found.

Conditions for existence, uniqueness and continuous dependence on the initial data for problem (1.1), (1.2) can be found for instance in the monograph of A.D. Myškis [1].

Under the conditions involved, the infinite dimensionals space Ω of initial vector functions Φ, which are continuous on E_{t_0}, yields an, infinite dimensional space L^∞_Ω of solutions of the problem (1.1), (1.2).

We consider also the system of ordinary differential equations

$$(1.3)\qquad x'(t) = A(t)x(t)$$

for $t \geqslant t_0$, containing no retarded argument, with initial condition

$$(1.4)\qquad x(t_0) = \Phi(t_0).$$

Here $x \in R^n$, $A(t)$ is a continuous $n \times n$ matrix for $t \geqslant t_0$, $\Phi(t_0)$ is the initial vector. It would be very interesting to compare the properties of the solutions of (1.1) and (1.3). However, this is not natural in general, since the space of solutions of the problem (1.3), (1.4) is n-dimensional, while that of problem (1.1), (1.2) is infinite dimensional. There are two ways to get through this difficulty. The first one is to impose conditions to the system (1.1) – these are generally speaking very restrictive – assuring that the space of solutions of the problem (1.1), (1.2) would be asymptotically finite dimensional, while the second one is to decompose the space L^∞_Ω into finite dimensional subspaces.

The first direction was pursued in the conference lectures of I. Győri [2], J. Kurzweil, J. Jarník [3], E. Kozakiewicz [4]. Even the title of I. Győri's lecture was: "On asymptotically ordinary differential equations".

I shall deal with the second direction. (For a more detailed treatment see the monographs of S.B. Norkin [5] and L.E. El'sgol'c, S.B. Nozkin [6]).

Let e be the n-dimensional column-vector with entries equal to 1, $D(x_0, x_1, \ldots, x_{n-1})$ a diagonal matrix with arbitrary real entries in the diagonal. Let $\varphi(t)$ be a fixed vector-function, continuous on the initial set E_{t_0} and such that

(1.5) $$\varphi(t_0) = e.$$

We shall investigate the homogeneous initial value problem for equation (1.1): we look for the solution of the equation (1.1) determined by the initial conditions

(1.6) $$\begin{aligned} &x(t_0) = D(x_0, x_1, \ldots, x_{n-1})e, \\ &x(t-s) \equiv D(x_0, x_1, \ldots, x_{n-1})\varphi_j(t-s), \qquad t-s \in E_{t_0}, \quad t \geqslant t_0 \end{aligned}$$

Denote by L^n_φ the space of solutions of the homogeneous initial value problem (1.1), (1.6) belonging to a fixed initial function $\varphi(t)$ $(\varphi(t_0) = e)$. The space L^n_φ is n-dimensional, thus the problems (1.1), (1.6) and (1.3), (1.4) get into a natural correspondence.

A system $u_1, u_2, \ldots, u_n$ of solutions is said to be fundamental if its Wronskain is nonzero for $t = t_0$:

(1.7) $$W(u_1(t), \ldots, u_n(t))|_{t=t_0} \neq 0.$$

In this case an arbitrary solution $x(t) \in L^n_\varphi$ can be represented in the form

(1.8) $$x(t) = \sum_{k=1}^{n} C_k u_k(t)$$

where the C_k's $(k = 1, \ldots, n)$ are suitable constants.

The investigation of the finite dimensional spaces of the homogeneous initial value problem makes possible to deduce the fundamental intrinsic property of the solutions of equation (1.1), implied by the presence of retardation. Actually the following theorem holds:

Theorem 1.1. *The space* L^n_φ *of solutions of the homogeneous initial value problem* (1.1), (1.6) *belonging to a fixed initial function* $\varphi(t)$ $(\varphi(t_0) = e)$ *coincides with the space of solutions of a non-retarded equa-*

tion of the form (1.3) *with a continuous matrix* $A(t)$, *if and only if the nontrivial solutions in* L^n_φ *have no zeros for* $t \geqslant t_0$.

For the scalar equation

$$(1.9) \qquad x^{(n)}(t) + \sum_{k=0}^{n-1} a_k(t)x^{(k)}(t) + \int_0^\infty x^{(j)}(t-s)\,dr(t,s) = 0, \qquad t \geqslant t_0$$

of order n, where the functions $a_k(t)$ are continuous, the kernel $r(t,s)$ satisfies the conditions of Myškis [1] and $j \in \{0, 1, \ldots, n-1\}$ is fixed, the following homogeneous initial value problem will be considered. Let the scalar initial function $\varphi_j(t)$, continuous on E_{t_0} be fixed, and such that

$$(1.5') \qquad \varphi_j(t_0) = 1.$$

We look for the solution of the equation (1.9) satisfying the initial conditions

$$(1.10) \qquad \begin{aligned} & x^{(k)}(t_0) = x_0^{(k)} \qquad (k = 0, 1, \ldots, n-1) \\ & x^{(j)}(t-s) \equiv x_0^{(j)}\varphi_j(t-s), \qquad t-s \in E_{t_0},\ t \geqslant t_0 \end{aligned}$$

where $x_0^{(0)}, \ldots, x_0^{(n-1)}$ are arbitrary real numbers. The space $L^n_{\varphi j}$ of solutions of the problem (1.9), (1.10) is n-dimensional. A fundamental system of solutions in $L^n_{\varphi_j}$ is defined by the analogue of the condition (1.7), in this case the expansion (1.8) holds.

The fundamental intrinsic property of solutions of the equation (1.9), implied by the presence of retardation is the possibility of occurrence of zeros of multiplicity $s \geqslant n$ of the nontirival solutions belonging to $L^n_{\varphi j}$ (Theorem 1.1).

For equation (1.9) the decomposition structure of the infinite dimensional space of solutions of the fundamental initial value problem into finite-dimensional subspaces $L^n_{\varphi_j}$ turns out to be especially simple: all $L^n_{\varphi j}$ have a unique common part – this is a $(n-1)$-dimensional space of solutions, belonging to the fixed value $x_0^{(j)} = 0$ (in the case $n = 2$ for more details see [5]).

In the nonlinear case the infinite dimensional manifold of solutions of a retarded differential equatioh obviously also can be decomposed into finite dimensional ones by restricting the corresponding set of initial functions. Such a decomposition provides the possibility to observe properties of the solutions, which may not hold for the manifold of global solutions.

§2. OSCILLATION OF SOLUTIONS OF SECOND ORDER EQUATIONS

In this § we shall investigate systems, described for $t \geqslant t_0$ by second-order retarded neutral differential equations of the form

$$x''(t) = f(t, x(t), x'(t), x(t-\tau(t)), x'(t-\tau(t)), x''(t-\tau(t))) \tag{2.1}$$

where $\tau(t) \geqslant 0$ is continuous on $[t_0, \infty)$. Let G be a fixed set of initial functions and suppose that the function $f(t, x, y, u, v, w)$ is such that to arbitrary initial function $\varphi \in G$ there corresponds a continuously differentiable solution of the fundamental initial value problem, defined for $[t_0, \infty)$. We do not exclude the special cases, when f does not depend on its last argument – retarded differential equation – or on its last three arguments – non-retarded equation.

Investigating the oscillation properties of equations of the form (2.1) a global approach – i.e. G is the set of all continuously differentiable initial functions on E_{t_0} – is also possible. For equations of order n this problem was treated inthe lacture of P. Marusiak [7]. However, for assuring the oscillation of all solutions there was necessary to impose a very restrictive condition: the sign of the right hand side has to be determined by the sign of merely one of the functional arguments. A restriction of the set G of admissible initial functions yields oscillation properties under less restrictive assumptions.

For example, the notion of φ-oscillation turned to be useful when investigating oscillation properties of the linear equation

$$x''(t) + N(t)x(t) + M(t)x(t-\tau(t)) = 0 \tag{2.2}$$

(see [5]). This means, that all nontrivial solutions belonging to the two-dimensional solution space L^2_φ of the homogeneous initial value problem for the equation (2.2) with initial function $\varphi(t)$ ($\varphi(t_0) = 1$) are oscillatory. It turns out, that in certain cases the equation (2.2) has oscillatory solutions, but it is not φ-oscillatory for each initial function φ, continuous on E_{t_0} (by Theorem 1.1 this is a consequence of the existence of nontrivial solutions with multiple zeros in each solution space L^2_φ). There are also special cases when φ-oscillation takes place for certain $\varphi(t)$'s, while the equation is not globally oscillatory. It is possible to give conditions (of course very restrictive ones) for global oscillation.

We shall suppose in the sequel that the choice of the set G of admissible initial functions already provides the necessary restriction and all considerations will be carried out for this set.

The literature devoted to the oscillation of solutions of equations of the form (2.1) is fearly extensive, however, couriously enough, there exists no unified definition of oscillation. The following three definitions occur in the literature:

A solution $x(t)$ of equation (2.1) is said to be oscillating, if

A: $x(t)$ changes sign in $[a, \infty)$ for arbitrary $a \geqslant t_0$;

B: the set of (isolated – B_+) zeros of the function is not bounded from the right;

C: the corresponding $(x, \dot{x})$ on the phase plane goes round the origin infinitely often.

The historical pattern of solutions is the function $A \sin \omega t$, possessing all of these properties. Probably this is the reason of the coexistence of three different definitions. This was natural as long as linear oscillating systems (described without retarded argument) were investigated. This is also natural for nonlinear systems (with small nonlinearity) when using the methods of Krylov – Bogolyubov – Mitropolskii. Also the definitions are equivalent for continuously differentiable functions without multiple zeros (see Theorem 1.1). For an arbitrary $x(t)$, continuously differentiable on $[t_0, \infty)$, C does not imply B, and B does not imply A. The solutions of (even the simplest) retarded equations often have such properties. For example the equation

$$(2.3) \qquad x''(t) + \frac{1}{2}\, x(t) - \frac{1}{2}\, x(t-\pi) = 0$$

of the form (2.1) with initial function $\varphi(t) = 1 - \sin t \;\; (-\pi \leqslant t \leqslant 0)$, has the solution

$$(2.4) \qquad x(t) = 1 - \sin t$$

satisfying condition B (actually B_+) without satisfying condition A.

It is easy to construct a retarded differential equation possessing a solution of the form

$$x(t) = \begin{cases} (t - t_1)^3 \sin (t - t_*)^{-1}, & 0 \leqslant t < t_* \\ 0 & t_* \leqslant t < \infty \end{cases}$$

which satisfies conditions C, B, but A.

Condition B is also satisfied by all finally trivial solutions, often occurring in the case of retarded equations.

On the other hand, any function continuously differentiable on $[t_0, \infty)$ and satisfying condition A, also satisfies conditions B and C, and consequently, condition A is the most suitable one to express the intuitive notion of oscillation (of the type $A \sin \omega t$). However, it is also natural to call oscillating the process described by solutions of the form (2.4) (with respect to the equilibrium $x \equiv 1$). Nevertheless this does not justify the definition B_+ since the functions $x(t) = 1 - \sin t + \epsilon$ are not less oscillating then the functions (2.4) while the former do not satisfy any of the conditions A, B, C.

In connection with this we propose here a more general definition of oscillation, that includes cases similar to the one mentioned here.

Definition 2.1. A solution $x(t)$ of (2.1) is said to be k-oscillatory if the function $x(t) - k$ changes sign on $[a, \infty)$ for all $a \geqslant t_0$.

Denote by $\mathscr{K}_x$ the set of values k for which a given solution $x(t)$ is k-oscillatory, and by $\mathscr{K}$ the intersection of all sets $\mathscr{K}_x$ for solutions $x(t) = x(t, \varphi)$ of (2.1), $\varphi \in G$.

Definition 2.2. The equation (2.1) is said to be k-oscillatory (or more precisely, k-oscillatory with respect to G) if the corresponding $\mathscr{K}$ is non-void.

Denote by $\mathscr{K}^*$ the union of all sets $\mathscr{K}_x$ belonging to solutions $x(t) = x(t, \varphi)$ of (2.1), $\varphi \in G$.

Definition 2.3. The equation (2.1) is said to be k-nonoscillatory (with respect to G) if the corresponding $\mathscr{K}^*$ is void.

Obviously if the equation (2.1) is oscillatory in the sense of definition A then it is k-oscillatory (the set $\mathscr{K}$ contains at least the point $k = 0$). If the equation is k-nonoscillatory then it is also nonoscillatory while the converse does not hold in general. The investigation of the structure of the sets $\mathscr{K}$ and $\mathscr{K}_x$ provides more information about the asymptotic behaviour of solutions than merely the knowledge of the oscillatoricity of equation (2.1) or their solutions.

This shows the usefulness of the definitions proposed.

We present some simple theorems about k-oscillation.

The continuity of solutions of (2.1) implies.

Theorem 2.4. *If the equation* (2.1) *is* k*-oscillatory then the corresponding* $\mathscr{K}$ *is simply connected.*

Consider the equation

$$(2.5) \qquad x'' + 2\alpha x' + \omega^2 x = 0.$$

If $\alpha^2 - \omega^2 \geqslant 0$, then the set $\mathscr{K}^*$ is void – the equation (2.5) is k-nonoscillatory. If $\alpha^2 - \omega^2 < 0$ then the equation (2.5) is k-oscillatory $\mathscr{K} = (-\infty, +\infty)$ if $\alpha < 0$ and $\mathscr{K} = \{0\}$ if $\alpha \geqslant 0$. The two-dimensional space of solutions of the homogeneous initial value problem for the equation (2.3), with initial function $\varphi(t) = 1 - \sin t$ contains, together with the solution $x_1(t) = 1 - \sin t$, also the solution $x_2(t) = \sin t - 1$. Here $\mathcal{K}_{x_1} = (0, 2)$, $\mathcal{K}_{x_2} = (-2, 0)$. The set $\mathcal{K}$ is void and the equation (2.3) is not k-oscillatory with respect to the set $G = \{C(1 - \sin t)\}$ of initial functions, though there exist k-oscillatory solutions.

These examples can be generalized:

Theorem 2.5. *If for the equation* (2.1) *the set* $\mathcal{K}$ *is non-void and the function* f *is a first order homogeneous function of their variables except the first (i.e. together with a solution* $x(t)$ *of* (2.1) *also* $Cx(t)$ *is a solution for an arbitrary constant* C*) then the set* $\mathcal{K}$ *is either the point* $k = 0$ *or the whole real line.*

Theorem 2.6. *If, under the conditions of Theorem* 2.5, *the equation* (2.1) *has at least one bounded nontrivial solution, then the set* $\mathcal{K}$ *reduces to the point* $k = 0$.

Theorem 2.7. *If the equation* (2.1) *is nonoscillatory in the sense of definition A and together with any solution* $x(f)$ *also* $-x(t)$ *is a solution, then the equation cannot be* k*-oscillatory.*

Let $g(t)$ be a twice differentiable function. Consider the equation

$$x'' + x = g''(t) + g(t). \tag{2.6}$$

1. $g(t) \equiv A$. The set $\mathcal{K}$ consists of the single point $k = A$.

2. $g(t) = A + \sin 2t$. The set $\mathcal{K}$ is the interval $(A - 1, A + 1)$.

3. $g(t) = A + \dfrac{t^2 + 2}{t^2 + 1} \sin 2t$. The set $\mathcal{K}$ is the segment $[A - 1, A + 1]$.

4. Let $g(t)$ be a function oscillating in the sense A with constant frequency $\omega \neq 1$, $\{t_i\}$ and $\{t_j'\}$ the sequences of places of minima and maxima of the function $g(t)$ respectively. For all i, let $g(t_i) = \alpha$, $\lim_{j \to \infty} g(t_j') = \beta > \alpha$ and for all j, $g(t_j') > \beta$. Then the set $\mathcal{K}$ is the semi-closed interval $(\alpha, \beta]$.

Similarly can be represented any possible type of interval. In each case for (2.6) $\mathscr{K} = \mathscr{K}_g$.

These examples can be generalized:

Theorem 2.8. *Let* $g(t)$ *be twice differentiable and suppose that equation* (2.1) *is of the form*

$$(2.7) \qquad L(x) + D(x - g(t)) = L(g(t))$$

where L *and* G *are linear and nonlinear operators respectively, and* $D(Cx) = CD(x)$. *Suppose further that the equation*

$$(2.8) \qquad L(x) + D(x) = 0$$

is k-oscillatory and all solutions of (2.8) *are bounded. Then each of the conditions*

(a) $g(t) \equiv \alpha$ *or* $g(t)$ *is k-oscillatory and* $\mathscr{K}_g = \{\alpha\}$;

(b) $g(t)$ *is k-oscillatory and the set* $\mathscr{K}_g$ *is not bounded;*

(c) $g(t)$ *is k-oscillatory, the set* $\mathscr{K}_g$ *is bounded and for an arbitrary solution* $u(t)$ *of equation* (2.8) *the set* $\mathscr{K}_{u+g}$ *is not a proper subset of* $\mathscr{K}_g$ *implies that equation* (2.7) *is k-oscillatory and* $\mathscr{K} = \mathscr{K}_g$. *(*$\mathscr{K} = \{\alpha\}$, *if* $g(t)$ *is not k-oscillatory).*

Indeed, for an arbitrary solution $u(t)$ of the equation (2.8) and for arbitrary C the function

$$(2.9) \qquad x(t) = Cu(t) + g(t)$$

is a solution of the equation (2.7). On the other hand, let $z(t)$ be an arbitrary solution of the equation (2.7), and $y(t) = z(t) - g(t)$. Then by (2.7)

$$L(g(t)) = L(y(t) + g(t)) + D(y(t)) = L(y(t)) + D(y(t)) + L(g(t)).$$

Hence $y(t)$ is a solution of (2.8) and $z(t)$ is of the form (2.9). Now the assertion of the theorem follows from (2.9), since by Theorem 2.6 for the equation (2.8) $\mathscr{K} = \{0\}$.

Finally, Theorem 2.4 and the above examples yield

Theorem 2.9. *If for the equation* (2.1) *the set* $\mathcal{K}$ *is non-void, then it is either a single point, or an interval (bounded or not, closed, open, or semi-closed).*

A further generalization of the notion of oscillation would arise by substituting the constant k in Definition 2.1 by a function $k(t)$ (oscillation with respect to a curved axis).

However, this case of introduction of new variables can be reduce to k-oscillation.

REFERENCES

[1] A.D. Myškis, *Linear differential equations with retarded argument,* Moscow, Nauka, 1972 (in Russian).

[2] I. Győri, On asymptotically ordinary functional differential equations, *(in this volume).*

[3] J. Kurzweil – J. Jarník, On Ryabov's special solutions of functional differential equations, *(in this volume).*

[4] E. Kozakiewicz, Über das asymptotische Verhalten der positiven Lösungen einiger Differentialgleichungen mit nacheilenden Argument, *(in this volume).*

[5] S.B. Norkin, *Differential equations of the second order with retarded argument,* Providence, Amer. Math. Soc., 1972.

[6] L.E. El'sgol'c – S.B. Norkin, *Introduction to the theory and application of differential equations with deviating arguments,* New York, Acad. Press, 1973.

[7] P. Marusiak, Oscillation of solutions of delay differential equations, *(in this volume).*

S.B. Norkin,

Moscow, A-445, Leningradskoe sosse 118, 1, 106. USSR.

COLLOQUIA MATHEMATICA SOCIETATIS JÁNOS BOLYAI
15. DIFFERENTIAL EQUATIONS, KESZTHELY (HUNGARY), 1975.

OSCILLATORY PROPERTIES OF CERTAIN SECOND ORDER DIFFERENTIAL EQUATIONS

L. PINTÉR

0. The equations to be considered in this article are the followings:

(E_1) $\qquad x'' + f(t)g(x) = 0,$

(E_2) $\qquad x'' + h(t)x' + f(t)g(x) = 0,$

(E_3) $\qquad x'' + f(t)g(x(\sigma(t))) = 0.$

We shall restrict attention to solutions existing on some halfline. A solution will be called oscillatory if it has arbitrarily large zeros.

Recently I. V. Kamenev [1] investigating the equation (E_1) obtained an interesting sufficient condition for oscillation which contains some known theorems as a special case. In his proof among others, he makes use of the assumptions $g'(s) \geqslant \epsilon > 0$ on $-\infty < s < \infty$, and $\int_{\delta}^{\infty} \frac{ds}{g(s)} < \infty$, $\int_{-\delta}^{-\infty} \frac{ds}{g(s)} < \infty$ for $\delta > 0$.

At first we shall prove a theorem; our method will be similar to that of Kamenev's theorem, though the conditions of the two theorems are not compatible. Further we employ the method on the ordinary differential equation (E_2) and on the special functional differential equation (E_3).

1. Suppose that the function $f(t) \in C[t_0, \infty)$ $(t_0 > 0)$; $g(s) \in \in C(-\infty, \infty)$, $sg(s) > 0$ if $s \neq 0$. Assume further that $\alpha(t)$, $\beta_i(t)$ $(i = = 0, \dots, n-1$; n is an arbitrary, fixed, natural number) and $\varphi(t)$ are functions with the following properties: $\alpha(t) \in C^1[t_0, \infty)$ and $\alpha'(t)$ is a nonnegative, decreasing function; the functions $\beta_i(t)$ $(i = 0, \dots, n-1)$ are positive, continuous functions on $[t_0, \infty)$; finally $\varphi(t) \in C[0, \infty)$ is nonnegative and nondecreasing.

Introduce the notations:

$$\gamma_1(t) = \beta_0(t), \gamma_2(t) = \beta_1(t) \int_{t_0}^{t} \gamma_1(s)\, ds, \dots, \gamma_n(t) =$$

$$= \beta_{n-1}(t) \int_{t_0}^{t} \gamma_{n-1}(s)\, ds;$$

and

$$\delta_1(t) = \frac{\int_{t_0}^{t} \alpha(s)\beta_0^2(s)\, ds}{\beta_0(t)}, \delta_2(t) = \frac{\int_{t_0}^{t} \delta_1(s)\beta_1^2(s)\, ds}{\beta_1(t)}, \dots$$

$$\dots, \delta_n(t) = \frac{\int_{t_0}^{t} \delta_{n-1}(s)\beta_{n-1}^2(s)\, ds}{\beta_{n-1}(t)}.$$

Assume that

$$A_n(t) = \Big[\int_{t_0}^{t} \gamma_n(s)\, ds\Big]^{-1} \int_{t_0}^{t} \beta_{n-1} dt_{n-1} \dots$$

$$\dots \int_{t_0}^{t_1} \beta_0\, ds \int_{t_0}^{s} \alpha(u) f(u)\, du \to \infty$$

as $t \to \infty$ $(t > t_0)$ and

$$\int^{\infty} \frac{\varphi(t)}{t^2}\, dt < \infty, \quad \int^{\infty} \frac{\varphi\left(\int_{t_0}^{t} \beta_{n-1}(s)\, ds\right)}{\delta_n(t)}\, dt = \infty.$$

Suppose that the functions $f(t), g(t)$ are such that there exist positive constants ϵ_1, ϵ_2 and k such that

$$(1) \qquad 0 < \epsilon_1 \leqslant \frac{g(s)}{s} - k \leqslant \epsilon_2 \qquad (s \neq 0)$$

and for every $T \geqslant t_0$ there exists $T^* \geqslant T$ so that

$$(2) \qquad \epsilon_1 \int_{T^*}^{t} \alpha(s) f^+(s)\, ds > \epsilon_2 \int_{T^*}^{t} \alpha(s) f^-(s)\, ds$$

for $t > T^*$ (f^+ and f^- denotes the positive resp. negative part of f).

Proposition. *The bounded solutions of* (E_1) *are oscillatory.*

Proof. Suppose that there exists a bounded nonoscillatory solution $x(t)$, we may assume that $x(t) > 0$ if $t \geqslant t_0$. By the Riccati transformation $w = x'x^{-1}$ we have the Riccati-like equation

$$w' + w^2 + f(t) \frac{g(x(t))}{x(t)} = 0.$$

Multiplying by $\alpha(t)$ and integrating from t_0 to t we get

$$\alpha(t) w(t) - \int_{t_0}^{t} \alpha'(s) w(s)\, ds + \int_{t_0}^{t} \alpha(s) w^2(s)\, ds =$$

$$= k_1 - \int_{t_0}^{t} \alpha(s) f(s) \frac{g(x(s))}{x(s)}\, ds.$$

Now multiply both sides by β_0 and integrate, then multiply by β_1 and integrate and so on till β_{n-1}.

We obtain the equality

$$B_n(t) + C_n(t) = \int_{t_0}^{t} \gamma_n(s)\, ds\, [k_1 + D_n(t) - A_n^*(t)],$$

where

$$B_n(t) = \int_{t_0}^{t} \beta_{n-1} dt_{n-1} \ldots \int_{t_0}^{t_1} \beta_0(s) \alpha(s) w(s)\, ds,$$

$$C_n(t) = \int_{t_0}^{t} \beta_{n-1} dt_{n-1} \ldots \int_{t_0}^{t_1} \beta_0(s)\, ds \int_{t_0}^{s} \alpha(u) w^2(u)\, du,$$

$$D_n(t) = \Big[\int_{t_0}^{t} \gamma_n(s)\, ds \Big]^{-1} \int_{t_0}^{t} \beta_{n-1} dt_{n-1} \ldots$$

$$\ldots \int_{t_0}^{t_1} \beta_0(s)\, ds \int_{t_0}^{s} \alpha'(u) w(u)\, du,$$

$$A_n^*(t) = \Big[\int_{t_0}^{t} \gamma_n(s)\, ds \Big]^{-1} \int_{t_0}^{t} \beta_{n-1} dt_{n-1} \ldots$$

$$\ldots \int_{t_0}^{t_1} \beta_0(s)\, ds \int_{t_0}^{s} \alpha(u) f(u) \frac{g(x(u))}{x(u)}\, du.$$

From the assumptions there is a number $k_2 > 0$ such that $A_n^*(t) \geqslant \geqslant k_2 A_n(t)$, therefore $A_n^*(t) \to \infty$ as $t \to \infty$.

Since the solution $x(t)$ is bounded, we obtain

$$\int_{t_0}^{t} \alpha'(s) w(s)\, ds = \alpha'(t_0) \int_{t_0}^{\xi} \frac{x'(s)}{x(s)}\, ds = \alpha'(t_0) \int_{x(t_0)}^{x(\xi)} \frac{dz}{z} \leqslant k_3,$$

consequently $D_n(t)$ is bounded too, $D_n(t) \leqslant k_3$.

Therefore there exists $t^* \geqslant t_0$ such that $k_1 + D_n(t) - A_n^*(t) < 0$ for $t > t^*$, i.e. $0 < C_n(t) < - B_n(t)$ or $C_n^2(t) < B_n^2(t)$ for $t > t^*$.

By applications of the Schwarz inequality

$$B_n^2(t) = \Big[\int_{t_0}^{t} (\sqrt{\delta_{n-1}}\,\beta_{n-1})\Big(\frac{1}{\sqrt{\delta_{n-1}}} \int_{t_0}^{t_{n-1}} \beta_{n-2}\, dt_{n-2} \ldots$$

$$\ldots \int_{t_0}^{t_1} \beta_0 \alpha w\, ds\Big)\, dt_{n-1}\Big]^2 \leqslant$$

$$\leqslant \int_{t_0}^{t} \delta_{n-1}\beta_{n-1}^2\, ds \int_{t_0}^{t} \frac{1}{\delta_{n-1}} \Big(\int_{t_0}^{t_{n-1}} \beta_{n-2}\, dt_{n-2} \ldots$$

$$\ldots \int_{t_0}^{t} \beta_0 \alpha w\, ds\Big)^2\, dt_{n-1} \leqslant \ldots \leqslant \delta_n(t)\, C_n'(t).$$

Hence

$$\frac{1}{\delta_n(t)} \leqslant \frac{C_n'(t)}{C_n(t)} \quad \text{for} \quad t \geqslant t^*.$$

On the other hand

$$C_n(t) \geqslant k_4 \int_{t^*}^{t} \beta_{n-1}(s)\, ds,$$

where

$$k_4 = \int_{t_0}^{t^*} \gamma_{n-2}\, dt_{n-2} \ldots \int_{t_0}^{s} \beta(u) w^2(u)\, du.$$

Therefore we obtain

$$\varphi\Big(k_4 \int_{t^*}^{t} \beta_{n-1}(s)\, ds\Big) \leqslant \varphi(C_n(t)) \quad \text{for} \quad t \geqslant t^*,$$

hence

$$\int_{t^*}^{t} \frac{\varphi\Big(k_4 \int_{t^*}^{s} \beta_{n-1}(u)\, du\Big)}{\delta_n(s)}\, ds \leqslant \int_{t^*}^{t} \frac{C_n'(s)\varphi(C_n(s))}{C_n^2(s)}\, ds =$$

$$= \int_{C_n(t^*)}^{C_n(t)} \frac{\varphi(s)}{s^2}\, ds \leqslant \int_{C_n(t^*)}^{\infty} \frac{\varphi(s)}{s^2}\, ds < \infty,$$

which is a contradiction, since as it can be easily seen, from

$$\int^{\infty} \frac{\varphi\left(\int_{t_0}^{t} \beta_{n-1}\, ds\right)}{\delta_n(t)}\, dt = \infty$$

it follows that also

$$\int^{\infty} \frac{\varphi\left(k_4 \int_{t^*}^{t} \beta_{n-1}\, ds\right)}{\delta_n(t)}\, dt = \infty.$$

In the proof the main problem was to assure the nonnegativity of the integral $\int_{t_1}^{t} \alpha(s) f(s) \left(\frac{g(x(s))}{x(s)} - k\right) ds$. I should like to mention that no assumption was made on the sign of $f(t)$.

As applications we give examples.

Example 1. Put $\alpha(t) = t$, $\beta_0(t) = t^{-1}$, $\varphi(t) \equiv 1$ and $t_0 > 0$. Then $\gamma_1(t) = t^{-1}$, hence according to the Proposition, the assumption

$$A_1(t) = \frac{1}{\ln t} \int_{t_0}^{t} \frac{1}{s} \int_{t_0}^{s} u f(u)\, du\, ds \to \infty$$

as $t \to \infty$, with (1) and (2) is sufficient for the bounded solutions to be oscillatory. It is easy to find function $f(t) \geq 0$, $\int^{\infty} t f(t)\, dt = \infty$ which assures obviously also $A_1(t) \to \infty$, the assumptions of the Proposition are valid, the equation (E_1) has nonoscillatory solution, but naturally this solution fails to be bounded.

Example 2. Put $\alpha(t) = t$, $\beta_0(t) = t^{-1}$, $\beta_1(t) = (t \ln t)^{-1}$, $\varphi(t) \equiv 1$. Then

$$A_2(t) = \frac{1}{\ln t} \int_{t_0}^{t} \frac{1}{s \ln s} \int_{t_0}^{s} \frac{1}{u} \int_{t_0}^{u} v f(v)\, dv\, du\, ds.$$

Example 3. Put

$$\alpha(t) = t^{\alpha}, \quad \alpha \in [0, 1), \quad \beta_0(t) = t^{-\frac{1+\alpha}{1}}, \qquad \varphi(t) \equiv 1.$$

Then

$$A_1(t) = t^{\frac{\alpha - 1}{2}} \int_{t_0}^{t} u^{-\frac{1+\alpha}{2}} \int_{t_0}^{u} s^{\alpha} f(s)\, ds\, du.$$

2. Although equation (E_2) may be transformed into an equation of the form (E_1) the change of variables applied makes more difficult to determine the effect of the separate behavior of $h(t)$ and $f(t)$.

Repeating the foregoing method we have the Riccati-like equation

$$w' + w^2 + h(t) + f(t)\frac{g(x(t))}{x(t)} = 0$$

and if beyond the above conditions we have

$$h(t) \in C[t_0, \infty)$$

and

$$H_n(t) = \Big[\int_{t_0}^{t} \gamma_n(s)\, ds\Big]^{-1} \int_{t_0}^{t} \beta_{n-1}\, dt_{n-1} \dots$$

$$\dots \int_{t_0}^{s} \gamma_0\, ds \int_{t_0}^{s} \alpha(u) h(u)\, du \geqslant H > -\infty$$

then our Proposition remains valid for (E_2) too.

3. Trying to apply the method for the more complicated special functional differential equation (E_3), after transformation we have the Riccati-like equation

$$w' + w^2 + f(t)\frac{g(x(\sigma(t)))}{x(t)} = 0.$$

Then the foregoing procedure gives that if $\sigma(t) \in C[t_0, \infty)$, $|t - \sigma(t)| < M$ for $t \in [t_0, \infty)$, M an appropriate constant; and instead of (2) (the other assumptions are unchanged)

$$(2') \qquad \int^{\infty} \alpha(s) f^{+}(s)\, ds = \infty, \quad \int^{\infty} \alpha(s) f^{-}(s)\, ds < \infty,$$

then every bounded solution of (E_3) is either oscillatory, or

$$\underline{\lim}_{t\to\infty} x(t) = 0 \quad \text{if} \quad x(t) > 0 \quad \text{for} \quad t > t_1,$$

$$\overline{\lim}_{t\to\infty} x(t) = 0 \quad \text{if} \quad x(t) < 0 \quad \text{for} \quad t > t_1$$

(t_1 may depend on the solution $x(t)$.)

Condition (2′) seems to be a strong. However, consider the following example (see [4])

$$x'' + \frac{\sin t}{2 - \sin t}\, x(t - \pi) = 0,$$

where $x = 2 + \sin t$ is obviously a solution which is neither oscillatory, nor $\underline{\lim}_{t\to\infty} x(t) = 0$. Every assumption is satisfied except (2′) since here $\int^{\infty} \alpha(s) f^{-}(s)\, ds = \infty$.

REFERENCES

[1] I.V. Kamenev, Oscillation criteria for solutions of second order ordinary differential equations, *Diff. Urav.*, 10 (1974), 246-253 (in Russian).

[2] L. Erbe, Oscillation criteria for second order nonlinear differential equations, *Annali di Mat.*, 94 (1972), 257-268.

[3] J. Hale, *Functional differential equations,* Springer-Verlag, New York – Heidelberg – Berlin, 1971.

[4] C.C. Travis, Oscillation theorems for second order differential equations with functional arguments, *Proc. Amer. Math. Soc.*, 31 (1972), 199-202.

L. Pintér

JATE Bolyai Intézet, 6720 Szeged, Aradi vértanúk tere 1, Hungary.

COLLOQUIA MATHEMATICA SOCIETATIS JÁNOS BOLYAI
15. DIFFERENTIAL EQUATIONS, KESZTHELY (HUNGARY), 1975.

LOCAL TOPOLOGICAL EQUIVALENCE OF DIFFERENTIAL EQUATIONS

L.È. REĬZIN'S

Statement of the problem. Consider the differential equation

(1) $\dot{x} = P(x),$

where $P\colon R^n \to R^n$, $P(\lambda x) = \lambda^k P(x)$, $\lambda \geqslant 0$, $k > 1$, $P \in C^2$, $P = (P_1, P_2)$, $P_i\colon R^{n_i} \to R^{n_i}$, $n_1 + n_2 = n$, $x_i \in R^{n_i}$, $x = (x_1, x_2)$, the trivial solution of the equation

(2_i) $\dot{x}_i = P_i(x_i)$

being asymptotically stable for $t \to (-1)^{i+1}\infty$, and the differential equation

(3) $\dot{x} = P(x) + p(x),$

where $p \in C^1$,

(4) $\dfrac{\partial p}{\partial x}(x) = o(|x|^{k-1+\alpha}),$ as $|x| \to 0,$ $\alpha > 0.$

In the paper, in terms of the mapping p, sufficient conditions are obtained for the local topological equivalence of the equations (1) and (3).

As compared to the well-known results of the works [1], [2], [3], the conditions are obtained by different methods. The comparison of the present result with those mentioned above is not fully accomplished. The scope of a strengthening of the results is analysed.

It is shown that the result can be generalized to equations homogeneous in the extended sense.

Lemma 1. *Equation* (2_i) *has a global Ljapunov – Krasovskiĭ function satisfying the relations*

$$V_i(\lambda x_i) = \lambda^{r-k+1} V_i(x_i),$$
$$\dot{V}_i(x_i) = |x_i|^r,$$

where $r > k$ *is a positive number.*

Proof. Set

$$V_i(x_i) = - \int_0^{(-1)^{i+1}\infty} |\varphi_i(\tau, x_i)|^r d\tau,$$

where $\varphi_i(\tau, x_i)$ is the solution of equation (2_i). Cf. [4] for details.

Lemma 2. *The transformation*

$$(5) \qquad x_i = y_i\left((-1)^i V_i\left(\frac{x_i}{|x_i|}\right)\right)^{-\frac{1}{r-k+1}}; \qquad i = 1, 2$$

turns equation (2_i) *into the equation*

$$(6_i) \qquad \dot{y}_i = Q_i(y_i)$$

with global Ljapunov – Krasovskiĭ function $W_i(y_i) = (-1)^i |y_i|^2$ *and the property* $Q_i(\lambda y_i) = \lambda^k Q_i(y_i)$.

Proof. Straightforward evaluation yields

$$Q_i(y_i) = \left(E - \frac{y_i y_i^T}{|y_i|^2}\right) P_i(y_i) \frac{(-1)^i V_i(y_i)}{|y_i|^{r-k+1}} + (-1)^i \frac{|y_i|^{k-1} y_i}{r-k+1},$$

whence

$$\frac{\partial W_i}{\partial y_i}(y_i) y_i = \frac{k}{r-k+1} |y_i|^{k+1} > 0.$$

Lemma 3. *The transformation*

$$(7_i) \qquad y_i = |\xi_i|^{-1} \psi_i\left((-1)^i \frac{|\xi_i|^{1-k} - 1}{(k-1)|\xi_i|^{k-1}}, \xi_i\right),$$

where $\psi_i(t, y_i)$ *is the solution of equation* (6_i), *takes* (2_i) *into the equation*

$$(8_i) \qquad \dot{\xi}_i = (-1)^i |\xi_i|^{k-1} \xi_i.$$

Proof. According to Lemma 2, each solution $\psi_i(t, y_i)$ of equation (2_i) intersects the sphere $|y_i| = 1$ in one and only one point. Let this point correspond to time $t_i(y_i)$. The solution $\chi_i(t, \xi_i)$ of equation (8_i) that passes through the point ξ_i also intersects the sphere $|\xi_i| = 1$ in one and only one point $\frac{\xi_i}{|\xi_i|}$. Moreover,

$$(9_i) \qquad \chi_i(t, \xi_i) = \frac{\xi_i}{|\xi_i|} (1 - (-1)^i (k-1) t)^{\frac{1}{1-k}}.$$

We construct a transformation turning (6_i) into (8_i) as follows.

With the point y_i we associate the point $\psi_i(t_i(y_i), y_i)$ lying on the sphere $|y_i| = 1$. We map the sphere $S_1^{n_i-1} = \{\tilde{y}_i : |\tilde{y}_i| = 1\}$ onto the sphere $S^{n_i-1} = \{\tilde{\xi}_i : |\tilde{\xi}_i| = 1\}$ identically. Let

$$(10_i) \qquad \Psi_i : S_1^{n_i-1} \to S^{n_i-1}, \quad \Psi_i(\tilde{y}_i) = \tilde{y}_i.$$

Further, with the point $\Psi_i(\psi_i(t_i(y_i), y_i))$ we associate the point

$$(11_i) \qquad \xi_i = \chi_i(-t_i(y_i), \Psi_i(t_i(y_i), y_i))).$$

From (9_i), (10_i) and (11_i) we obtain

$$\xi_i = \psi_i(t_i(y_i), y_i)(1 + (-1)^i(k-1)t_i(y_i))^{\frac{1}{1-k}}.$$

Hence

$$|\xi_i| = (1 + (-1)^i(k-1)t_i(y_i))^{\frac{1}{1-k}}$$

and

$$(12_i) \qquad t_i(y_i) = (-1)^i \frac{|\xi_i|^{1-k} - 1}{k-1}.$$

As

$$\frac{\xi_i}{|\xi_i|} = \psi_i(t_i(y_i), y_i),$$

we have

$$(13_i) \qquad y_i = \psi_i\left(-t_i(y_i), \frac{\xi_i}{|\xi_i|}\right).$$

Substituting the value (12_i) into (13_i) and making use of the equation

$$\psi_i(t\lambda^{1-k}, \lambda\xi_i) = \lambda\psi_i(t, \xi_i),$$

which follows from the homogenety of Q_i, we obtain equation (6_i).

Theorem. *In equation* (3), *let* $p = (p_1, p_2)$, $p_i\colon R^n \to R^{n_i}$ *and*

$$(14) \qquad \begin{aligned} &p_i^j(x) = x_i^j \widetilde{p}_i^j(x), \quad \widetilde{p}_i \in C^1, \quad \widetilde{p}_i(0) = 0; \\ &i = 1, 2; \; j = 1, \dots, n_i. \end{aligned}$$

Then the differential equation (3) *is locally topologically equivalent to the differential equation* (1).

Proof. Applying the transformations (5) and (7_1), (7_2) to the differential equation (3), we obtain the system of differential equations

$$(15) \qquad \dot{\xi}_i = (-1)^i |\xi_i|^{k-1}\xi_i + \pi_i(\xi); \qquad i = 1, 2,$$

where $\xi = (\xi_1, \xi_2)$, $\pi_i\colon R^n \to R^{n_i}$. The system (15) satisfies the conditions of Theorem 6.3.7 of the work [1] with $\beta_1 = \beta_2 = \frac{k}{1-k}$, so the result follows from that theorem. We only have to check that the mappings π_i satisfy the conditions of the theorem. Thus the relation

$$(16) \qquad \frac{\partial \pi_i}{\partial \xi}(\xi) = \sigma(|\xi|^{k-1+\alpha}); \qquad i = 1, 2$$

must be valid with some $\alpha > 0$.

For brevity, let the transformation carrying equation (3) into equation (15) be denoted by

$$\xi_i = \Phi_i(x_i); \qquad i = 1, 2.$$

then $\pi(\xi) = \frac{\partial \Phi}{\partial x}(\Phi^{-1}(\xi))p(\xi))p(\Phi^{-1}(\xi))$ and

$$\frac{\partial \pi_i}{\partial \xi}(\xi) =$$

$$= \Big(\frac{\partial^2 \Phi_i}{\partial x_i^2}(\Phi_i^{-1}(\xi_i))\frac{\partial \Phi_i^{-1}}{\partial \xi_i}(\xi_i)p_i(\Phi_1^{-1}(\xi_1), \Phi_2^{-1}(\xi_2)), 0\Big) +$$

$$(17) \qquad + \frac{\partial \Phi_i}{\partial x_i}(\Phi_i^{-1}(\xi_i))\frac{\partial p_i}{\partial x}(\Phi_1^{-1}(\xi_1), \Phi_2^{-1}(\xi_2)) \times$$

$$\times \Big(\frac{\partial \Phi_1^{-1}}{\partial \xi_1}(\xi_1) \dotplus \frac{\partial \Phi_2^{-1}}{\partial \xi_2}(\xi_2)\Big).$$

Further,

$$\frac{\partial \Phi_i}{\partial x_i}(\lambda x_i) = \frac{\partial \Phi_i}{\partial x_i}(x_i)$$

and

$$(18) \qquad \frac{\partial^2 \Phi_i}{\partial x_i^2}(\lambda x_i) = \frac{1}{\lambda}\frac{\partial^2 \Phi_i}{\partial x_i^2}(x_i).$$

Therefore (13), (18) and (17) imply (16).

Example. The constant α in relations (4) or (16) cannot be replaced by zero. This can be seen from the following example due to L.A. Bieza.

Consider the system of equations

$$\dot{x} = x(x^2 + y^2 - z^2) - y(x^2 + y^2) + \frac{3z^2 x}{2 \ln |z|},$$

$$\dot{y} = y(x^2 + y^2 - z^2) + x(x^2 + y^2) + \frac{3z^2 y}{2 \ln |z|},$$

$$\dot{z} = - z^3.$$

Here x, y, z are scalars. If one omits the last terms in the first two equations, then along the positive z semi-axis there is only one trajectory coming into the origin: the semi-axis itself.

On the other hand,

$$\frac{d}{dt} \left(x^2 + y^2 + \frac{z^2}{\ln |z|}\right) \Bigg|_{x^2 + y^3 + \frac{x^2}{\ln |z|} = 1} = 0.$$

Thus the whole surface $x^2 + y^2 + \dfrac{z^2}{\ln |z|} = 0$ is filled with trajectories coming into the origin along the positive z semi-axis. The same is true for all trajectories of the region $x^2 + y^2 + \dfrac{z^2}{\ln |z|} < 0$, $|z| < 1$. Therefore, naturally, the entire system is not topologically equivalent to its homogeneous truncation.

Remark. Condition (14) is, at the first glance, very restrictive. In many cases, however, one can get rid of it. It can be shown that, under certain assumptions on p, the equation can be brought to a form for which condition (14) is satisfied. In particular, the following lemma is valid.

Lemma 4. *If the equations*

$$\begin{aligned} &P_1(\theta_1(x_2)) + p_1(\theta_1(x_2), x_2) = \\ &\quad = \frac{\partial \theta_1}{\partial x_2}(x_2)(P_2(x_2) + p_2(\theta_1(x_2), x_2)), \end{aligned} \tag{19}$$

$$\begin{aligned} &P_2(\theta_2(x_1)) + p_2(x_1, \theta_2(x_1)) = \\ &\quad = \frac{\partial \theta_2}{\partial x_1}(x_1)(P_1(x_1) + p_1) + p_1(x_1, \theta_2(x_1))), \end{aligned}$$

define a twice differentiable functions $\theta_1 : R^{n2} \to R^{n1}$ *and* $\theta_2 : R^{n1} \to$ $\to R^{n2}$, *then equation* (3) *can be transformed to a form such that condition* (14) *is satisfied.*

Proof. The equation

$$x_1 = \theta_1(x_2)$$

and

$$x_2 = \theta_2(x_1)$$

define asymptotically invariant manifolds for equation (3). Indeed, in order that these manifolds be invariant, the identities

$$\varphi_1(t, \theta_1(x_2), x_2) = \theta_1(\varphi_2(t, \theta_1(x_2), x_2)),$$

$$\varphi_2(t, x_1, \theta_2(x_1)) = \theta_2(\varphi_1(t, x_1, \theta_2(x_1)))$$

must hold. Hence, differentiating and setting $t = 0$, we obtain the identities (19). Therefore the initial data of trajectories belonging to invariant manifolds satisfy the equations (19). But, equations (3) being autonomous, the same will be true then also for the coordinates of any point of the trajectories belonging to asymptotically invariant manifolds.

Let us perform now the transformation

$$\begin{aligned} \bar{x}_1 &= x_1 - \theta_1(x_2), \\ \bar{x}_2 &= x_2 - \theta_2(x_1). \end{aligned} \tag{20}$$

Then from equation (3) we obtain the system

$$\dot{\bar{x}}_1 = P_1(x_1) + p_1(x_1, x_2) - \frac{\partial \theta_1}{\partial x_2}(x_2)(P_2(x_2) + p_2(x_1, x_2)),$$

$$\dot{\bar{x}}_2 = P_2(x_2) + p_2(x_1, x_2) - \frac{\partial \theta_2}{\partial x_1}(x_1)(P_1(x_1) + p_1(x_1, x_2))$$

or, applying the transformation inverse to (20), the system

$$\begin{aligned} \dot{\bar{x}} &= P_1(\bar{x}_1) + \bar{p}_1(\bar{x}_1, \bar{x}_2), \\ \dot{\bar{x}} &= P_2(\bar{x}_2) + \bar{p}_2(\bar{x}_1, \bar{x}_2). \end{aligned} \tag{21}$$

where on account of relations (19) we have

$$\bar{p}_1(0, \bar{x}_2) = 0,$$

$$\bar{p}_2(\bar{x}_1, 0) = 0.$$

Thus, indeed, system (21) has the property (14), as required by the lemma.

Lemma 5. *Let* P *be a generalized homogeneous mapping of class* $k_1, \ldots, k_n$ *an order* k, *i.e. for any* $\lambda > 0$ *the equations*

$$P^i(\lambda^{k_1}x^1, \ldots, \lambda^{k_n}x^n) = \lambda^{k+k_i}P^i(x^1, \ldots, x^n); \quad i = 1, \ldots, n \tag{22}$$

are valid (see [5]*). Then the transformation*

$$x^i = |y^i|^{k_i c} \operatorname{sgn} y^i; \qquad i = 1, \ldots, n, \tag{23}$$

where $c \leqslant \min_i \frac{1}{k_i}$, *carries equation* (1) *into a homogeneous equation* $\dot{y} = Q(y)$, *where*

$$Q(\lambda y) = \lambda^{kc+1} Q(y). \tag{24}$$

Proof. We have

$$y^i = |x^i|^{\frac{1}{k_i c}} \operatorname{sgn} x^i.$$

Therefore

$$\dot{y}^i = \frac{1}{k_i c} |x^i|^{\frac{1}{k_i c} - 1} \dot{x}^i,$$

whence

$$Q^i(y) = \frac{1}{k_i c} |y^i|^{1 - k_i c} P^i(|y^1|^{k_1 c} \operatorname{sgn} y^1, \ldots, |y^n|^{k_n c} \operatorname{sgn} y^n).$$

By the aid of (22) we obtain (24).

Lemma 6. *Let the mapping* p *satisfy the conditions*

(25) $\qquad \dfrac{\partial p^i}{\partial x^j}(x) = \sigma(|x^j|^{\frac{1}{k_j c}}).$

Then transformation (23) turns equation (3) into the equation

(26) $\qquad \dot{y} = Q(y) + q(y),$

where q *satisfies the condition*

(27) $\qquad \dfrac{\partial q}{\partial y}(y) = \sigma(|y|^{kc + \alpha}).$

Proof. Applying transformation (23), we find

$$q^i(y) = \frac{1}{k_i c}|y^i|^{1-k_i c}p^i(|y^1|^{k_1 c}\,\mathrm{sgn}\,y^1, \ldots, |y^n|^{k_n c}\,\mathrm{sgn}\,y^n).$$

Hence the statement follows by differentiating and making use of relation (25).

REFERENCES

[1] L.È. Reĭzin's, *Local equivalence of differential equations.* Riga, 1971 (in Russian).

[2] N.N. Ladis, On the structural stability of some systems of differential equations. Differencial'nye Uravnenija, 7 (1971), 419-423 (in Russian).

[3] A.A. Reĭnfeld, The reduction theorem. *Differential'nye Uravnenija,* 10 (1974), 838-843 (in Russian).

[4] A.Ja. Kanevskiĭ – L.È. Reĭzin's, The construction of Ljapunov – Krasovskiĭ functions. Differencial'nye Uravnenija, 9 (1973), 251-259 (in Russian).

[5] V.I. Zubov, *A.M. Ljapunov's methods and their application.* Leningrad, 1957 (in Russian).

L.E. Reizin's

Phys. Inst. A.N. Latvian SSR Riga, Salaspils, USSR.

COLLOQUIA MATHEMATICA SOCIETATIS JÁNOS BOLYAI
15. DIFFERENTIAL EQUATIONS, KESZTHELY (HUNGARY), 1975.

LINEAR DIFFERENTIAL EQUATIONS WITH CONSTANT COEFFICIENTS IN TWO INDEPENDENT VARIABLES

A. SCHMIDT

For certain linear partial differential equations all solutions* or a part of them – can be determined explicitely. If we look for solutions in the field of distributions, we have no difficulties in connection with continuity and differentiability of the solutions. Wishing to emphasize the elementary character of the following investigations, we shall use a definition of distributions which in the course of the usual treatment arises only from the representation theorem.**

*Cf. L. Lammel, Über eine zur Differentialgleichung $(a_0 \frac{\partial^n}{\partial x^n} + a_1 \frac{\partial^n}{\partial x^{n-1}\partial y} + \dots$ $\dots + a_n \frac{\partial^n}{\partial y^n})u(x, y) = 0$ gehörige Funktionentheorie. *Math. Ann.* 122 (1950), 109-126.

A. Schmidt, Lineare partielle Differentialgleichungen mit konstanten Koeffizienten. *J. reine u. angew. Math.*, 189 (1951), 160-167.

I. Fenyő – A. Schmidt, Über gewisse partielle Differentialgleichungen im Bereich der Distributionen. *J. reine u. angew. Math.*, 232 (1970), 215-220.

**We introduce here the notations used in the sequel. All theorems regarding distributions used here are consequences of (1) without the usual estimations. Cf. A. Schmidt, Zur "Differentialgleichung" $\frac{\partial}{\partial y} S = 0$ in the field of distributions, *Wiss. Zeitschr. d. Univ. Rostock*, (to appear).

Let Ω be an open connected domain of the x, y plane E. The real valued Schwartzian test functions $\{\varphi(x, y) \subset \mathfrak{C}_E^\infty,\ \mathrm{supp}\,\varphi \subset \Omega\} = \mathscr{D}_\Omega$ form an Abelian group with respect to the usual addition of functions. $\mathscr{D}_\Delta$ can also be defined for a compact set Δ. A mapping S from $\mathscr{D}_\Omega$ into the additive group of complex numbers is said to be a distribution if for every compact subset Δ of Ω there exist a finite set of continuous functions $f_{m,n}(x, y)$ such that for all $\varphi(x, y) \in \mathscr{D}_\Delta$

$$(1) \qquad S\varphi(x, y) = \int_\Delta \sum_{m,n} (-1)^{m+n} f_{m,n}(x, y) \frac{\partial^{m+n}}{\partial x^m \partial y^n} \varphi(x, y)\, dx\, dy$$

holds. By partial integration it is easy to deduce a representation with only one function $f(x, y)$, moreover this can be supposed to be continuously differentiable. Defining $(S_1 + S_2)\varphi = S_1\varphi + S_2\varphi$, the distributions form an Abelian group denoted by $\mathscr{D}'_\Omega$. For all $f(x, y) \in \mathfrak{C}_\Omega^\infty$ a left multiplication can be defined by $(f(x, y)S)\varphi = S(f(x, y)\varphi)$. We introduce $D_y S$ by $(D_y S)\varphi = -S\left(\frac{\partial}{\partial y}\varphi\right)$, and D_x similarly. Then for the above addition and multiplication, the usual laws of the calculus hold. D_y and D_x are homomorphisms of $\mathscr{D}'_\Omega$ into itself. The equation $D_y S = T$ has solutions for $T \in \mathscr{D}'_\Omega$ arbitrary. Denote by $S = \check{D}_y T$ a special solution. For an interval $\mathfrak{i}$ the definition of $\mathscr{D}_\mathfrak{i}$ and $\mathscr{D}'_\mathfrak{i}$ is obvious. Suppose that the intersection of Ω with any line $x = \mathrm{const}$ is either void or it is an interval. If

$$\mathfrak{i} = \left(\inf_{(x,y)\in\Omega} x,\ \sup_{(x,y)\in\Omega} x\right)$$

then we have

$$(S \in \mathscr{D}'_\Omega \ \&\ D_y S = 0) \Leftrightarrow$$

$$\Leftrightarrow \left(\exists R \in \mathscr{D}'_\mathfrak{i}\ \forall\varphi \in \mathscr{D}_\Omega\ \ S\varphi = R \int_{-\infty}^{+\infty} \varphi(x, y)\, dy\right).$$

Theorem. *If $g(x, y)$ is a polynomial with complex coefficients, and g is not divisible by $\alpha y - \beta x - \lambda$, then*

$$(g(D_x, D_y)(\alpha D_y - \beta D_x - \lambda)^k S = 0)$$

(2) $$\Updownarrow$$

$$(S = S_{\mathrm{I}} + S_{\mathrm{II}} \ \& \ g(D_x, D_y) S_{\mathrm{I}} = 0 \ \& \ (\alpha D_y - \beta D_x - \lambda)^k S_{\mathrm{II}} = 0).$$

Here $S, S_{\mathrm{I}}, S_{\mathrm{II}} \in \mathscr{D}'_{\Omega}$, *$\Omega$ be open and connected, if $\frac{\alpha}{\beta}$ is not real then in addition Ω be simply connected, if α, β are real, then any line* $\alpha x + \beta y =$ = const *intersect Ω in a connected set. In both cases S_{II} can be determined explicitely.**

It can be supposed that $\lambda = 0$. For choose complex numbers γ, δ so that $\alpha\delta - \beta\gamma \neq 0$ and put $S = e^{\lambda \frac{\gamma x + \delta y}{\alpha\delta - \beta\gamma}} T$. If $\lambda = 0$, we can rewrite the equation in (2) into the form

(3) $$\sum_{\mu, \nu} a_{\mu\nu} (-\gamma D_y + \delta D_x)^{\mu} (\alpha D_y - \beta D_x)^{\nu + k} S = 0$$

where at least one $a_{\mu\sigma}$ differs from 0. For $k = 1$ we have

$$(S \in \mathscr{D}'_{\Omega} \ \& \ (\alpha D_y - \beta D_x)^k S = 0)$$

$$\Updownarrow$$

(4) $$\Big(\exists \{S_1, S_2, \ldots, S_k\} \subset \mathscr{D}'_{\Omega},$$

$$S = \sum_{\kappa=1}^{k} \Big(\frac{\gamma x + \delta y}{\alpha\delta - \beta\gamma}\Big)^{k-\kappa} \frac{1}{(k-\kappa)!} S_{\kappa} \ \& \ (\alpha D_y - \beta D_x) S_{\kappa} = 0,$$

$$\kappa = 1, 2, \ldots, k\Big).$$

For $k > 1$ (4) follows easily by induction since

*Other representations of the solutions can be found in the paper I. Fenyő and A. Schmidt, referred to above. There was investigated the differential equation $\prod_{\kappa=1}^{k} (\alpha_{\kappa} D_y - \beta_{\kappa} D_x)^{m_s} S = 0$ in $\mathscr{D}'_E$. For $\alpha : \beta$ not real the equation $(\alpha D_y - \beta D_x) S = 0$ has solution, which for $\Omega \neq E$ cannot be extended to $\mathscr{D}'_E$ as solutions, since there exists functions holomorphic in $\Lambda\Omega$, which cannot be continued to any larger domain.

$$(S \in \mathscr{D}'_\Omega \;\;\&\;\; (\alpha D_y - \beta D_x)^{k+1} S = 0)$$

$$\Updownarrow$$

$$\Big(\exists \{S_1, S_2, \ldots, S_k\} \subset \mathscr{D}'_\Omega,$$

$$(\alpha D_y - \beta D_x) S = \sum_{\kappa=1}^{k} \left(\frac{\gamma x + \delta y}{\alpha\delta - \beta\gamma} \right)^{k-\kappa} \frac{1}{(k-\kappa)!} S_\kappa \;\;\&$$

$$\& \;\; (\alpha D_y - \beta D_x) S_\kappa = 0, \qquad \kappa = 1, 2, \ldots, k \Big).$$

Now we shall need a Lemma about Abelian groups.

Lemma. *Let G be an Abelian group, θ_1, θ_2 homomorphismus of G into itself, denote $\mathscr{N}_\nu$ their kernels, $\mathscr{N}_\nu = \{u \in G\colon \theta_\nu u = 0\}$, $\nu = 1, 2$. Moreover, suppose that θ_1 and θ_2 commute, and let $\mathscr{N} = \{u \in G, \theta_1 \theta_2 u = 0\}$. Then $\mathscr{N}_2 \subset \theta_1 \mathscr{N}_2$ implies $\mathscr{N} = \mathscr{N}_1 + \mathscr{N}_2$.*

By $\theta_1 \mathscr{N} \subset \mathscr{N}_2 \subset \theta_1 \mathscr{N}_2$, for each $u \in \mathscr{N}$ there exists a $v \in \mathscr{N}_2$ such that $\theta_1 (u - v) = 0$. By $u - v \in \mathscr{N}_1$, $u = (u - v) + v$ is the desired decomposition of u.

For the proof of statement (2) it is sufficient to show that if $S_1^*, S_2^*, \ldots, S_k^*$ are arbitrary distributions from $\mathscr{D}'_\Omega$, satisfying $(\alpha D_y - \beta D_x) S_\kappa^* = 0$ $(\kappa = 1, 2, \ldots, k)$ then there exist distributions $S_1, S_2, \ldots, S_k \in \mathscr{D}'_\Omega$ also satisfying $(\alpha D_y - \beta D_x) S_\kappa = 0$ $(\kappa = 1, 2, \ldots, k)$ such that

$$\begin{aligned} &\sum_\mu a_{\mu 0} (\delta D_x - \gamma D_y)^\mu S_\kappa \, +, \\ (5) \qquad &+ \sum_{0<\nu<\kappa} \sum_\mu a_{\mu\nu} (\delta D_x - \gamma D_y)^\mu S_{\kappa-\nu} = S_\kappa^* \\ &(\kappa = 1, 2, \ldots, k). \end{aligned}$$

For finding solutions to (5) in the domain $\{S \in \mathscr{D}'_\Omega\colon (\alpha D_y - \beta D_x) S = 0\}$ we introduce the affine mappings $\Lambda(x, y) = (ax + by, cx + dy)$, $(a, b, c, d$ real, $ad - bc \neq 0)$. Using the inverse $\check{\Lambda}(x, y) = \left(\frac{dx - by}{ad - bc}, \frac{ay - cx}{ad - bc} \right)$, Λ transforms Ω into $\Lambda\Omega = \{(x, y)\colon \check{\Lambda}(x, y) \in \Omega\}$. For a function f defined on $\Lambda\Omega$ we introduce the function $\Lambda f(x, y) = f(\Lambda(x, y))$, defined on Ω. For a distribution $T \in \mathscr{D}'_{\Lambda\Omega}$ we define $\Lambda T\varphi = |ad - bc|^{-1} T \Lambda \varphi$

according to $\int\limits_{\Delta}(\Delta f)\varphi\,dx\,dy = \int\limits_{\Lambda\Delta} f(\check{\Lambda}\varphi)\,|ad-bc|^{-1}d(\Lambda(x,y))$. For proving $\Lambda T \in \mathscr{D}'_{\Omega}$ it is sufficient to rewrite the integral representation of the form (1) for $T\check{\Lambda}\varphi$ to that for $\Lambda T\varphi$. It can be proved that $(aD_y - bD_x)\Lambda T = (ad-bc)\Lambda(D_y T)$. Indeed, $(aD_y - bD_x)\Lambda T\varphi =$
$= -(\Lambda T)(aD_y - bD_x)\varphi = \dfrac{-1}{|ad-bc|}\,T\check{\Lambda}((aD_y - bD_x)\varphi) =$
$= -\dfrac{(ad-bc)}{|ad-bc|}\,TD_y(\check{\Lambda}\varphi) = \dfrac{(ad-bc)}{|ad-bc|}\,(D_y T)(\check{\Lambda}\varphi) = (ad-bc)\Lambda(D_y T)\varphi.$

By analogy, $(\alpha D_x - cD_y)(\Lambda T) = (ad-bc)\Lambda(D_x T)$.

For proving (2) for α, β real, choose the real numbers γ, δ so that $\alpha\delta - \beta\gamma \neq 0$, and put $S = \Lambda T$, where Λ is determined by $(a,b,c,d) = (\alpha,\beta,\gamma,\delta)$. Thus $(\alpha D_y - \beta D_x)S = 0$ is transformed into $D_y T = 0$ for $T \in \mathscr{D}'_{\Lambda\Omega}$. Hence, $T\varphi = RI_{\mathfrak{i}}\varphi$ where $I_{\mathfrak{i}}\varphi = \int\limits_{-\infty}^{+\infty}\varphi(x,y)\,dy$, $R \in \mathscr{D}'_{\mathfrak{i}}$, $\mathfrak{i} = (\inf\limits_{(x,y)\in\Lambda\Omega} x, \sup\limits_{(x,y)\in\Lambda\Omega} x)$ since the intersection of $\Lambda\Omega$ with any line $x = \text{const}$ is an interval. Substituting $S_\kappa = \Lambda R_\kappa I_{\mathfrak{i}}$, $S^*_\kappa = \Lambda R^*_\pi I_{\mathfrak{i}}$ into (5) we get

$$\sum_{\mu} a_{\mu 0} D_x^{\mu} R_\kappa + \sum_{0<\nu<\kappa} \sum_{\mu} a_{\mu\nu} D_x^{\mu} R_{\kappa-\nu} = R^*_\kappa$$

$(\kappa = 1, 2, \ldots, k)$.

For each of these inhomogeneous linear differential equations a particular solution, can be determined successively starting from $\kappa = 1$, for example, by the scheme

$$\prod_{\mu=1}^{m}(D_x - \lambda_\mu)R =$$

$$= e^{\lambda x}\check{D}_x e^{-\lambda x}R^{**} \to \prod_{\mu=1}^{m}(D_x - \lambda_\mu)(D_x - \lambda)R = R^{**}.$$

Thus (2) is proved for α, β real.

Applying (2) for the rectangle $\mathfrak{P} = \mathfrak{i} \times \mathfrak{j}$ we get the following

Corollary.

$$(D_x^m D_y^n T = 0 \ \& \ T \in \mathscr{D}'_{\mathfrak{P}})$$

$$\Updownarrow$$

$$\Big(\exists\{Q_1, Q_2, \ldots, Q_m \subseteq \mathscr{D}'_{\mathfrak{j}},\ \exists\{R_1, R_2, \ldots, R_n\} \subseteq \mathscr{D}'_{\mathfrak{i}},$$

$$T = \sum_{\mu=1}^{m} \frac{x^{m-\mu}}{(m-\mu)!} Q_\mu I_{\mathfrak{i}} + \sum_{\nu=1}^{n} \frac{y^{n-\nu}}{(n-\nu)!} R_\nu I_{\mathfrak{j}}\Big)$$

where for arbitrary intervals $\mathfrak{i}$ *and* $\mathfrak{j}$

$$I_{\mathfrak{i}}\varphi = \int_{\mathfrak{i}} \varphi(x, y)\, dx, \quad I_{\mathfrak{j}}\varphi = \int_{\mathfrak{j}} \varphi(x, y)\, dy.$$

If $\alpha : \beta$ is not real, then for proving (2) put $S = \Lambda T$ with $(a, b, c, d) = (\alpha_1, \beta_1, \alpha_2, \beta_2)$, where $\alpha = \alpha_1 + i\alpha_2$, $\beta = \beta_1 + i\beta_2$. Thus $(\alpha D_y - \beta D_x)S = 0$ will be transformed into

$$(D_y - iD_x)T = 0.$$

This implies: $T\varphi = \int_{\Lambda\Omega} h(x + iy)\varphi(x, y)\, dx\, dy$ where $h(z)$ is holomorphic in $\Lambda\Omega$, $z = x + iy$.

That is what we are now going to prove on elementary way.

If T satisfies the above "Cauchy – Riemann" differential equation, then there exist a compact rectangle $\mathfrak{P} = \mathfrak{i} \times \mathfrak{j} \subset \Lambda\Omega$ and an $F \in \mathscr{D}'_{\mathfrak{P}}$ so that $T = D_x^m D_y^n F$ and $F\varphi = \int_{\mathfrak{P}} f(x, y)\varphi(x, y)\, dx\, dy$ where the function $f(x, y)$ is continuously differentiable on $\mathfrak{P}$.

By the above Corollary $D_x^m D_y^n (D_y - iD_x)F = 0$ implies

(6)
$$(D_y - iD_x)F = \sum_{\mu=1}^{m} \frac{x^{m-\mu}}{(m-\mu)!} Q_\mu I_{\mathfrak{i}} + \sum_{\nu=1}^{n} \frac{y^{n-\nu}}{(n-\nu)!} R_\nu I_{\mathfrak{j}},$$

$$Q_\mu \in \mathscr{D}'_{\mathfrak{j}}, \quad R_\nu \in \mathscr{D}'_{\mathfrak{i}}.$$

Let $\psi(y) \in \mathscr{D}_{\mathfrak{j}}$ so that $\int_{\mathfrak{j}} \psi(y)\, dy = 1$, $\chi(x) \in \mathscr{D}_{\mathfrak{i}}$ and apply the distribution (6) to $\chi\psi^{(k)}$

$$\int_{\mathfrak{i}} \Big(\int_{\mathfrak{j}} (f_y(x,y) - i f_x(x,y) \quad \psi^{(k)}(y)\, dy \Big) \chi(x)\, dx =$$

$$= \sum_{\mu=1}^{m} Q_\mu \psi^{(k)}(y) \int_{\mathfrak{i}} \frac{x^{m-\mu}}{(m-\mu)!} \chi(x)\, dx +$$

$$+ \sum_{\nu=1}^{n-k-1} (-1)^k \int_{\mathfrak{j}} \frac{y^{n-k-\nu}}{(n-k-\nu)!} \psi(y)\, dy \cdot R_\nu \chi(x) +$$

$$+ (-1)^k R_{n-k} \chi(x)$$

for $k = 0, 1, \ldots, n-1$. Here we have used that

$$I_{\mathfrak{j}} \frac{y^{n-\nu}}{(n-\nu)!} \psi^{(\kappa)}(y) \chi(x) = \int_{\mathfrak{j}} (-1)^{n-\nu} \psi^{(k-n+\nu)}(y)\, dy \chi(x) =$$

$$= \begin{cases} 0 & \text{if} \quad \nu > n-k \\ (-1)^k \chi(x) & \text{if} \quad \nu = n-k. \end{cases}$$

Thus we obtain successively for $R_1, R_2, \ldots$ the representation $R_{n-k}\chi(x) = \int_{\mathfrak{i}} r_{n-k}(x) \chi(x)\, dx$ where the functions $r_{n-k}(x)$ are continuous on $\mathfrak{i}$. The same method yields $Q_{m-k}\psi(y) = \int_{\mathfrak{j}} q_{n-k}(y) \psi(y)\, dy$, where the functions $q_{m-k}(y)$ are continuous on $\mathfrak{j}$. Thus instead of (6) we can write

$$(D_y - iD_x) f(x,y) = \sum_{\mu=1}^{m} \frac{x^{m-\mu}}{(m-\mu)!} q_\mu(y) + \sum_{\nu=1}^{n} \frac{y^{n-\nu}}{(n-\nu)!} r_\nu(x).$$

This differential equation can be solved easily for the unknown function

$$f(x,y) = \sum_{\mu=1}^{m} \Big(\frac{x^{m-\mu}}{(m-\mu)!} \check{D}_y + i \frac{x^{m-\mu-1}}{(m-\mu-1)!} \check{D}_y^2 + \ldots$$

$$(7) \qquad \ldots + i^{m-\mu} \frac{x^\sigma}{\sigma!} \check{D}_y^{m-\mu+1} \Big) q_\mu(y) -$$

$$- i \sum_{\nu=1}^{n} \Big(\frac{y^{n-\nu}}{(n-\nu)!} \check{D}_x + (-i) \frac{y^{n-\nu-1}}{(n-\nu-1)!} \check{D}_x + \ldots$$

$$\ldots + (-i)^{n-\nu} \frac{y^\sigma}{\sigma!} \check{D}_x^{n-\nu+1} \Big) r_\nu(x) + f_0(x,y).$$

Here $f_0(x, y)$ denotes a solution of the homogeneous equation $(D_y - iD_x)u = 0$, i.e. the real and imaginary parts of f_0 satisfy the Cauchy – Riemann equation. Since all other terms in (7) are differentiable, so is $f_0(x, y)$, whence $f_0(x, y) = h_0(x + iy)$ coincides with the holomorphic function $h_0(z)$, $z = x + iy$, $(x, y) \in \mathfrak{P}$. The operator $D_x^m D_y^n$ can be applied to the right-hand side of (7), this yields $D_x^m D_y^n f(x, y) = D_x^m D_y^n h_0(x + iy) = h(x + iy)$ where $h(z)$ is holomorphic in $\mathfrak{P}$. We have proved that for any rectangle $\mathfrak{P} \subset \Lambda\Omega$ a representation $T\varphi = \int_{\mathfrak{P}} h(x + iy)\varphi(x, y)\, dx\, dy$ is valid with $h(z)$ holomorphic on $\mathfrak{P}$. The functions $h(z)$, belonging to different rectangles with non-void intersection are analytic continuations of each other. Hence for Ω simply connected the representation $T\varphi = \int_{\Lambda\Omega} h(x + iy)\varphi(x, y)\, dx\, dy$ holds as well, with $h(z)$ holomorphic in $\Lambda\Omega$. Conditions (5) is transformed into

$$\sum_{\mu} a_{\mu 0} D_z^{\mu} h_{\kappa}(z) + \sum_{0<\nu<\kappa} \sum_{\mu} a_{\mu\nu} D_z^{\mu} D_{\kappa-\nu} h(z) = h_{\kappa}^{*}(z)$$

$$(\kappa = 1, 2, \ldots, k)$$

having holomorphic solutions $h_{\kappa}(z)$ if the $h_{\kappa}^{*}(z)$'s are holomorphic in $\Lambda\Omega$.

A. Schmidt

25 Rostock, Schliemannstr. 40., GDR.

COLLOQUIA MATHEMATICA SOCIETATIS JÁNOS BOLYAI
15. DIFFERENTIAL EQUATIONS, KESZTHELY (HUNGARY), 1975.

ON THE EXISTENCE OF THE SOLUTIONS OF FUNCTIONAL-DIFFERENTIAL EQUATIONS

J. TERJÉKI

It is known that in a general Banach space the analogue of the Peano existence theorem for the initial value problem

$$\dot{x}(t) = f(t, x(t)), \qquad x(0) = 0 \tag{1}$$

(see e.g. [1], [2], [6] and [7]) does not hold. On the other hand it is easy to see by the step-by-step method that the initial value problem

$$\dot{x}(t) = f(t, x(t - \tau)), \quad x(t) = 0 \quad \text{for} \quad t \leqslant 0 \tag{2}$$

$\tau > 0$ constant, has solution for every continuous function $f(t, u)$ in an arbitrary Banach space.

In this paper we consider initial value problems which have no solutions in certain Banach spaces. Assume that $r \geqslant 0$ and $a > 0$ are real numbers, σ is a function, $\sigma \in C([0, a], R)$ with $\sigma(t) \leqslant t$, B is a real Banach space.

Introduce the notations $p = \inf_{t \in [0,a]} \sigma(t) - r$, $C = C([-r, 0], B)$, $C' = C([-r, 0], R)$.

If $x \in C([p, a], B)$, $\hat{\sigma} \in [p + r, a]$ define $x_{\hat{\sigma}} \in C$ as follows:

$$x_{\hat{\sigma}}(s) = x(\hat{\sigma} + s), \qquad s \in [-r, 0].$$

If $f \in C([0, a] \times C, B)$ then the equation

$$(3) \qquad \dot{x}(t) = f(t, x_{\sigma(t)})$$

will be called functional differential equation. The solution of (3) is a function $x(t)$ such that for some suitable $0 < h \leqslant a$, $x \in C([p, h], B)$, $x \in C^1([0, h], B)$ and for $t \in [0, h]$ $x(t)$ satisfies (3).

The function $x(t)$ is a solution of the initial value problem

$$(4) \qquad \dot{x}(t) = f(t, x_{\sigma(t)}), \quad x(t) = \varphi(t) \quad \text{for} \quad t \in [p, 0]$$

if $\varphi \in C([p, 0], B)$ and $x(t)$ is a solution of (3) with $x(t) = \varphi(t)$ for $t \in [p, 0]$.

Equation (3) is very general. For $r = 0$ we have the retarded equation $\dot{x}(t) = f(t, x(t - \sigma(t)))$ and for $\sigma(t) = t$ the functional equation $\dot{x}(t) = f(t, x_t)$ (see e.g. [3]). But also the integro-differential equation $\dot{x}(t) = \int_{\frac{t}{2}-1}^{\frac{t}{2}} x(s)\, ds$ belongs to (3) without belonging to the two previously mentioned types.

If there exists a Schauder basis $\{\varphi^i\}$ in B, then a function $f \in C([0, a] \times C, B)$ can be written in the form $f(t, u) = \sum_{i=1}^{\infty} f^i(t, u)\varphi^i$, the function $u \in C([-r, 0], B)$ can be written in the form $u(t) = \sum_{i=1}^{\infty} u^i(t)\varphi^i$ where $f^i \in C([0, a] \times C, R)$, $u^i \in C([-r, 0], R)$.

Theorem. *Suppose that there exists a Schauder basis in* B *and a mapping* $g \in C([0, a] \times C', R)$ *with the properties*

1. $g(t, \psi)$ *is nondecreasing with respect to* ψ,

2. $f^i(t, u) > g(t, u^i)$ *if* $(t, u) \in [0, a] \times C$, $i = 1, 2, \ldots$

3. *the initial value problem*

(5) $\dot{u}(t) = g(t, u_{\sigma(t)}), \quad u(t) = 0, \quad t \in [p, 0]$

has positive solution on $(0, a]$,

then the initial value problem

(6) $\dot{x}(t) = f(t, x_{\sigma(t)}), \quad x(t) = 0, \quad t \in [p, 0]$

does not have solution on $[p, h]$ *where* h *is an arbitrary positive number.*

Proof. Suppose on the contrary that $x(t)$ is a solution of (6) on $[p, h]$, where $0 < h \leqslant a$. Then for arbitrary fixed $i = 1, 2, \ldots,$ $\dot{x}^i(t) = f^i(t, x_{\sigma(t)}) > g(t, x^i_{\sigma(t)})$ is valid. According to 3 we have $\dot{x}^i(0) > g(0, x^i_{\sigma(0)}) = g(0, 0) = \dot{u}(0)$ hence

(7) $x^i(t) > u(t)$

for sufficiently small t, $t > 0$. If (7) did not hold on $(0, a]$, for example $x^i(t) > u(t)$ in $t \in (0, h^*)$, $x^i(h^*) = u(h^*)$ $(h^* \in (0, a])$ then from

$$\frac{u(h^* + H) - u(h^*)}{H} > \frac{x^i(h^* + H) - x^i(h^*)}{H}, \qquad H > 0$$

we have for $H \to 0$ the inequality $\dot{u}(h^*) \geqslant \dot{x}^i(h^*)$ contradicting to $\dot{x}^i(h^*) = f^i(h^*, x_{\sigma(h^*)}) > g(h^*, x^i_{\sigma(h^*)}) \geqslant g(h^*, u_{\sigma(h^*)}) = \dot{u}(h^*)$. Therefore (7) is valid on $(0, a]$, $i = 1, 2, \ldots$. For t fixed, $\lim_{i \to \infty} x^i(t) = 0$, hence according to (7) $u(t) \leqslant 0$ which is a contradiction.

Remarks. Condition 3 of our Theorem is strong and difficult to apply. For example there must exist a sequence $\{t_n\}$ such that $t_n \to 0$, $t_n > 0$, $\sigma(t_n) > 0$, further $g(t, 0) \leqslant 0$. Thus the initial value problem (5) fails to have the uniqueness property.

In the case $\sigma(t) \equiv t$, $r = 0$ and $g(t, u) \geqslant d(t) b(u)$, with $d, b \in C([0, a], R)$ and

(8) $\int_0^t d(s)\, ds > 0 \quad \text{for} \quad t > 0,$

(9) $\quad b(t) > 0 \quad \text{for} \quad t > 0,$

(10) $$\int_0^a \frac{ds}{b(s)} < \infty,$$

the initial value problem

(11) $$\dot{u}(t) = g(t, u(t)), \quad u(0) = 0$$

admits a positive solution.

If $g(t, u) = d(t)b(u)$, and (9) holds, then (8) and (10) are necessary for (11) to have positive solution.

If $b(u)$ is an increasing function then (9) and (10) are necessary for the existence of a positive solution of the problem

$$\dot{u}(t) = b(u(\sigma(t))), \quad u(t) = 0 \quad \text{for} \quad t \in [p, a].$$

It is not known whether the above statement is valid or not for other $b(t)$'s.

An example. Suppose that $B = c_0$ where c_0 is the Banach space of real sequences tending to zero, with the supremum norm. The mapping $f \in C(C, c_0)$ is given by $f(\varphi) = \{f^i(\varphi)\} = \left\{\sqrt[\alpha]{\|\varphi^i\|^{\alpha-1}} + \frac{1}{i}\right\}$ where $\alpha > 1$. If $0 < \beta \leqslant 1$, then the initial value problem

$$\dot{x}(t) = f(x_{\beta t}), \quad x(t) = 0, \quad t \leqslant 0$$

has no solution. This follows from the fact that the initial value problem

$$\dot{u}(t) = \sqrt[\alpha]{\|u_{\beta t}\|^{\alpha-1}}, \quad u(t) = 0, \quad t \leqslant 0$$

has the solution $u(t) = \beta^{\alpha(\alpha-1)}\left(\frac{t}{\alpha}\right)^{\alpha}$ for $t \geqslant 0$. It is natural to ask for further conditions on f and σ, which assure the existence of the solution of the problem. Using the step-by-step method it is easy to see that if there exists a number τ $0 < \tau \leqslant -r-p$ such that $\sigma(t) \leqslant t - \tau$ for $t \in [0, a]$, then in every Banach space B there exists a solution of the initial value problem (3), defined on $[p, a]$.

If f is compact and uniformly continuous on bounded closed subsets of $[0, a] \times C$, $\sigma(t)$ is continuous and $\sigma(t) \leqslant t$, then for a suitable $h > 0$ (3) has a solution defined on $[p, h]$. One can prove this latter statement by modification of the proof of the Peano theorem.

REFERENCES

[1] J. Dieudonné, Deux examples singuliers d'équations différentielles, *Acta Sci. Math.*, 12 (1950), Leopoldo Fejér et Frederico Riesz LXX annos natis dedicatus, pars B, 38-40.

[2] J. Dieudonné, *Foundations of Modern Analysis,* New York and London, 1960.

[3] J. Hale, *Functional Differential Equations,* New York, Heidelberg and Berlin, 1971.

[4] A.N. Kolmogorov – S.V. Fomin, *Elements of the theory of functions and functional analysis,* Moscow, 1972.

[5] Ya.A. Mamedov, *One-sided estimations in the conditions of investigating solutions of differential equations in Banach-spaces,* Baku, 1972.

[6] J.A. Yorke, A Continuous Differential Equation in Hilbert Space without Existence, *Funkcialaj Ekvacioj,* 13 (1970), 19-21.

[7] A. Cellina, On the Nonexistence of Solutions of Differential Equations in Nonreflexive Spaces, *Bull. Amer. Math. Soc.*, 78 (1972), 1069-1072.

J. Terjéki

JATE Bolyai Intézet, 6720 Szeged, Aradi vértanúk tere 1, Hungary.

COLLOQUIA MATHEMATICA SOCIETATIS JÁNOS BOLYAI
15. DIFFERENTIAL EQUATIONS, KESZTHELY (HUNGARY), 1975.

LINEAR FUNCTIONAL-DIFFERENTIAL OPERATORS: NORMAL SOLVABILITY AND ADJOINTS

M. TVRDÝ

1. PRELIMINARIES

Given a real $p \times q$-matrix $A = (a_{i,j})_{\substack{i=1,\dots,p \\ j=1,\dots,q}}$, A' denotes its transpose and

$$|A| = \max_{i=1,\dots,p} \sum_{j=1}^{q} |a_{i,j}|,$$

The space of real column n-vectors ($n \times 1$-matrices) is denoted by $R_n \cdot R_n^d$ is the space of real row n-vectors ($1 \times n$-matrices). (Generally, vectors are assumed to be column. Row vectors are written as transposes of column vectors.)

Given $c, d \in R_1$, $(c < d)$, $[c, d]$ denotes the closed interval $c \leqslant t \leqslant d$, (c, d) is its interior $c < t < d$ and $[c, d)$, $(c, d]$ are the corresponding half-open intervals.

We assume throughout $-\infty < a < b < +\infty$, $0 < r \leqslant b - a$ and denote $J = [a, b]$ and $J_0 = [a - r, b]$.

$C_n(J)$ is the Banach space (B-space) of continuous functions $J \to R_n$ ($x \in C_n(J) \to \|x\|_C = \sup_J |x(t)|$). $L_n^1(J)$ is the B-space of functions $J \to R_n$ which are Lebesgue integrable on J ($x \in L_n^1(J) \to \|x\|_L = \int_J |x(t)|\,dt$). $AC_n(J)$ is the B-space of absolutely continuous functions $J \to R_n$ ($x \in AC_n(J) \to \|x\|_{AC} = |x(a)| + \mathrm{var}_J\, x$). Given $x \in AC_n(J)$, its derivative is denoted by $\dot{x}$ ($\dot{x} \in L_n^1(J)$) and D is the linear bounded operator: $x \in AC_n(J) \to \dot{x} \in L_n^1(J)$.

Given a B-space X, its dual space is denoted by X^d and $(x, y)_X$ denotes the value of the functional $y \in X^d$ on $x \in X$. Accordingly, $C_n^d(J_0)$ denotes the B-space of row n-vector functions of bounded variation on $J_0 = [a - r, a]$, left-continuous on $(a - r, a)$ and vanishing at $t = a$. ($C_n^d(J_0)$ is isometrically isomorphic with the dual space of $C_n(J_0)$. For $u \in C_n(J_0)$ and $z' \in C_n^d(J_0)$, $(u, z')_C = \int_{a-r}^{a} [dz'(t)]u(t)$ holds.) $L_n^\infty(J)$ is the B-space of row n-vector functions essentially bounded on J. $L_n^\infty(J)$ and $AC_n^d(J) = L_n^\infty(J) \times R_n^d$ are isometrically isomorphic with the dual spaces of $L_n^1(J)$ and $AC_n(J)$, respectively. (Given $x \in AC_n(J)$, $f \in L_n^1(J)$, $y' \in L_n^\infty(J)$ and $w = (y', c') \in AC_n^d(J)$, $(f, y')_L = \int_a^b y'(t)f(t)\,dt$ and $(x, w)_{AC} = c'x(a) + (Dx, y')_L$.)

Let X, Y be B-spaces, $Z \subset Y$ and let the linear bounded operator A with values in Y be defined on the whole space X ($A\colon X \to Y$). Then $\mathrm{Im}\,(A)$ denotes the range of A, $\mathrm{Ker}\,(A)$ is its kernel (null space), $A^d\colon Y^d \to X^d$ its adjoint ($(Ax, y)_Y = (x, A^d y)_X$ for all $x \in X$ and $y \in Y^d$) and $A_{-1}(Z)$ is the set of all $x \in X$ such that $Ax \in Z$. The operator $A\colon X \to Y$ is normally solvable if $\mathrm{Im}\,(A)$ is closed in Y. For any closed linear subspace Z in Y, $\dim Z$ denotes its dimension and $\mathrm{codim}_Y Z = \dim \frac{Y}{Z}$, where $\frac{Y}{Z}$, is the corresponding quotient space.

All integrals used are the Perron – Stieltjes integrals. (The σ-Young – Stieltjes integral [4] and the Lebesgue integral are sufficient for our purposes.)

2. FREDHOLM – STIELTJES INTEGRO-DIFFERENTIAL OPERATOR

2.1. Assumptions. Let the $n \times n$-matrix function $P(t, s)$ be defined and measurable in (t, s) on $J \times J$, while

$$p(t) = |P(t, b)| + \operatorname{var}_J P(t, \cdot) < \infty \quad \text{for any} \quad t \in J$$

$$\text{and} \quad \int_a^b p(t)\, dt < \infty .$$

Given $x \in AC_n(J)$, $y(t) = \int_a^b [d_s P(t, s)] x(s) \in L_n^1(J)$ and P: $x \in AC_n(J) \to$ $\to \int_a^b [d_s P(t, s)] x(s) \in L_n^1(J)$ is a compact and continuous (completely continuous) linear operator (cf. [11] Lemma 5.4.1 or [7] Theorem 1).

Let us summarize the fundamental properties of the operator

$$L = D - P.$$

2.2. Theorem.

(a) *L is a linear, bounded and normally solvable (*$\operatorname{Im}(L)$ *is closed) operator* $AC_n(J) \to L_n^1(J)$.

(b) $n \leqslant \dim \operatorname{Ker}(L) < \infty$, *while* $\dim \operatorname{Ker}(L) = n$ *iff the equation* $Lx = f$ *possesses a solution in* $AC_n(J)$ *for any* $f \in L_n^1(J)$.

Proof.

(a) Obviously L is linear and bounded. Put

$$S\colon f \in L_n^1(J) \to \int_a^t f(s)\, ds \in AC_n(J),$$

$$R\colon x \in AC_n(J) \to g(t) \equiv x(a) \in AC_n(J)$$

and

$$K = R + SP\colon AC_n(J) \to AC_n(J).$$

Then $f \in L_n^1(J)$ belongs to $\operatorname{Im}(L)$ iff $Sf \in \operatorname{Im}(I-K)$, where I denotes the identity operator on $AC_n(J)$. Obviously R is completely continuous and since S is continuous and P is completely continuous, K is also completely continuous and $\operatorname{Im}(I-K)$ is closed. The closedness of $\operatorname{Im}(L) = S_{-1}(\operatorname{Im}(I-K))$ follows from the continuity of S.

(b) Since

$$\operatorname{Im}(I-K) \subset Z = \{g \in AC_n(J);\ g(a) = 0\},$$

$$\dim \operatorname{Ker}(L) = \dim \operatorname{Ker}(I-K) =$$

$$= \operatorname{codim}_{AC_n(J)} \operatorname{Im}(I-K) \geqslant \operatorname{codim}_{AC_n(J)} Z = n$$

and

$$\dim \operatorname{Ker}(L) = n \quad \text{iff} \quad \operatorname{Im}(I-K) = Z,$$

i.e. iff the equation $Lx = f$ has a solution $x \in AC_n(J)$ for any $f \in L_n^1(J)$,

2.3. Remark. The assertion (b) of Theorem 2.2 was proved formerly in a less straightforward manner in [6]. The normal solvability of L follows also from the following theorem.

2.4. Theorem. *Let the operator* $N\colon AC_n(J) \to R_n$ $(m < \infty)$ *be linear and bounded. Then the operator*

$$A\colon x \in AC_n(J) \to \begin{bmatrix} Lx \\ Nx \end{bmatrix} \in Y = L_n^1(J) \times R_m$$

is normally solvable, $\dim \operatorname{Ker}(A) < \infty$ *and* $\operatorname{ind}(A) = \dim \operatorname{Ker}(A) - \operatorname{codim}_Y \operatorname{Im}(A) = n - m$.

(The first and second assertion of the theorem were proved e.g. in [11], Theorem 5.5.1. The last assertion could be proved analogously to Lemma 4.2 in [10].)

2.5. Remark. Similar questions has been treated recently by several authors (cf. e.g. [5], [6], [7], [8], [10]-[12]). In particular, [7] and [10] contain also the proof of Theorem 2.4. The normal solvability of the operator L can be derived also from Theorem 3.1 of [8].

3. VOLTERRA – STIELTJES INTEGRO-DIFFERENTIAL OPERATOR

Let the matrix function $P(t, s)$ satisfy Assumptions 2.1. In the special Volterra-type case $(P(t, s) = P(t, t)$ for $s > t)$ we have

3.1. Theorem. *Let* $P(t, s) = P(t, t)$ *for* $s > t$. *Given* $c \in R_n$ *and* $f \in L_n^1(J)$, *there exists a unique solution* $x \in AC_n(J)$ *to the equation* $Lx = f$ *such that* $x(a) = c$. *This solution may be written in the form* $x = Fc + Gf$, *where* $F\colon R_n \to AC_n(J)$ *and* $G\colon L_n^1(J) \to AC_n(J)$ *are some linear and bounded operators.*

Proof. Put $A(t) = P(t, t) - P(t, t-0)$, for $t \in J$ and $G(t, s) = P(t, s-0)$ for $s < t$, $G(t, s) = G(t, t) = P(t, t-0)$ for $s \geqslant t$ $(P(t, a-0) = P(t, a))$. $G(t, \cdot)$ is left-continuous on $(a, b]$ for any $t \in J$ and the equation $Lx = f$ takes the form

$$\dot{x}(x) - A(t)x(t) - \int_a^t [d_s G(t, s)]x(s) = f(t) \quad \text{a.e. on } J. \tag{3.1}$$

Let $X(t)$ denote the fundamental matrix solution corresponding to $A(t)$. The initial value problem $Lx = f$, $x(a) = c$ reduces to the Volterra – – Stieltjes integral equation

$$x(t) - \int_a^t [d_s K(t, s)]x(s) = X(t)c + \int_a^t X(t)X^{-1}(s)f(s)\,ds \quad \text{on } J, \tag{3.2}$$

where

$$K(t, s) = X(t)\Big(\int_a^s X^{-1}(r)G(r, r)\,dr +$$

$$+ \int_s^t X^{-1}(r)G(r, s)\,dr\Big) \quad \text{if} \quad t \geqslant s,$$

$K(s, t) = K(t, t)$ if $t < s$. It is easy to show that $K(t, s)$ is of strongly bounded variation on $J \times J$, i.e. that for given $t \in J$ and $s \in J$, $\operatorname{var}_J K(t, \cdot) + \operatorname{var}_J K(\cdot, s) + v(K) < \infty$, where $v(K)$ denotes the two-dimensional Vitali-variation of K on $J \times J$ (cf. [4], III §4.) (For the proof of a similar assertion see e.g. [10], Lemma 5.1.) Moreover, for any $t \in (a, b]$ $K(t, s)$ is left-continuous on $(a, b]$. Hence by [9], Theorem 2.1 the equation (3.2) possesses a unique solution for any $c \in R_n$ and $f \in L_n^1(J)$ and there exists an $n \times n$-matrix function $R(t,s)$ of strongly bounded variation on $J \times J$ such that this solution is given on J by

$$x(t) = \left\{X(t) + \int_a^t [d_s R(t,s)] X(s)\right\} c +$$

$$+ \int_a^t \left\{X(t) + \int_s^t [d_r R(t,r)] X(r)\right\} X^{-1}(s) f(s)\, ds.$$

3.2. Corollary. *Let* U *be a B-space and let the operator* $Q\colon U \to$ $\to L_n^1(J)$ *be linear and bounded. Then for any* $f \in L_n^1(J)$, $u \in U$ *and* $c \in R_n$, *there exists a unique solution* $x \in Fc + GQu + Gf$ *to the equation*

$$Dx - Px = Qu + f$$

such that $x(a) = c$.

3.3. Remark. In particular, putting $U = C_n(J_0)$ and $c = u(a)$, we could obtain in this way the variation-of-constants formula $x = Fu(a) +$ $+ GQu + Gf = Hu + Gf$ for functional-differential equations of the retarded type

$$\dot{x}(t) - \int_{t-r}^t [d_s P(t,s)] x(s) = f(t) \quad \text{a.e. on } J,$$

$$x(t) = u(t) \quad \text{on } J_0.$$

(See also e.g. [2], Theorem 16.3.) As a consequence we get furthermore that the boundary value type problem (V is a B-space)

$$Dx - Px - Qu = f, \quad x(a) - u(a) = 0, \quad Mu + Nx = v \in V,$$

where $M\colon C_n(J_0) = U \to V$ and $N\colon AC_n(J) \to V$ are linear bounded operators, $f \in L_n^1(J)$ and $v \in V$ are known and $x \in AC_n(J)$ and $u \in C_n(J_0)$ are sought, is normally solvable iff $\operatorname{Im}(M + NH)$ is closed. This happens to be the case e.g. if $r \leq b - a$, M admits a bounded inverse and $N =$ $= N_b T_b$ where $T_b\colon x \in AC_n(J) \to x_b = x|_{[b-r,b]} \in C_n[b-r,b]$ and N_b is a linear bounded operator $C_n[b-r,b] \to V$ (cf. e.g. [2], [3], [11].)

3.4. Corollary. *Let* U, V *be B-spaces. Given a linear bounded operator* $N\colon AC_n(J) \to V$, *the operator*

$$A\colon x \in AC_n(J) \to \begin{bmatrix} Lx \\ Nx \end{bmatrix} \in Y = L_n^1(J) \times V$$

is normally solvable. If, in addition $\operatorname{codim}_Y \operatorname{Im}(A) < \infty$, *then the operator*

$$C\colon \binom{x}{u} \in AC_n(J) \times U \to Ax + Bu \in Y$$

is normally solvable for any linear and bounded operator $B\colon U \to Y$.

Proof.

(a) $\operatorname{Im}(A) = S_{-1}(\operatorname{Im}(NF))$, where $S\colon \binom{f}{v} \in Y \to v - NGf \in V$ and $\dim \operatorname{Im}(NF) < \infty$.

(b) The second assertion follows from the following

Proposition ([5]). *If the operator* $A\colon X \to Y$ *is normally solvable and* $\operatorname{codim}_Y \operatorname{Im}(A) < \infty$, *then* $C\colon \binom{x}{u} \in X \times U \to Ax + Bu \in Y$ *is normally solvable and* $\operatorname{codim}_Y \operatorname{Im}(C) < \infty$ *for any linear bounded operator* $B\colon U \to Y$.

4. BOUNDARY VALUE TYPE PROBLEMS FOR FUNCTIONAL-DIFFERENTIAL EQUATIONS AND THEIR ADJOINTS

4.1. Assumptions. Let the $n \times n$-matrix function $P(t, s)$ be measurable in (t, s) on $J \times [a - r, b]$ and such that $p(t) = \operatorname{var}_{a-r}^{b} P(t, \cdot) < \infty$ for any $t \in J$ and $p(t)$ is integrable on J. Without loss of generality we can assume that $P(t, \cdot)$ is left-continuous on $(a - r, b)$ for any $t \in J$, while $P(t, b) \equiv 0$ on J. Let V be a B-space and let the linear operators $M\colon C_n(J_0) \to V$ and $N\colon AC_n(J) \to V$ be bounded. Let $f \in L_n^1(J)$ and $v \in V$. ($J = [a, b]$, $J_0 = [a - r, a]$.)

In the sequel we shall investigate the following boundary value type problem.

4.2. Problem (*P*). Determine $x \in AC_n(J)$ and $u \in C_n(J_0)$ such that

$$\dot{x}(t) - \int_a^b [d_s P(t,s)]x(s) - \int_{a-r}^a [d_s P(t,s)]u(s) = f(t) \quad \text{a.e. on } J, \tag{4.1}$$

$$Mu + Nx = v, \tag{4.2}$$

$$u(a) - x(a) = 0. \tag{4.3}$$

Let us denote $X = AC_n(J) \times C_n(J_0)$ and $Y = L_n^1(J) \times V \times R_n$. We shall write the problem (P) briefly as the operator equation

$$C\binom{x}{u} = \begin{bmatrix} f \\ v \\ 0 \end{bmatrix}, \tag{4.4}$$

where the linear bounded operator $C\colon X \to Y$ is defined in the obvious manner.

4.3. Remark. If $w \in V^d$, then there exist (uniquely determined) functions $(M^d w)(t)$ and $(N^d w)(t)$ and a constant $(\tilde{N}^d w) \in \boldsymbol{R}_n^d$ such that $(M^d w)(t) + (N^d w) \in C_n^d(J_0)$, $(N^d w)(t) \in L_n^\infty(J)$,

$$(Mu, w)_V = \int_{a-r}^a [d(M^d w)(t)]u(t)$$

and

$$(Nx, w)_V = (\tilde{N}^d w)x(a) + \int_a^b (N^d w)(t)\dot{x}(t)\,dt$$

for all $x \in AC_n(J)$ and $u \in C_n(J_0)$.

4.4. Theorem. *Let* $y' \in L_n^\infty(J)$, $w \in V^d$ *and* $c' \in \boldsymbol{R}_n^d$. *Then* $(y', w, c) \in \operatorname{Ker}(C^d)$ *iff*

$$\int_a^b y'(s)P(s,t)\,ds - (M^d w)(t) = 0 \quad \textit{for} \quad t \in [a-r, a), \tag{4.5}$$

$$y'(t) + \int_a^b y'(s)P(s,t)\,ds + (N^d w)(t) = 0 \quad \textit{a.e. on } J, \tag{4.6}$$

and

$$c' = -\int_a^b y'(s)P(s,a)\,ds - (\tilde{N}^d w).$$

Proof. Let $y' \in L_n^\infty(J)$, $w \in V^d$ and $c' \in R_n^d$, then $(y', w, c') \in$ $\in \operatorname{Ker}(C^d)$ iff for any $\binom{x}{u} \in X$,

$$0 = \left(C\binom{x}{u}, (y', w, c')\right)_Y = p'x(a) + \int_a^b g'(t)\dot{x}(t)\,dt +$$

$$+ \int_{a-r}^{a} [dh'(t)]u(t),$$

where $p' = c' + \int_a^b y'(s)P(s,a)\,ds + (\tilde{N}^d w) \in R_n^d$,

$$g'(t) = y'(t) + \int_a^b y'(s)P(s,t)\,ds + (N^d w)(t) \in L_n^\infty(J)$$

and

$$h'(t) = \int_a^b y'(s)(P(s,a) - P(s,t))\,ds + (M^d w)(t) + (\tilde{N}^d w) +$$

$$+ \begin{Bmatrix} c', & t < a \\ 0, & t = a \end{Bmatrix} \in C_n^d(J_0).$$

Choosing $x(t) \equiv 0$ on J and then $u(t) \equiv 0$ on J_0 and $x(t) \equiv \text{const} \neq 0$ on J, we obtain successively that $(y', w, c') \in \operatorname{Ker}(C^d)$ iff $h'(t) \equiv 0$ on J_0, $p' = 0$ and $g'(t) = 0$ a.e. on J. The assertion of the theorem easily follows.

4.5. Remark. Theorem 4.4 shows that the problem of finding a couple $(y', w) \in L_n^\infty(J) \times V^d$ satisfying the system (4.5), (4.6) is a "well defined" adjoint problem to (P).

4.6. Remark. The ordinary Fredholm – Stieltjes integro-differential equation is a special case of the equation $Lx = f$ and functional-differential equation of the retarded type is a special case of the system (4.1), (4.3), where $P(t,s) = P(t,t)$ for $t < s$. Hence most of the results from [1],

[3], [6] and [12] concerning boundary value type problems for such equations may be derived from the results presented above. For further references see e.g. [6], [11] and [12]. The problem (P) and its various special cases are also investigated in the paper [11]. In [11] the problem (P) is studied also in other possible settings (e.g. with initial functions u of bounded variation or square integrable on J_0).

REFERENCES

[1] A. Halanay, On a boundary-value problem for linear systems with time lag, *J. Diff. Eq.*, 2 (1966), 47-56.

[2] J.K. Hale, *Functional Differential Equations,* Springer-Verlag, New York, 1971.

[3] D. Henry, The adjoint of a linear functional-differential equation and boundary value problems, *J. Diff. Eq.*, 9 (1971), 55-66.

[4] T.H. Hildebrandt. *Introduction to the Theory of Integration,* Academic Press, New York-London, 1963.

[5] Ju.K. Lando, On controlled operators, *Diff. Urav.*, 10 (1974), 531-536 (in Russian).

[6] V.P. Maksimov – L.F. Rachmatullina, Linear functional-differential equation solved with respect to the derivative, *Diff. Urav.,* 9 (1973), 2231-2240 (in Russian).

[7] V.P. Maksimov, General boundary value problem for linear functional-differential equation is noetherian, *Diff. Urav.,* 10 (1974), 2288-2291 (in Russian).

[8] St. Schwabik, On an integral operator in the space of functions with bounded variation, *Cas. pest. mat.,* 97 (1972), 297-330.

[9] St. Schwabik, On Volterra – Stieltjes integral equations, *Cas. pest. mat.,* 99 (1974), 255-278.

[10] M. Tvrdý, Boundary value problems for linear generalized differential equations and their adjoints, *Czech. Math. J.*, 23, 98 (1973), 183-217.

[11] M. Tvrdý, Linear boundary value type problems for functional-differential equations and their adjoints, *Czech. Math. J.*, 25, 100 (1975), 37-66.

[12] M. Tvrdý – O. Vejvoda, General boundary value problem for an integro-differential system and its adjoint, *Cas. pest. mat.*, 97 (1972), 399-419 and 98 (1973), 26-42.

M. Tvrdý

ČSAV Matematicky Ustav 11567 Praha 1, Žitna 25, Czechoslovakia.

COLLOQUIA MATHEMATICA SOCIETATIS JÁNOS BOLYAI
15. DIFFERENTIAL EQUATIONS, KESZTHELY (HUNGARY), 1975.

AN ASYMPTOTIC PROBLEM FOR DIFFERENTIAL EQUATIONS WITH RETARDED ARGUMENT

JU.A. VED'

The problem of power asymptotics for the solutions of differential equations with retarded argument is considered.

For the differential equation

$$x^{(n)}(t) = \sum_{k=0}^{n-1} \sum_{j=0}^{m} a_{kj}(t)x^{(k)}(\sigma_{kj}(t)) + F(t, x^{(k)}(t), x^{(k)}(\sigma_{kj}(t))), \quad t \geqslant t_0 > 0 \tag{1}$$

of order $n \geqslant 2$ with normed initial conditions

$$x^{(k)}(t) = x^{(k)}(t_0)\varphi_k(t) \quad (k = 0, 1, \dots, n-1), \qquad t \in E_{t_0} \tag{1'}$$

we investigate the existence of a unique solution $x(t)$ satisfying the boundary conditions

$$\lim_{t\to\infty} \frac{1}{k!} \sum_{i=k}^{n-1} \frac{(-1)^{i-k}}{(i-k)!} t^{i-k} x^{(i)}(t) = c_k \quad (k = 0, 1, \dots, n-1), \tag{2}$$

where

$$F(t, x^{(k)}, x^{(k)}(\sigma_{kj})) \equiv$$

$$\equiv F(t, x, x', \ldots, x^{(n-1)}, x(\sigma_{01}), \ldots, x(\sigma_{0m}),$$

$$x'(\sigma_{11}), \ldots, x'(\sigma_{1m}), \ldots, x^{(n-1)}(\sigma_{n-11}), \ldots$$

$$\ldots, x^{(n-1)}(\sigma_{n-1m}));$$

$$\sigma_{k0}(t) \equiv t \qquad (k = 0, 1, \ldots, n-1);$$

$a_{k0}(t), a_{kj}(t)$ and $\sigma_{kj}(t)$ $(k = 0, 1, \ldots, n-1;\ j = 1, \ldots, m)$ are continuous functions on the semi-interval $J = [t_0, \infty)$;

$$\sigma_{kj}(t) \leqslant t \quad \text{for} \quad t \in J;$$

$F(t, x_k, x_{kj})$ is a continuous function in the domain

$$D = \{t_0 \leqslant t < \infty,\ |x_k| < \infty,\ |x_{kj}| < \infty\}$$

and there satisfies the Lipschitz condition

$$|F(t, x_k^{(1)}, x_{kj}^{(1)}) - F(t, x_k^{(2)}, x_{kj}^{(2)})| \leqslant$$

$$\leqslant \sum_{k=0}^{n-1} \Big[g_{k0}(t)|x_k^{(1)} - x_k^{(2)}| + \sum_{j=1}^{m} g_{kj}(t)|x_{kj}^{(1)} - x_{kj}^{(2)}|\Big]$$

with non-negative continuous functions $g_{kj}(t)$ $(k = 0, 1, \ldots, n-1;\ j = 0, 1, \ldots, m)$ on the semi-interval J; $\varphi_k(t)$ $(k = 0, 1, \ldots, n-1)$ are continuous functions on the initial set E_{t_0} consisting of the point t_0 and of the values $\sigma_{kj}(t) < t_0$ $(k = 0, 1, \ldots, n-1;\ j = 1, \ldots, m)$ for $t \in J$, $\varphi_k(t_0) = 1$; c_k $(k = 0, 1, \ldots, n-1)$ are prescribed constants.

The problem under consideration requires us to establish the existence of a unique solution $x(t)$ of the problem (1)-(1′) for which the given parabole

$$x = \sum_{k=0}^{n-1} c_k t^k$$

serves as an asymptotic parabola (i.e., as a parabola which is tangent of order $n-1$ to the curve $x = x(t)$ at infinity). From (2) it follows that

$$\lim_{t\to\infty} \left[x(t) - \sum_{k=0}^{n-1} c_k t^k\right]^{(i)} = 0 \qquad (i = 0, 1, \ldots, n-1).$$

By a method based on the use of the criterion [1], [2] for the representation of a function $x(t)$ satisfying the boundary conditions (2) as well as on the construction of special successive approximations, we establish sufficient conditions for the existence of a unique solution of problem (1), (1′), (2) in the class $C^n[t_0, \infty)$ of all functions $x(t)$ defined and n times continuously differentiable on the semi-interval J. For $n = 2$ the problem (1), (1′), (2) was investigated by the method of constructing the usual successive approximations in the works [3], [4], the results of which are special cases, for $n = 2$, of the result of the present paper; a special result concerning the existence of a unique solution of problem (1), (1′), (2) appears in [5]. We clarify the influence of terms with retardation on the behaviour of the solutions. We also show that if the retardations in equation (1) tend to zero, and at the same time the conditions assuring that problem (1), (1′), (2) has a unique solution are fulfilled, then the resulting differential equation with unretarded argument need not have solutions satisfying conditions (2). Further, we exhibit a specific property of the differential equation with retarded argument (1) as compared to the corresponding differential equation with unretarded argument ($m = 0$ in (1)).

For the differential equation

$$(1_1) \qquad x^{(n)}(t) = L(t; x) + \sum_{k=0}^{n-1} \sum_{j=0}^{m} b_{kj}(t)(x^{(k)}(\sigma_{kj}(t)))^r, \qquad t \geqslant t_0 > 0$$

of order $n \geqslant 2$, where $L(t; x)$ stands for the right-hand side of equation (1), $b_{kj}(t)$ ($k = 0, 1, \ldots, n-1$; $j = 0, 1, \ldots, m$) are continuous functions on the semi-interval J and r denotes a positive integer, greater than 1, for normed initial conditions (1′) we investigate the existence of a unique solution of the boundary value problem (2) with $c_0 \neq 0$, $c_k = 0$ ($k = 1, \ldots, n-1$). A particular case was treated in [5].

Making use of the criterion [1], [2] for the representation of a function $x(t)$ satisfying the boundary conditions (2) we obtain that in order to establish, in the class $C^n[t_0, \infty)$, the existence of a unique solution of the problem (1), (1'), (2) it is sufficient to prove the existence, in the class $\Gamma[t_0, \infty)$ of all functions $\psi(t)$ which are continuous on J and tend to the constant c_0 as $t \to \infty$, of a unique solution of the integral equation

$$\text{(3)} \qquad \psi(t) = c_0 + \int_t^{\infty} [S(s) + T(s; \psi) + V(s; \psi) + W(s; \psi, \psi)]\, ds, \quad t \in J,$$

where

$$S(t) \equiv -\frac{1}{\delta} t^{n-1} \Big[\sum_{i=1}^{n-1} i! c_i \sum_{j=0}^{m} A_{ij}(t) + F(t, C_K^0(t), D_k(\sigma_{kj}(t)) \Big];$$

$$T(t; \psi) \equiv t^{n-1} \sum_{i=1}^{n-1} \frac{(-1)^{n-1-i_i}}{(n-1-i)!} A_{i0}(t) \times$$

$$\times\, \psi_i^0(t; \psi) - a_{n-10}(t)\, \psi(t);$$

$$V(t; \psi) \equiv t^{n-1} \times$$

$$\times \sum_{j=1}^{m} \Big[\sum_{i=1}^{n-1} \frac{(-1)^{n-1-i_i}}{(n-1-i)!} \sum_{k=0}^{i} \frac{1}{(i-k)!} a_{kj}(t) \varphi_{ik}^0(\sigma_{kj}(t)) \times$$

$$\times \begin{Bmatrix} \psi_i^0(\sigma_{kj}(t), \psi), \sigma_{kj}(t) \geqslant t_0 \\ \psi_i^0(t_0; \psi), \sigma_{kj}(t) \leqslant t_0 \end{Bmatrix} - a_{n-1j}(t) \varphi_{0n-1}^0(\sigma_{n-1j}(t)) \times$$

$$\times \begin{Bmatrix} \psi(\sigma_{n-1j}(t)), \sigma_{n-1j}(t) \geqslant t_0 \\ \psi(t_0), \sigma_{n-1j}(t) \leqslant t_0 \end{Bmatrix} \Big];$$

$$W(t; \psi, \varphi) \equiv -\frac{1}{\delta} t^{n-1} [F(t, C_k^0(t) + \Psi_k^0(t; \psi), D_k(\sigma_{kj}(t)) +$$

$$+\, \Phi_k(\sigma_{kj}(t); \varphi)) - F(t, C_k^0(t), D_k(\sigma_{kj}(t)))];$$

$$\delta = (-1)^{n-1}(n-1)!;$$

$$A_{ij}(t) \equiv \sum_{k=0}^{i} \frac{1}{(i-k)!} \varphi_{ik}^0(\sigma_{kj}(t)) a_{kj}(t)$$

$$(i = 1, \ldots, n-1;\ j = 0, 1, \ldots, m);$$

$$\varphi_{ik}^0(t) \equiv \begin{cases} t^{i-k}, & t \in J \\ t_0^{i-k}\varphi_k(t), & t \in E_{t_0} \end{cases} \qquad (i, k = 0, 1, \ldots, n-1);$$

$$C_k^0(t) \equiv \sum_{i=k}^{n-1} i\,\frac{(i-1)!}{(i-k)!}\,c_i t^{i-k} \qquad (k = 0, 1, \ldots, n-1);$$

$$D_k(t) \equiv \begin{cases} C_k^0(t), & t \in J \\ C_k^0(t_0)\varphi_k(t), & t \in E_{t_0} \end{cases} \qquad (k = 0, 1, \ldots, n-1);$$

$$\psi_i^0(t;\psi) \equiv \int_t^\infty \eta^{-1-i}\psi(\eta)\,d\eta \qquad (i = 1, \ldots, n-1);$$

$$\Psi_k^0(t;\psi) \equiv -\delta \sum_{i=k}^{n-1} \frac{(-1)^{n-i-1}i}{(n-1-i)!\,(i-k)!}\,t^{i-k}\psi_i^0(t;\psi) +$$

$$+\,\delta\delta_k t^{1-n}\psi(t); \quad \delta_k = 0 \ (k = 0, 1, \ldots, n-2); \ \delta_{n-1} = 1;$$

$$\Phi_k(t;\psi) \equiv \begin{cases} \Psi_k^0(t;\psi), & t \in J \\ \Psi_k^0(t_0;\psi)\varphi_k(t), & t \in E_{t_0} \end{cases} \qquad (k = 0, 1, \ldots, n-1).$$

The solution $x(t)$ of problem (1), (1′), (2) is connected with the solution $\psi(t)$ of integral equation (3) by the relation

$$\psi(t) = \sum_{i=0}^{n-1} \frac{(-1)^i}{i!}\,t^i x^{(i)}(t), \qquad t \in J.$$

Theorem 1. *If*

$$(\mathrm{A}_1) \qquad \left|\int_{t_0}^\infty t^{n-1} \sum_{j=0}^m A_{ij}(t)\,dt\right| < \infty \qquad (i = 1, \ldots, n-1),$$

$$(\mathrm{A}_2) \qquad \left|\int_{t_0}^\infty t^{n-1}F(t, 0, 0)\,dt\right| < \infty,$$

$$(\mathrm{A}_3) \qquad \int_{t_0}^\infty t^{2n-2-k}g_{k0}(t)\,dt < \infty \qquad (k = 0, 1, \ldots, n-1),$$

$$\text{(A}_4\text{)}\qquad \int_{t_0}^{\infty} t^{n-1} \sum_{k=0}^{i} |\varphi_{ik}^{0}(\sigma_{kj}(t))|\, g_{kj}(t)\, dt < \infty$$

$$(i = 1, \ldots, n-1;\ j = 1, \ldots, m),$$

$$\text{(B}_1\text{)}\qquad \int_{t_0}^{\infty} t^{n-1-k} |a_{k0}(t)|\, dt < \infty \qquad (k = 0, 1, \ldots, n-1),$$

$$\text{(C}_1\text{)}\qquad \int_{t_0}^{\infty} t^{n-1} |\varphi_{0k}^{0}(\sigma_{kj}(t))|\, [|a_{kj}(t)| + g_{kj}(t)]\, dt < \infty$$

$$(k = 0, 1, \ldots, n-1;\ j = 1, \ldots, m),$$

$$\text{(q}_1\text{)}\qquad q_1 = \int_{t_0}^{\infty} (E(t))^{-1} \sum_{j=1}^{m} \left\{ G_j(t) \begin{Bmatrix} E(\sigma_{n-1j}(t)),\ \sigma_{n-1j}(t) \geqslant t_0 \\ E(t_0),\ \sigma_{n-1j}(t) \leqslant t_0 \end{Bmatrix} + \right.$$

$$+ t^{n-1} \sum_{i=1}^{n-1} \frac{i}{(n-1-i)!} \sum_{k=0}^{i} \frac{1}{(i-k)!} |\varphi_{ik}^{0}(\sigma_{kj}(t))|\, [|a_{kj}(t)| +$$

$$\left. + g_{kj}(t)] \begin{Bmatrix} \psi_i^0(\sigma_{kj}(t); E),\ \sigma_{kj}(t) \geqslant t_0 \\ \psi_i^0(t_0; E),\ \sigma_{kj}(t) \leqslant t_0 \end{Bmatrix} \right\} dt < 1,$$

where

$$E(t) \equiv \exp\left\{ \int_t^{\infty} \left[G_0(s) + \int_s^{\infty} H_0(s, \eta) \exp\left(-\int_s^{\eta} G_0(\theta)\, d\theta \right) d\eta \right] ds \right\};$$

$$G_j(t) \equiv t^{n-1} |\varphi_{0n-1}^{0}(\sigma_{n-1j}(t))|\, [|a_{n-1j}(t)| + g_{n-1j}(t)]$$

$$(j = 0, 1, \ldots, m);$$

$$H_0(t, \eta) \equiv t^{n-1} \sum_{i=1}^{n-1} \frac{i}{(n-1-i)!} [|A_{i0}(t)| +$$

$$+ \sum_{k=0}^{i} \frac{1}{(i-k)!} t^{i-k} g_{k0}(t)] \eta^{-1-i},$$

then the problem (1), (1′), (2) *has, in the class* $C^n[t_0, \infty)$, *a unique solution* $x(t)$ *satisfying the inequality*

$$(4) \qquad \left| \sum_{i=0}^{n-1} \frac{(-1)^i}{i!} t^i x^{(i)}(t) \right| \leqslant \frac{E(t)}{1-q_1} \sup_J |f(t)|. \qquad t \in J,$$

where

$$f(t) \equiv c_0 + \int_t^\infty S(s)\, ds.$$

Proof. For $t \in J$ we find

$$(5) \qquad \begin{aligned} &|F(t, C_k^0(t), D_k(\sigma_{kj}(t))) - F(t, 0, 0)| \leqslant \\ &\leqslant \sum_{i=1}^{n-1} i!\, |c_i| \sum_{k=0}^{i} \frac{1}{(i-k)!} \sum_{j=0}^{m} |\varphi_{ik}^0(\sigma_{kj}(t))|\, g_{kj}(t). \end{aligned}$$

In view of conditions (A_1), (A_2), (A_3), (A_4) from the relations (5) and

$$(6) \qquad \begin{aligned} &F(t, C_k^0(t), D_k(\sigma_{kj}(t))) \equiv \\ &\equiv F(t, 0, 0) + [F(t, C_k^0(t), D_k(\sigma_{kj}(t))) - F[t, 0, 0)] \end{aligned}$$

it follows that

$$(7) \qquad \left| \int_{t_0}^\infty S(s)\, ds \right| < \infty,$$

and therefore $f(t) \in \Gamma[t_0, \infty)$.

For any function $\psi(t)$ which is continuous and bounded on J, the integrals

$$(8) \qquad \int_t^\infty \eta^{-1-i} \psi(\eta)\, d\eta \qquad (i = 1, \ldots, n-1)$$

are absolutely and uniformly convergent with respect to t on J. For $t \in J$ we obtain

$$(9) \qquad |T(t; \psi)| \leqslant M\Big[|a_{n-10}(t)| + \sum_{i=1}^{n-1} \frac{t^{n-1-i}}{(n-1-i)!} |A_{i0}(t)|\Big],$$

$$(10) \qquad |V(t, \psi)| \leqslant M t^{n-1} \sum_{j=1}^{m} \big[|\varphi_{0n-1}^0(\sigma_{n-1j}(t))|\, |a_{n-1j}(t)| +$$

$$+\sum_{i=1}^{n-1} \frac{1}{(n-1-i)!} \sum_{k=0}^{i} \frac{1}{(i-k)!} |\varphi_{0k}^{0}(\sigma_{kj}(t))|\,|a_{kj}(t)|\Big],$$

$$(11)\qquad \begin{aligned} |W(t;\psi,\psi)| \leqslant M t^{n-1} \sum_{j=0}^{m} \Big[|\varphi_{0n-1}^{0}(\sigma_{n-1j}(t))|\, g_{n-1j}(t) + \\ + \sum_{i=1}^{n-1} \frac{1}{(n-1-i)!} \sum_{k=0}^{i} \frac{1}{(i-k)!} |\varphi_{0k}^{0}(\sigma_{kj}(t))|\, g_{kj}(t)\Big], \end{aligned}$$

where

$$M = \sup_J |\psi(t)| < \infty.$$

Owing to the conditions (B_1), from (9) it follows that

$$(12)\qquad \int_{t_0}^{\infty} |T(t;\psi)|\, dt < \infty.$$

Owing to conditions (C_1), from (10) it follows that

$$(13)\qquad \int_{t_0}^{\infty} |V(t;\psi)|\, dt < \infty.$$

Also, by the conditions (A_3) and (C_1), from (11) we find

$$(14)\qquad \int_{t_0}^{\infty} |W(t;\psi,\psi)|\, dt < \infty.$$

Every function of the class $\Gamma[t_0,\infty)$ is bounded on J. By (7), (12), (13) and (14), any solution $\psi(t)$ of the integral equation (3) which is continuous and bounded on J belongs to the class $\Gamma[t_0,\infty)$. Therefore it is sufficient to prove that the integral equation (3) has a unique solution in the class $O[t_0,\infty)$ of all functions $\psi(t)$ which are continuous and bounded on J.

For integral equation (3) we construct the successive approximations

$$\psi_0(t) = 0,$$

$$(15)\qquad \psi_p(t) = f(t) + \int_t^{\infty} [T(s;\psi_p) + V(s;\psi_{p-1}) + W(s;\psi_p,\psi_{p-1})]\, ds$$

$$(p = 1, 2, \ldots), \qquad t \in J.$$

Using conditions (B_1), (A_3) and (C_1), by the method of successive approximations it can be shown that the integral equation

$$z(t) = f(t) + u(t) + \int_t^\infty [T(s; z) + W(s; z, v)]\, ds, \qquad t \in J, \tag{16}$$

where $u(t)$ and $v(t)$ are given functions of the class $O[t_0, \infty)$, has a unique solution in the class $O[t_0, \infty)$.

On account of conditions (C_1), for each natural number p relation (15) is an integral equation of the form (16). Consequently, the recursive integral equations (15) uniquely define the functions $\psi_p(t) \in O[t_0, \infty)$ $(p = 1, 2, \ldots)$.

From (15) we obtain

$$\begin{gathered} |z_p(t)| \leqslant d_p(t) + \int_t^\infty \Big[G_0(s)|z_p(s)| + \int_s^\infty H_0(s, \eta)|z_p(\eta)|\, d\eta\Big]\, ds \\ (p = 1, 2, \ldots), \qquad t \in J, \end{gathered} \tag{17}$$

where

$$z_p(t) \equiv \psi_p(t) - \psi_{p-1}(t) \qquad (p = 1, 2, \ldots);$$

$$d_1(t) \equiv f_0 = \sup_J |f(t)|;$$

$$d_p(t) \equiv \int_t^\infty \Big[|V(s; z_{p-1})| + \frac{1}{(n-1)!} s^{n-1} \sum_{k=0}^{n-1} \sum_{j=1}^{m} g_{kj}(s)|\Phi_k(\sigma_{kj}(s); z_{p-1})|\Big]\, ds$$

$$(p = 2, 3, \ldots).$$

The integral inequalities (17) yield

$$Z_1(t) \leqslant f_0, \qquad t \in J, \tag{18}$$

$$Z_p(t) \leqslant -\int_t^\infty (E(s))^{-1} d_p'(s)\, ds \qquad (p = 2, 3, \ldots), \qquad t \in J, \tag{19}$$

where

$$Z_p(t) \equiv (E(t))^{-1} |z_p(t)| \qquad (p = 1, 2, \ldots).$$

From (18) and (19) we find

(20) $$M_p \leqslant f_0 q_1^{p-1} \qquad (p = 1, 2, \ldots),$$

where

$$M_p = \sup_J Z_p(t) < \infty \qquad (p = 1, 2, \ldots).$$

Owing to condition (q_1), from (20) it follows that the series

(21) $$\sum_{p=1}^{\infty} z_p(t)$$

is absolutely and uniformly convergent on J. The sum, say $\psi(t)$, of the series (21) belongs to the class $O[t_0, \infty)$, and the estimate (4) is valid. Passing in (15) to the limit $p \to \infty$, we obtain that $\psi(t)$ satisfies integral equation (3).

Let $\psi_1(t), \psi_2(t)$ be any pair of solutions of integral equation (3) in the class $O[t_0, \infty)$. Then, setting $z(t) \equiv \psi_1(t) - \psi_2(t)$, we obtain

(22) $$|z(t)| \leqslant d(t) + \int_t^{\infty} \left[G_0(s) |z(s)| + \int_s^{\infty} H_0(s, \eta) |z(\eta)| \, d\eta \right] ds, \quad t \in J,$$

where

$$d(t) \equiv \int_t^{\infty} \Big[|V(s; z)| +$$

$$+ \frac{1}{(n-1)!} s^{n-1} \sum_{k=0}^{n-1} \sum_{j=1}^{m} g_{kj}(s) |\Phi_k(\sigma_{kj}(s); z)| \Big] ds.$$

By (22) we have

(23) $$M_0 \leqslant q_1 M_0,$$

where

$$M_0 = \sup_J (E(t))^{-1} |z(t)|.$$

In view of condition (q_1), from (23) it follows that $\psi_2(t) \equiv \psi_1(t)$, $t \in J$.

In the case $m = 0$ Theorem 1 yields Theorem 3.1 of [2] for the differential equation with unretarded argument

$$(1_0) \qquad x^{(n)}(t) = \sum_{k=0}^{n-1} a_{k0}(t) x^{(k)}(t) + F(t, x^{(k)}(t)), \qquad t \geq t_0 > 0.$$

Remark 1. In the case

$$(\sigma_\nu) \qquad \sigma_{k\nu}(t) \equiv \sigma_\nu(t) \qquad (k = 0, 1, \ldots, n-1;\ 1 \leq \nu \leq m), \quad t \in J,$$

in condition (q_1) the expression

$$\sum_{k=0}^{i} \frac{1}{(i-k)!} |a_{k\nu}(t)|\, |\varphi_{ik}^0(\sigma_\nu(t))|$$

can be replaced by $|A_{i\nu}(t)|$ $(i = 1, \ldots, n-1;\ 1 \leq \nu \leq m)$.

Remark 2. If in the boundary conditions (2) we have $c_\nu = 0$ $(1 \leq \nu \leq n-1)$, then conditions (A_1) and (A_4) with $i = \nu$ may be omitted. If in the conditions (2) we have $c_\nu = 0$ $(\nu = k_0, \ldots, n-1;\ k_0 \geq 1)$, then conditions (A_3) can be replaced by the conditions

$$\int_{t_0}^{\infty} t^{n-2+k_0-k} g_{k0}(t)\, dt < \infty \qquad (k = 0, 1, \ldots, k_0 - 1),$$

$$\int_{t_0}^{\infty} t^{n-1-k} g_{k0}(t)\, dt < \infty \qquad (k = k_0, \ldots, n-1).$$

Remark 3. Conditions (A_1), (A_4) and (C_1) will be fulfilled if

$$\left| \int_{t_0}^{\infty} t^{n-1} A_{i0}(t)\, dt \right| < \infty \qquad (i = 1, \ldots, n-1),$$

$$\int_{t_0}^{\infty} t^{n-1} |\sigma_{kj}(t)|^{n-1-k} [|a_{kj}(t)| + g_{kj}(t)]\, dt < \infty$$

$$(k = 0, 1, \ldots, n-1;\ j = 1, \ldots, m),$$

$$\varphi_k(t) = O(t^{n-1-k}) \qquad (k = 0, 1, \ldots, n-1), \qquad t \in E_{t_0}.$$

Condition (q_1) is satisfied if

$$(q_1^0) \qquad q_1^0 = \int_{t_0}^{\infty} \Big\{ \sum_{j=0}^{m} G_j(t) + t^{n-1} \sum_{i=1}^{n-1} \frac{1}{(n-1-i)!} \Big[t^{-i} | A_{i0}(t) | + \sum_{k=0}^{i} \frac{1}{(i-k)!} \Big(t^{-k} g_{k0}(t) + \sum_{j=1}^{m} | \varphi_{0k}^0(\sigma_{kj}(t)) | (| a_{kj}(t) | + g_{kj}(t)) \Big) \Big] \Big\} \, dt < 1.$$

Remark 4. If the conditions of Theorem 1 are satisfied then the corresponding differential equation with unretarded argument (1_0) need not have solutions satisfying the boundary conditions (2) with $c_{n-1} \neq 0$. Indeed for the equation

$$x''(t) = \frac{2(t+3)}{(t+1)(14t^2 + 51t + 35)} \Big[x'(t) - \frac{2}{t+1} x\Big(\frac{t+1}{2}\Big) \Big],$$

$$t \geqslant 1,$$

(the initial set consists of the single initial point $t_0 = 1$) all conditions of Theorem 1 are fulfilled, and it has the unique solution

$$x(t) = \frac{7t+6}{7(t+1)} c_0 + c_1, \qquad t \geqslant 1,$$

satisfying the boundary conditions (2) (for $n = 2$); the general solution of this equation has the form

$$x(t) = d_1 t + \frac{7t+6}{t+1} d_2, \qquad t \geqslant 1,$$

where d_1, d_2 are arbitrary constants. The corresponding unretarded equation

$$x''(t) = \frac{2(t+3)}{(t+1)(14t^2 + 51t + 35)} x'(t), \qquad t \geqslant 1,$$

has no solutions satisfying conditions (2) (for $n = 2$) with $c_1 \neq 0$.

Remark 5. If, the conditions of Theorem 1 being fulfilled, in equation (1) the retardations tend to zero, then the resulting differential equation with unretarded argument need not have solutions satisfying the boundary conditions (2) with $c_{n-1} \neq 0$. Indeed, for the equation

$$x'''(t) = -\frac{1}{3t^5} x(t) + \frac{1}{6t^3} x''(t) + \frac{1}{3t^5} x(t^\epsilon),$$

$$0 < \epsilon < 1, \quad t \geqslant 1,$$

all conditions of Theorem 1 are fulfilled. Letting here the retardation tend to zero, we obtain the equation

$$x'''(t) = \frac{1}{6t^3} x''(t), \qquad t \geqslant 1,$$

which has no solutions satisfying the conditions (2) (for $n = 3$) with $c_2 \neq 0$.

Remark 6. If condition (q_1) is violated, the problem (1), (1'), (2) with $c_{n-1} \neq 0$ or $c_0 \neq 0$ need not have solutions. Indeed, for the equation

$$x'''(t) = -\frac{4}{t^4} x''(t^{\frac{1}{4}}), \qquad t \geqslant 1,$$

all conditions, excepting (q_1), of Theorem 1 are fulfilled, we have $q_1 = \frac{16}{3}$, and none of the solutions

$$x(t) = d_1 + d_2 t + \frac{d_3}{t^2}, \qquad t \geqslant 1,$$

where d_1, d_2, d_3 are arbitrary constants, satisfies the boundary conditions (2) (for $n = 3$) with $c_2 \neq 0$. For the equation

$$x'''(t) = -\frac{1}{t^{4-\epsilon}} [2x(t^\epsilon) - 2t^\epsilon x'(t^\epsilon) + t^{2\epsilon} x''(t^\epsilon)],$$

$$0 < \epsilon < 1, \qquad t \geqslant 1,$$

all conditions, excepting (q_1), of Theorem 1 are fulfilled, on account of Remark 1 we have $q_1 = \frac{1}{1-\epsilon} > 1$, and none of the solutions

$$x(t) = d_1 t^2 + d_2 t + \frac{d_3}{t}, \qquad t \geqslant 1,$$

(d_1, d_2, d_3 are arbitrary constants) satisfies conditions (2) (for $n = 3$) with $c_0 \neq 0$.

According to Theorem 3.1 of [2] and Theorem 1, the differential equation with unretarded argument (1_0), where the conditions (A_1) for $m = 0$, (A_2), (A_3) and (B_1) are fulfilled, has, in the class $C^n[t_0, \infty)$, a unique solution satisfying the boundary conditions (2) with arbitrary fixed values c_k $(k = 0, 1, \ldots, n-1)$.

Thus, as to the solvability of boundary problem (2), the differential equation with retarded argument (1) behaves, in general, differently from the corresponding differential equation with unretarded argument (1_0).

Let us turn to the investigation of the problem (1_1), (1), (2) with $c_0 \neq 0$, $c_k = 0$ $(k = 1, \ldots, n-1)$. In equation (1_1) $F(t, 0, 0) \equiv 0$. For this problem we have, instead of (3), the equation

$$(3_1) \qquad \psi(t) = c_0 + \int_t^{\infty} [T(s; \psi) + V(s; \psi) + W(s; \psi, \psi) + W_1(s; \psi)]\, ds, \qquad t \in J,$$

where

$$W_1(t; \psi) \equiv -\frac{1}{\delta} t^{n-1} \sum_{k=0}^{n-1} [b_{k0}(t)(\Psi_k^0(t; \psi))^r + \sum_{j=1}^{m} b_{kj}(t)(\Phi_k(\sigma_{kj}(t)))^r].$$

Theorem 2. *If conditions* (B_1), (C_1), (q_1) *are fulfilled, and also*

$$(A_3^0) \qquad \int_{t_0}^{\infty} t^{n-1-k} g_{k0}(t)\, dt < \infty \qquad (k = 0, 1, \ldots, n-1),$$

$$(D_1) \qquad \int_{t_0}^{\infty} t^{n-1} |\varphi_{0k}^0(\sigma_{kj}(t))|^r |b_{kj}(t)|\, dt < \infty \qquad (k = 0, 1, \ldots, n-1;\ j = 0, 1, \ldots, m),$$

$$q_2 \overset{*}{=} ((n-1)!)^{r-1} \int_{t_0}^{\infty} t^{n-1}(E(t))^{-1} \sum_{k=0}^{n-1} \sum_{j=0}^{m} |b_{kj}(t)| \times$$

$$\times \Big[\sum_{i=k}^{n-1} \frac{i}{(n-1-i)!\,(i-k)!} |\varphi_{ik}^0(\sigma_{kj}(t))| \times$$

$$(q_2) \qquad \times \begin{Bmatrix} \psi_i^0(\sigma_{kj}(t); E), \sigma_{kj}(t) \geq t_0 \\ \psi_i^0(t_0; E), \sigma_{kj}(t) \leq t_0 \end{Bmatrix} + \delta_k |\varphi_{0n-1}^0(\sigma_{n-1j}(t))| \times$$

$$\times \begin{Bmatrix} E(\sigma_{n-1j}(t)), \sigma_{n-1j}(t) \geq t_0 \\ E(t_0), \sigma_{n-1j}(t) \leq t_0 \end{Bmatrix} \Big]^r dt <$$

$$< \frac{1}{r^r} \left(\frac{r-1}{|c_0|} \right)^{r-1} (1-q_1)^r,$$

then the problem (1_1), $(1')$, (2) *with* $c_0 \neq 0$, $c_k = 0$ $(k = 1, \ldots, n-1)$ *has a unique solution in the class* $C^n[t_0, \infty)$ *of all functions* $x(t)$ *for which*

$$(24) \qquad \left| \sum_{i=0}^{n-1} \frac{(-1)^i}{i!} t^i x^{(i)}(t) \right| \leq RE(t), \qquad t \in J,$$

where

$$R = \frac{r|c_0|}{(r-1)(1-q_1)}.$$

Proof. By conditions (D_1), for any function $\psi(t) \in O[t_0, \infty)$ we have

$$(25) \qquad \int_{t_0}^{\infty} |W_1(t; \psi)| \, dt < \infty.$$

For integral equation (3_1) we construct the successive approximations

$$\psi_0(t) = 0,$$

$$(26) \qquad \psi_p(t) = c_0 + \int_t^{\infty} [T(s; \psi_p) + V(s; \psi_{p-1}) +$$

$$+ W(s; \psi_p, \psi_{p-1}) + W_1(s; \psi_{p-1})] \, ds \quad (p = 1, 2, \ldots), \quad t \in J.$$

Proceeding as in the proof of Theorem 1, we obtain that $\psi_p(t) \in O[t_0, \infty)$ $(p = 1, 2, \ldots)$,

$$(E(t))^{-1} |\psi_p(t)| \leqslant |c_0| + \int_t^\infty (E(s))^{-1} \Big[|V(s; \psi_{p-1})| +$$

$$\text{(27)} \qquad + \frac{1}{(n-1)!} s^{n-1} \sum_{k=0}^{n-1} \sum_{j=1}^{m} g_{kj}(s) |\Phi_k(\sigma_{kj}(s); \psi_{p-1})| +$$

$$+ |W_1(s; \psi_{p-1})| \Big] ds \quad (p = 1, 2, \ldots), \quad t \in J,$$

$$(E(t))^{-1} |z_p(t)| \leqslant \int_t^\infty (E(s))^{-1} \Big[|V(s; z_{p-1})| +$$

$$\text{(28)} \qquad + \frac{1}{(n-1)!} s^{n-1} \sum_{k=0}^{n-1} \sum_{j=1}^{m} g_{kj}(s) |\Phi_k(\sigma_{kj}(s); z_{p-1})| +$$

$$+ |W_1(s; \psi_{p-1}) - W_1(s; \psi_{p-2})| \Big] ds \quad (p = 1, 2, \ldots), \quad t \in J,$$

where $z_p(t) \equiv \psi_p(t) - \psi_{p-1}(t)$ $(p = 1, 2, \ldots)$.

Making use of conditions (q_1) and (q_2), from (27) we obtain by induction

$$\text{(29)} \qquad |\psi_p(t)| \leqslant RE(t) \qquad (p = 1, 2, \ldots), \quad t \in J.$$

Taking (29) into account, (28) yields

$$\text{(30)} \qquad M_p \leqslant [q_1 + rR^{r-1} q_2]^{p-1} M_1 \qquad (p = 1, 2, \ldots),$$

where

$$M_p = \sup_J (E(t))^{-1} |z_p(t)| \qquad (p = 1, 2, \ldots).$$

Since $q_1 + rR^{r-1} q_2 < 1$, on the basis of (30) we conclude that there exists a solution $\psi(t) \in O[t_0, \infty)$ of the integral equation (3_1); by (29) we have (24).

Let $\psi_1(t)$, $\psi_2(t)$ be any pair of solutions of the integral equation (3_1) in the class $O[t_0, \infty)$ such that

(31) $\quad |\psi_i(t)| \leqslant RE(t) \qquad (i = 1, 2), \quad t \in J.$

Then similarly to the proof of Theorem 1 we obtain

(32) $\quad M_0 \leqslant [q_1 + rR^{r-1}q_2]M_0,$

where

$$M_0 = \sup_J (E(t))^{-1} |\psi_1(t) - \psi_2(t)| < \infty.$$

Since $q_1 + rR^{r-1}q_2 < 1$, from (32) it follows that $\psi_2(t) \equiv \psi_1(t)$, $t \in J$.

REFERENCES

[1] O. Haupt, Über Asymptoten ebener Kurven. *J. Reine Angew. Math.*, 152 (1922), 6-10.

[2] Ju.A. Ved', On the question of the existence of solutions for differential and integro-differential equations having asymptotic parabolas. *Differencial'nye Uravnenija,* 2 (1966), 1594-1610 (in Russian).

[3] Ju.A. Ved' – L.N. Kitaeva, On a boundary problem for a second order differential equation with retarded argument. *Materials of the 6-th inter-university physical and mathematical scientific conference of the Far East (Habarovsk,* 1966*), Differential and integral equations, vol.* 3, *Habarovsk,* 1967, *pp.* 53-59 *(in Russian).*

[4] Ju.A. Ved' – L.N. Kitaeva, On a boundary problem for second order differential equations with retarded argument. *Sibirskiĭ Matematičeskiĭ Žurnal,* 9 (1968), 773-782 (in Russian).

[5] Ju.A. Ved', On a boundary problem for differential equations with retarded argument. *Summaries of reports of the 4-th scientific conference, Faculty of physics, mathematics and natural sciences, Section of mathematics, P. Lumumba University of People's Friendship, Moscow,* 1968, pp. 14-17 (in Russian).

Yu.A. Ved'

720071 Frunze 71 ul 22 Partsiezda 265 a Inst. Fiz. i Mat. A.N. Kirg. SSR, USSR.

COLLOQUIA MATHEMATICA SOCIETATIS JÁNOS BOLYAI
15. DIFFERENTIAL EQUATIONS, KESZTHELY (HUNGARY), 1975.

METHODICAL REMARKS ON THE NOTION OF A SATURATED SOLUTION OF THE DIFFERENTIAL EQUATION $x' = f(t, x)$

J. WALTER

1. INTRODUCTION

Let $n \in N$, $t \in R$, $x \in R^n$, f a function of t and x with $f(t, x) \in \in R^n$, and $(t_0, x_0) \in D(f)$. We consider the initial value problem

(1) $$x' = f(t, x), x(t_0) = x_0.$$

No further assumptions on f are made at this stage.

Definition 1. A function g is called a solution of (1) if $D(g)$ is an interval with $\inf D(g) < \sup D(g)$ and if $g \in (C^1(D(g)))^n$, $g \subset D(f)$, $(t_0, x_0) \in g$ and $g'(t) = f(t, g(t))$ for $t \in D(g)$. If t is a boundary point of $D(g)$ then $g'(t)$ is to be considered as a one-sided derivative. The set of all solutions of (1) is denoted by S.

Definition 2. A solution g of (1) is called saturated* if

(2) $$\bigwedge_{\gamma \in S} \{\gamma \supset g \Rightarrow \gamma = g\}.$$

*We have adopted this name from [2; p. 40].

Theorem 1. *Let* g *be any solution of* (1). *Then a saturated solution* $\tilde{g}$ *exists such that* $\tilde{g} \supset g$.

In case the initial value problem (1) is uniquely solvable, i.e., if for every pair g_1, g_2 of solutions of (1) we have $g_1(t) = g_2(t)$ for $t \in D(g_1) \cap \cap D(g_2)$ * the proof of Theorem 1 is trivial since

$$\tilde{g} := \bigcup_{g \subset \gamma \in S} \gamma$$

has already the desired properties.

In the case of nonuniqueness the assertion of Theorem 1 looks evident too, but its proof is considerably more difficult, cf. the proofs in [6; p. 99], [8; p. 24]**.

This might have induced some authors to restrict themselves to the proof of special cases of Theorem 1. Under the assumption

(3) $\qquad D(f)$ open, $\quad f \in C^0(D(f))$

Theorem 1 is proved in [3; p. 17], [4; p. 13]*** and under the assumption

(4) $\qquad D(f)$ open, $\quad f \in C^1(D(f))$

(which already guarantees uniqueness) it is proved in [1; p. 132], [2; p. 43]. All these authors, however, weld together into one theorem the assertion of Theorem 1 with that of the following Theorem 2.

Theorem 2. *Under the assumption of* (3) *every saturated solution* g *of* (1) *tends to the boundary of* $D(f)$, *i.e.*,

(5) $\qquad (t, g(t)) \to \partial D(f)$ *as* $t \to \inf D(g)$ *or* $t \to \sup D(g)$.

It appears to us, however, that welding together two notions which are originally independent of each other, viz. (2) and (5), is likely to cause

*Using the notion of a saturated solution the uniqueness assertion takes the more familiar form: If $g_1 g_2$ are two saturated solutions of (1) then $g_1 = g_2$.

**Very short proofs are possible at the price of using Zorn's lemma.

***(3) is assumed in [8; p. 23], although it is not used in the proof of that part of the theorem that corresponds to our Theorem 1.

misunderstandings among uninitiated readers. This applies particularly to the presentation given in [2; p. 43-45]. Conforming to our point of view the assertions of Theorems 1 and 2 are placed into two separate theorems (loc. cit. Theorem 3.2 and 3.3) both, however, under the general assumption (4). Since the proof of Theorem 3.2 becomes trivial in this case – as was mentioned above already – the question arises as to what is actually shown in the rather long proof of this theorem. In fact it turns out that a lot more is proved than is asserted. Taking the existence of the saturated solution for granted, the nontrivial part of information given amounts essentially to the following: Under the *additional* assumption (4) the saturated solution has the *additional* property (5).

The last assertion is now important as the prototype of a series of formally similar theorems. The characteristic feature of these theorems is that they concern only a special type of solutions, viz., saturated solutions. By way of compensation they give information about the size not only of the values $g(t)$ of g but also of the domain of definition $D(g)$ of g, i.e., they provide an answer to the two main questions of the qualitative theory*

In the next section we shall give a simplified version of the proofs in [6] and [8] hoping to reduce some of the difficulties which seem to be connected with the notion of a saturated solution.

2. SIMPLIFIED PROOF OF THEOREM 1

Let $s \in S$. By recursion we define numbers a_n, b_n and functions $g_n \in S$ such that $a_n < b_n$, $D(g_n) = (a_n, b_n)$ and $g_n \subset g_{n+1}$ for $n \in N$.

First step. $a_1 := \inf D(s)$, $b_1 := \sup D(s)$, $g_1 := s|_{(a_1, b_1)}$.

Second step. Let n be a natural number for which a_n, b_n, g_n are already defined. Put

*Information on the size of $D(g)$ relating to *all* solutions (in the sense of Definition 1) cannot be obtained for the simple reason that every restriction of a solution is again a solution. The notion of a saturated solution is thus somewhat closer to the intuitive idea of a solution than the usual one. Maybe one should therefore call a "saturated solution" simply a "solution" as is done in [10; p. 249].

$$M_n := \{g \in S \mid g \supset g_n\},$$

$$(6) \qquad \alpha_n := \inf_{g \in M_n} \inf D(g),$$

$$\beta_n := \sup_{g \in M_n} \sup D(g),$$

$$(7) \qquad a_{n+1} := \begin{cases} \dfrac{\alpha_n + a_n}{2} & \text{for} \quad \alpha_n > -\infty \\ a_n - 1 & \text{for} \quad \alpha_n = -\infty, \end{cases}$$

$$b_{n+1} := \begin{cases} \dfrac{b_n + \beta_n}{2} & \text{for} \quad \beta_n < \infty \\ b_n + 1 & \text{for} \quad \beta_n = \infty. \end{cases}$$

This implies

$$(8) \qquad \alpha_n \leqslant \alpha_{n+1} \leqslant a_{n+1} \leqslant a_n < b_n \leqslant b_{n+1} \leqslant \beta_{n+1} \leqslant \beta_n.$$

By definition of M_n there exist $\hat{g}, \hat{\hat{g}} \in M_n$ with

$$\epsilon := \inf D(\hat{g}) \leqslant a_{n+1}, \quad b_{n+1} \leqslant \sup D(\hat{\hat{g}}) =: \delta.$$

Define

$$g^* := (\hat{g} \mid_{(\epsilon, b_n)}) \cup (\hat{\hat{g}} \mid_{(a_n, \delta)}).$$

Then $g^* \in S$, $g^* \supset g_n$, $D(g^*) \supset (a_{n+1}, b_{n+1})$.

Finally let

$$g_{n+1} := g^* \mid_{(a_{n+1}; b_{n+1})}.$$

This completes the definition by recursion.

We now introduce the following abbreviations (the limits exist – possibly as improper ones – because of (8))

$$\alpha := \lim_{k \to \infty} \alpha_k, \quad \beta := \lim_{k \to \infty} \beta_k,$$

$$a := \lim_{k \to \infty} a_k, \quad b := \lim_{k \to \infty} b_k,$$

$$\gamma := \bigcup_{k=1}^{\infty} g_k .$$

Clearly one has $\inf D(\gamma) = a$ and $\sup D(\gamma) = b$.

Let us now show that no $g \in S$ exists with $g \supset \gamma$ and $\inf D(g) < a$ or $\sup D(g) > b$. For, let g be an element of S such that, e.g.,

(9) $\inf D(g) < a$.

In particular this means $a > -\infty$. Then $\alpha_k > -\infty$ for sufficiently large k since otherwise (compare (7)) the sequence of a_k could not be bounded from below. For sufficiently large k we therefore have

$$\frac{\alpha_k + a_k}{2} = a_{k+1}$$

which is equivalent to

$$2(a_{k+1} - a_k) = \alpha_k - a_k .$$

Since the left hand side tends to zero by hypothesis we have $\alpha = a$. Now $g \in M_k$ for $k \in N$ and (6) imply $\alpha_k \leqslant \inf D(g)$ for $k \in N$. Because of (9) this is not compatible with $\alpha = a$.

Now, if γ itself is not saturated yet then $\tilde{g} := \bigcup_{\gamma \subset g \in S} g$ fulfills the requirements of Theorem 1.

3. A TYPICAL APPLICATION

The answer given by Theorem 2 to the two main questions mentioned in the introduction is not very precise yet. The reason is that firstly one does not know *where* precisely the saturated solutions tend to the boundary of $D(f)$ and that secondly the only information on the size of $g(t)$ consists in the fact that g must be a subset of $D(f)$.

The following theorem should give an idea how one can arrive at more specific results of the same kind by introducing additional assumptions.

Theorem 3. ([8; p. 22, Th. 3.4]). *Under the assumption of* (3) *let* a, b *be positive numbers such that*

$$G := \{(t,x) \mid t_0 \leqslant t \leqslant t_0 + a, \mid x - x_0 \mid \leqslant b\} \subset D(f).$$

Define $M := \sup_{(t,x) \in G} |f(t,x)|$ *and* $d := \min \left\{ a, \frac{b}{M} \right\}$.

For every saturated solution g *of* (1) *we then have*

$$\sup D(g) \geqslant t_0 + d \quad \text{and} \quad |g(t) - x_0| \leqslant b$$

$$\text{for} \quad t \in [t_0, t_0 + d].$$

In order to proceed to more specific theorems it seems to be appropriate (following [5; p. 35], [10; p. 250, Cor. 8.4]) to set out from the following theorem which is, in fact, equivalent to Theorem 2.

Definition 3. A set is called a test domain if it is a compact subset of $D(f)$ *.

Theorem 4. *Assume* (3), *and let* g *be a saturated solution of* (1). *For every test domain* G *there exists a number* $t_1 < \sup D(g)$ *such that* $(t, g(t)) \notin G$ *for* $t \in [t_1, \infty) \cap D(g)$. *(An analogous statement holds with respect to the left hand side of* g.*)*

With the help of this theorem various answers to the two main questions can be given by suitable choices of the test domains if in addition to the assertion of Theorem 4, namely *that* every saturated solution ultimately leaves G, one is able to procure information about *where* this takes place.

For an adequate explication of this heuristic ideā Ważewski's [9] concept of an ingress (egress) point suggests itself as the appropriate tool. Since Ważewski's definition is restricted to the case that every initial value problem associated with the differential equation

(10) $\quad x' = f(t,x)$

is uniquely solvable we adopt here the definitions given in [4; p. 37] which are valid in the general case.

*The name "test domain" has been suggested to us by U.-W. Schmincke, Aachen.

Definition 4. Let G be a test domain. A point $(t_1, x_1) \in \partial G$ is called an ingress point of G with respect to the differential equation (10) if, for every saturated solution g of (10) through (t_1, x_1), a number $\epsilon > 0$ exists such that $[t_1, t_1 + \epsilon) \subset D(g)$ and $(t, g(t)) \in G^0$ for $t \in \in (t_1, t_1 + \epsilon)$ *.

Definition 5. A point $(t_1, x_1) \in \partial G$ is called an egress point if the following holds: There exists** a solution g of (10) through (t_1, x_1) such that for some $\epsilon > 0$ we have $[t_1 - \epsilon, t_1) \subset D(g)$ and $(t, g(t)) \in G^0$ for $t \in [t_1 - \epsilon, t_1)$.

Because we want to include cases where $D(f)$ is not necessarily open we use a modification of Theorem 4 (cf. [7; *p.* 74] for a similar modification). (Let G_E denote the set of all ingress point of G and define $\widetilde{G} := = G_E \cup G^0$.)

Theorem 5. *Assume* $f \in C^0(D(f))$ *with* $D(f)$ *not necessarily open. Let* G *be a test domain,* $(t_0, x_0) \in \widetilde{G}$ *and* g *a saturated solution of* (1). *Then a number* $t_1 > t_0$ *with* $t_1 \in D(g)$ *exists such that* $(t_1, g(t_1))$ *is an egress point of* G *and* $(t, g(t)) \in G^0$ *holds for* $t \in (t_0, t_1)$.

Proof. Because of Definition 4 there is a number $\epsilon > 0$ with $[t_0, t_0 + \epsilon) \subset D(g)$ and $(t, g(t)) \in G^0$ for $t \in (t_0, t_0 + \epsilon)$. Let ϵ_1 be the largest number with this property (obviously such a number exists) and define $t_1 := t_0 + \epsilon_1$. Since G is compact, g as well as g' are uniformly continuous in the interval $[t_0, t_1)$. Since g is saturated, we have $t_1 \in \in D(g)$ and (together with Theorem 3) $(t_1, g(t_1)) \notin G^0$. This implies $(t_1, g(t_1)) \in \partial G$, q.e.d.

In the following theorem test domains are constructed which are very useful in applications.

Theorem 6. *a, b be real numbers with* $a < b$, $D(f) = [a, b] \times \boldsymbol{R}^n$, *and* f *continuous. Let* ω *be a continuous increasing function with* $D(\omega) = [0, \infty)$, $\omega(0) = 0$ *and* $\omega(u) > 0$ *for* $u > 0$. *Suppose*

*G^0 in the interior of G.

**Instead of the quantifier "there exists" [4; p. 37] has the quantifier "for all" but needs our version in the proof of his Lemma 8.1.

$$|f(t, x_1) - f(t, x_2)| \leqslant \omega(|x_1 - x_2|) \quad \textit{for} \quad (t, x_i) \in D(f), \tag{11}$$
$$i = 1, 2.$$

Take $h \in (C^1([a, b]))^n$. *Define*

$$C := \sup_{t \in [a,b]} |f(t, h(t)) - h'(t)|. \tag{12}$$

Let $r_0 > 0$ *and* r *be a saturated solution of*

$$r' = \omega(r) + C, \quad r(a) = r_0 \tag{13}$$

and $I := [a, b] \cap [a, \sup D(r))$. *For every* $\beta \in I$ *the "quasicylindrical" set*

$$G_\beta := \{(t, x) \mid a \leqslant t \leqslant \beta, \; |h(t) - x| \leqslant r(t)\}$$

is clearly well defined and a test domain. Assertion: The surface

$$H := \{(t, x) \mid a < t \leqslant \beta, \; |h(t) - x| = r(t)\}$$

does not contain any egress points.

Proof. Let (t_1, x_1) be an egress point on H. By Definition 5 this means that a solution g through (t_1, x_1) and a number $t_0 \in (a, t_1)$ exist such that $[t_0, t_1] \subset D(g)$ and $(g(t) - h(t))^2 - r^2(t)$ is smaller than or equal to zero for $t \in [t_0, t_1)$ or $t = t_1$, respectively. This implies

$$0 < \int_{t_0}^{t_1} \frac{d}{d\tau} [(g - h)^2 - r^2] \, d\tau. \tag{14}$$

Using (10), (13) this integral can be written in the form

$$2 \int_{t_0}^{t_1} \{(g - h)([f(\tau, h) - h'] + [f(\tau, g) - f(\tau, h)]) -$$
$$- r[\omega(r) + C]\} \, d\tau.$$

The scalar product appearing in the integrand can be estimated from above by $r[\omega(r) + C]$ with the help of Schwarz's inequality, (11), (12) and the monotony of ω. Hence the integral is nonpositive contrary to (14).

Corollary. *Let* $|x_0 - h(a)| < r_0$ *and* g *be a saturated solution of*

$x' = f(t, x)$, $x(a) = x_0$. *Then for every* $\beta \in I$

$$\sup D(g) \geqslant \beta \quad \textit{and} \quad |h(t) - g(t)| \leqslant r(t) \quad \textit{for} \quad t \in [a, \beta].$$

Proof. Clearly (a, x_0) is an ingress point of G_β, i.e., $(a, x_0) \in \widetilde{G}_\beta$. On the other hand, according to Theorem 6 the set of egress points coincides with the right face of G_β, i.e., with $\{(t, x) \mid t = \beta, \ |x - h(\beta)| \leqslant r(\beta)\}$. Now the Corollary is an immediate consequence of Theorem 5.

Remark. The assertion of the Corollary is all the more precise the larger β and the smaller $r(t)$ is. Two questions naturally arise:*

1. Under which conditions we even have $D(g) = [a, b]$? Obviously this is the case for $b \in I$ or, what is equivalent, for $\sup D(r) > b$. The last condition, however, can easily be checked because of $\sup D(r) = = a + \int_{r_0}^{\infty} \frac{du}{\omega(u) + C}$. An important special case (which, e.g., applies if $f(t, x)$ is linear in x) is $\int^{\infty} \frac{du}{\omega(u)} = \infty$. In this case $D(g) = [a, b]$ for *all* solutions (Theorem of Wintner [4; p. 29]).

2. When does the test domain G_β shrink down to its middle line h as $r_0 \to 0$? Or equivalently: When do we have $r(\beta) \to 0$ as $r_0 \to 0$? A necessary condition for this to occur is $C = 0$, which means that h is a solution of (10). But the desired effect only takes place if in addition $\int_0 \frac{du}{\omega(u)} = \infty$. Now the usual assertions on uniqueness and continuous dependence of the solutions upon their initial data can be read off.

REFERENCES

[1] F. Brauer – J.A. Nohel, *Qualitative theory of ordinary differential equations.* New York: W.A. Benjamin 1968.

[2] C. Corduneanu, *Principles of differential and integral equations* Boston: Allyn and Bacon 1971.

[3] J.K. Hale, *Ordinary differential equations.* New York: Wiley Interscience 1969.

[4] P. Hartman, *Ordinary differential equations.* New York: John Wiley & Sons 1964.

[5] F. John, *Ordinary differential equations.* New York: Courant Institute, New York University 1965.

[6] E. Kamke, *Differentialgleichungen,* I. Leipzig: Geest & Portig 1969.

[7] A. Peyerimhoff, *Gewöhnliche Differentialgleichungen,* I und II. Frankfurt a. M.: Akademische Verlagsgesellschaft 1970.

[8] W.T. Reid, *Ordinary differential equations.* New York: John Wiley & Sons 1971.

[9] T. Ważewski, Sur un principe topologique de l'examen de l'allure asymptotique des intégrales des équations différentielles ordinaires. *Ann. Soc. Polon. Math.,* 20 (1947), 81-125.

[10] H.K. Wilson, *Ordinary differential equations.* Reading, Mass.: Addison – Wesley 1971.

J. Walter
51 Aachen, Tempelgraben 55, GFR.